Accession no.
01000409

Fundamentals of
Mathematical
Analysis

WITHDRAWN

Fundamentals of
Mathematical
Analysis

SECOND EDITION

CHESTER COLLEGE	
ACC. No. 01000409	DEPT.
CLASS No. SIS HAG	
LIBRARY	

ROD HAGGARTY
Oxford Brookes University

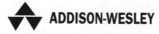
ADDISON-WESLEY

WOKINGHAM, ENGLAND • READING, MASSACHUSETTS • MENLO PARK, CALIFORNIA • NEW YORK
DON MILLS, ONTARIO • AMSTERDAM • BONN • SYDNEY • SINGAPORE
TOKYO • MADRID • SAN JUAN • MILAN • PARIS • MEXICO CITY • SEOUL • TAIPEI

To Linda

© 1993 Addison-Wesley Publishers Ltd.
© 1993 Addison-Wesley Publishing Company Inc.

All rights reserved. No part of this publication may be reproduced, stored in a retrieval system, or transmitted in any form or by any means, electronic, mechanical, photocopying, recording or otherwise, without prior written permission of the publisher.

Many of the designations used by manufacturers and sellers to distinguish their products are claimed as trademarks. Addison-Wesley has made every attempt to supply trademark information about manufacturers and their products mentioned in this book.

Cover designed by Designers & Partners, Oxford
and printed by The Riverside Printing Co. (Reading) Ltd.
Typeset by Keytec Typesetting, Bridport, Dorset.
Printed and bound in Great Britain by The University Press, Cambridge.

First printed 1992.   Reprinted 1993 and 1994.

British Library Cataloguing in Publication Data
A catalogue record for this book is available from the British Library.

ISBN 0-201-63197-0

Library of Congress Cataloging-in-Publication Data is available

Preface to the second edition

As with the first edition, this textbook is intended to give an introduction to mathematical analysis for students in their first or second year of an undergraduate course in mathematics. Its main concern is the analysis of real valued functions of one real variable and the limiting processes underlying this analysis. Since analysis is one of the cornerstones of twentieth-century mathematics, the element of 'proof' is of fundamental importance. A proof of a theorem is a carefully reasoned argument that validates the stated theorem relative to a set of basic assumptions. The basic assumptions in analysis are the axioms of the system of real numbers. These form the substance of Chapter 2, and subsequent chapters develop the limiting processes necessary to discuss the convergence of sequences and series, and ultimately to define the notion of a continuous function. Once these fundamental ideas are in place, the twin concepts of differentiation and integration are covered.

This second edition draws on the suggestions of many users of the first edition. My thanks go to them, and I trust that they will be pleased to see the new features of this edition, one which now deals exclusively with single-variable functions. There is a new Chapter 1, containing preliminary material on logic, methods of proof, sets and functions, and the material on sequences and series has been expanded and divided into two separate chapters. In addition, there are many improvements in exposition, including in each chapter a brief introductory overview. Another new feature is a prologue entitled 'What is analysis?', which sets mathematical analysis in its historical context. This is meant not only to provide the reader with increased motivation but also to highlight the fact that in mathematics the order of presentation of topics rarely follows the original chronological development. Also included in the text are brief biographical sketches of some of the many mathematicians who have contributed to the subject.

The aim throughout is to convey the fundamental concepts of analysis in as painless a manner as possible. The key definitions are well motivated, and proofs of central results are written in a sympathetic style to demonstrate clearly how the definitions are used to develop the theory. Important definitions and results are prominently displayed and the main theorems are given meaningful names. The import of each definition and the content of each theorem are further reinforced by examples. Many straightforward worked examples are included, and each section of each chapter ends with a short set of exercises designed to test the reader's grasp of the concepts involved and to provide some practice in the construction of proofs. These exercises again reinforce the main subject matter, and full solutions are included at the end of the book. In addition, each chapter ends with a set of problems designed for class use. Where appropriate, answers to these problems are given at the end of the book.

Prerequisites are a working knowledge of the techniques of calculus and a familiarity with elementary functions. The latter are used to motivate and illustrate the theoretical results, although the logical development of the theory is independent of their particular properties. The rigorous definitions of these elementary functions are given at the end of Section 4.3 and their analytic properties are derived in the Appendix. An informal flavour is maintained throughout, but with due attention to the rigour required in mathematics, with the overall aim of putting the reader's previous knowledge of calculus in its proper context. Mathematical analysis courses often have a rather negative impact on students, and I hope that this book will encourage its readers to tackle with confidence more advanced texts in the subject.

I should like to thank past mathematics students at Oxford Polytechnic who have suffered earlier versions of this material and whose positive reaction to its style of presentation encouraged me to write this book. My thanks also go to the reviewers of draft material for both editions who provided many helpful comments, and to the staff at Addison-Wesley in Wokingham for eliciting so many useful suggestions for improvement. Finally, my thanks go to my family for their unfailing support and encouragement.

Oxford   *Rod Haggarty*
July 1992

Contents

What is Analysis?

Introduction

This prologue seeks to answer the question 'What is analysis?' In a sentence, mathematical analysis† may be regarded as the study of infinite processes. Historically the subject saw its genesis in the work of the eminent Swiss mathematician, Leonard Euler (1707–1783). Euler took the calculus of Newton and Leibniz and, by giving the notion of a function central place, converted calculus from an essentially geometrical field of study into one where formulae and their relations were 'analysed'. The calculus itself was the greatest mathematical tool discovered in the seventeenth century, and it proved so powerful and capable of attacking problems that had been intractable in earlier times that its discovery heralded a new era in mathematics.

As with many branches of mathematics, calculus developed through an interplay between problems and theories, and is best understood through its applications. Differentiation is used to describe the way in which things change, move or grow, and most problems can be reduced to a geometric model of a curve in which a tangent is required at some point of the curve. If, for example, the curve represents the path of a moving body, the tangent gives the direction of motion at any particular time. See Figure P.1(a). Other types of problem require the determination of maximum and minimum values of some quantity, and this too can be reduced to a problem of tangents and hence may be solved by differential calculus. Integration was developed for finding the areas bounded by curves, called the **quadrature** of curves. If, for example, the curve is a graph of the velocity of a moving object, plotted against time, then the 'area under the curve' gives the total distance covered in a given time. See Figure P.1(b). The visual imagery present in Figure P.1 reflects the central role played by geometric models in the historical development of differentiation and

†The word analysis comes from the Greek word *analyein* meaning untie or unravel.

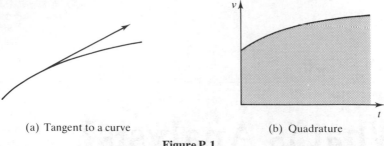

(a) Tangent to a curve   (b) Quadrature

Figure P.1

integration. The geometric problems of tangent and quadrature were themselves separate subjects of study for centuries before the advent of calculus, and, although it was suspected that the problems were in some sense inverse to each other, they were not formally linked until the seventeenth century. This linkage, the fundamental theorem of calculus, appears in Newton's work on the calculus.

As the seventeenth century unfolded, new notations were introduced that enabled the geometric notions of curve, tangent and quadrature to be superseded by the analytic notions of function, derivative and integral. More and more applications of this new analytic calculus were generated, and it might be supposed that mathematicians everywhere would have eagerly embraced the subject. However, there was much resistance to, and criticism of, calculus, principally because of its reliance on 'infinitely small quantities' or 'infinitesimals'. This was a notion that, along with the concept of infinity, had plagued mathematics from the time of Ancient Greece. The struggle to handle infinitesimals in the context of calculus so as to avoid contradictions or absurdities arising was eventually successful when, in the nineteenth century, infinitesimals were abandoned and calculus became based on the fundamental concept of a limit.

The remainder of this prologue traces the main aspects of the history of analysis beginning with the problematic nature of infinity and infinitesimals as perceived by the Greeks. Then precalculus attempts to solve the problems of tangent and quadrature are examined, followed by a description of the calculi of Newton and Leibniz and their inherent deficiencies. Finally, attention is paid to the way in which mathematical analysis grew out of the need to provide a satisfactory foundation for the calculus and, in turn, why mathematicians were forced to examine the foundations of analysis itself.

Precalculus developments in Ancient Greece

Archimedes (287–212 BC) is regarded by most commentators as having anticipated the integral calculus in his treatise, *The Method*, lost for a millennium, which came to light in Constantinople in 1906. In this work, Archimedes uses the rigorous approach developed by the Greeks since the founding of the Pythagorean School in 550 BC. This school placed the concept of number, by which was meant whole number, at the centre of their philosophy. In common with scientific thought at that time, the Pythagoreans also embraced the idea that all things were made up of finite indivisible elements. Other works of Archimedes, which were readily available to later mathematicians, contain ingenious techniques for calculating areas bounded by curves such as parabolas, and provide evidence of the Greeks' anticipation of the integral calculus.

The Greeks' reluctance to use any kind of infinite process is perhaps best exemplified by considering two of the famous paradoxes of Zeno of Elea (*c.* 460 BC). The first paradox, *Achilles and the tortoise*, begins with the assertion that space and time are infinitely divisible. If the tortoise is at B and Achilles is at A (see Figure P.2) then Achilles can never catch the tortoise since by the time Achilles reaches B, the tortoise will be at some further point C, and by the time Achilles reaches C, the tortoise will be further ahead at D, and so on *ad infinitum*: the tortoise will always be ahead!

The second paradox, the *Arrow*, adopts the alternative hypothesis that space and time are not infinitely divisible: hence there is an indivisible smallest unit of space (a point) and of time (an instant). Zeno now asserts that an arrow must be at a given point at a given instant. Since it cannot be in two places at the same instant it cannot move in that instant and so it is at rest in that instant. But this argument applies to all instants: the arrow cannot move at all! The dilemma raised by these two paradoxes, whereby alternative hypotheses both lead to conclusions that contradict common sense, is one of the reasons why Euclid's *Elements* never invoke the infinite.

The *Elements* appeared in the third century BC and used a single deductive system based upon a set of initial postulates, definitions and axioms to develop, in a purely geometric form, the mathematical wisdoms of previous generations. The whole of Pythagorean number theory was included, all couched in geometric terms. For example, consideration of the areas of the squares and rectangles in the dissection in Figure P.3(a) gives the well-known algebraic result

$$(x + y)^2 = x^2 + 2xy + y^2$$

Concepts that were not expressible in geometric terms were rejected, as were methods of proof that did not conform to the strict deductive

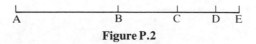

Figure P.2

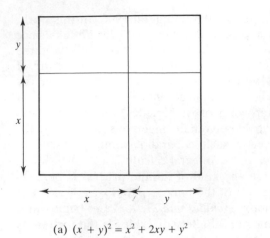

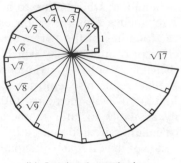

(b) Irrational magnitudes

(a) $(x + y)^2 = x^2 + 2xy + y^2$

Figure P.3

requirements of the *Elements*. This manic adherence to all things geometric was due to the fact that geometry could accommodate irrational
magnitudes as well as the whole numbers and ratios of whole numbers
(fractions) that formed the substance of Pythagorean mathematics. Irrationals such as $\sqrt{2}$, $\sqrt{3}$, $\sqrt{5}$ and so on cannot be expressed as fractions
but can be represented geometrically as indicated in Figure P.3(b).

The Greek approach to the problem of quadrature was thus geometric
in origin. Any rectilinear figure was reduced by geometric transforma-
tions to a square of the same area. In the sequence of diagrams in Figure
P.4 a triangle is transformed into a square with the same area by a
succession of geometric constructions involving parallel lines, congruent
triangles, perpendiculars and properties of the circle. For figures bounded
by curves this approach runs into difficulties, but the genius of Archimedes succeeded in obtaining the quadrature of several curves. The
quadrature of the parabola is one example: in Figure P.5 the area, T, of
triangle ABC is four times the sum of the areas of triangles APB and

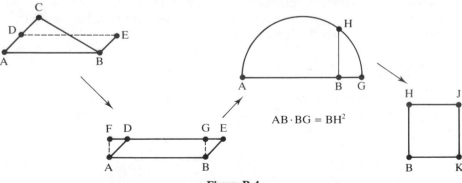

$AB \cdot BG = BH^2$

Figure P.4

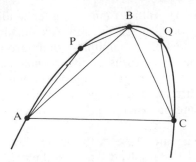

Figure P.5

BQC. Further triangles can then be inscribed within the parabolic segment with bases AP, PB, BQ and QC, and the same relativity of areas invoked. It is clear that, in modern parlance, the area K of the parabolic segment ABC is the sum of the infinite series

$$T + \frac{T}{4} + \frac{T}{4^2} + \frac{T}{4^3} + \ldots + \frac{T}{4^n} + \ldots$$

which gives $\frac{4}{3}T$. Of course, Archimedes did not use this 'infinite process', but instead showed that the conditions $K > \frac{4}{3}T$ and $K < \frac{4}{3}T$ both gave rise to absurd conclusions. The Greek reluctance to contemplate the logical validity of any process implemented *ad infinitum* and the lack of a notion of a function meant that no generally applicable method of quadrature was developed. In addition, Greek attention was focused on relatively few curves. This paucity of curves, allied with an unsatisfactory notion of angle (which included 'angles' between curves), also meant that there was little interest or progress in the problem of tangents. However, the properties of tangents and normals to parabolas, ellipses and hyperbolas were known, and Archimedes succeeded in constructing tangents to spirals.

The territorial conquests of the Arabs and the rise of the Roman Empire led to the fall of the Ancient Greek civilization and cut off Greek mathematical learning from European scholars. It was not until the twelfth century AD that elements of Greek scientific understanding began to filter through Arab sources and fragments of Euclid's *Elements* were studied. In medieval Europe there was no identifiable discipline called mathematics, although within philosophy debate raged about whether or not infinity existed and whether or not space and time were infinitely divisible.

Precalculus developments in Western Europe

During the fifteenth and sixteenth centuries in Europe, mathematics was applied over a wide range of practical subjects, and interest was centred on how to use geometry, rather than on understanding the proofs. Latin translations of the works of Euclid and Archimedes aroused much interest, and work began on streamlining their proof methods. In arguments involving repetitive processes the formal Greek arguments that avoided passage to the limit were abandoned. Although work was careful and soundly based, it remained geometric in nature, and no real attempt was made to formulate precise conditions under which limiting processes could be invoked.

Problems of quadrature arose in the many new practical applications of mathematics, and were solved by appealing to arguments involving infinitesimals. For instance, the astronomer Johann Kepler (1571–1630), who discovered that planets move around the sun in elliptical orbits, used infinitesimals to determine the area of an ellipse. He regarded the circumference of a circle as a regular polygon with an infinite number of sides. Hence the circle is made up of an infinite number of thin triangles (see Figure P.6(a)), each of height equal to the radius r of the circle. Since the area of any triangle is half its base times its height, the area of the circle is half its circumference times its radius (that is, $\pi r \cdot r = \pi r^2$). The definition of π used here is that it is the ratio of the circumference C of a circle to its diameter $2r$. Since all circles are similar, this makes $C = 2\pi r$, and hence the consequent formula for the area of a circle just described involves the same ratio π. The area of an ellipse whose axes are in the ratio $b : a$ is now derived by first inscribing the ellipse in a circle of radius a. The ellipse is obtained from the circle by a geometrical transformation that shortens the vertical component of any point on the circle in the ratio $b : a$. Then both the circle and the ellipse are thought of as consisting of infinitely many, infinitesimally thin vertical strips (see Figure P.6(b)) whose areas are, by construction, also in the ratio $b : a$. Hence the areas of the circle and the ellipse are in this same ratio and so the area of the ellipse $x^2/a^2 + y^2/b^2 = 1$ is $(\pi a^2)(b/a) = \pi ab$. However, Kepler was unable to determine the circumference of the ellipse, a problem that turned out a century later to be yet another application of quadrature, albeit one that merely expresses the circumference as an integral; this **elliptical integral** cannot, however, be evaluated in closed form.

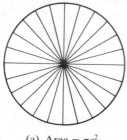

(a) Area $= \pi a^2$

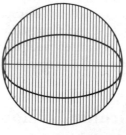

(b) Area $= \pi ab$

Figure P.6

From the time of Kepler, powerful methods were developed, primarily by Bonaventura Cavalieri (1598–1647), for calculating areas and volumes. Central to Cavalieri's methods is the notion that a planar area is composed of an infinite number of parallel chords, and that a solid is composed of an infinite number of parallel sections. By sliding these 'indivisibles' along their axes, new areas and volumes are formed that have the same area or volume as the originals. Although opposition to these new methods was widespread, numerous mathematicians used them and were able to produce results equivalent to the modern-day integration of expressions such as x^n, $\sin x$ and so on.

New tangent methods also appeared in the seventeenth century, considerably aided by the introduction into geometry of the developing language and symbolism of algebra. Foremost among the proponents of these techniques, which relied on the use of infinitesimals, was Pierre de Fermat (1601–1665). For instance, to determine a maximum or minimum value of an expression, suppose, using modern notation, that $f(x)$ attains an extreme value at x. If e is very small then $f(x - e)$ is approximately equal to $f(x)$, and so, tentatively, set $f(x - e)$ equal to $f(x)$, simplify and then set e equal to zero to determine those x for which $f(x)$ is a maximum or a minimum. As an illustration, let $f(x) = x^2 - 3x$, so that $f(x - e) = (x - e)^2 - 3(x - e)$. Then

$$x^2 - 3x = (x - e)^2 - 3(x - e)$$

leading to

$$0 = -2xe + e^2 + 3e$$

which simplifies to

$$2x = e + 3$$

Putting $e = 0$ gives $x = \frac{3}{2}$.

Fermat used similar techniques to find the tangent at a point on a curve whose equation is given in Cartesian coordinates. His methods are equivalent to the modern use of differential calculus, although his use of the infinitesimal e is clearly open to logical objections.

By the middle of the seventeenth century, quadrature and tangent methods had advanced and been applied to a wide variety of curves. Infinitesimals and indivisibles were accepted and the idea of a limit had been conceived. In addition, the problem of finding the length of a curve had been shown to be equivalent to finding the quadrature of a related curve, whose Cartesian equation involved tangents: if $y = f(x)$, the arc length s between two points (x_1, y_1) and (x_2, y_2) on the curve is given by

$$s = \int_{x_1}^{x_2} \sqrt{1 + \left(\frac{dy}{dx}\right)^2} \, dx$$

A link had been established between quadrature and tangents, and it was recognized that the problems were inversely related. What was missing was a satisfactory symbolism and a set of formal analytic rules for finding tangents and quadrature, and, more importantly, a consistent and rigorous redevelopment of the fundamental concepts involved. The former was soon to be furnished with the independent discoveries by Newton and Leibniz of the differential and integral calculus.

The discovery of calculus

Although the emergence of differential and integral calculus neither started nor finished at any particular time, it was during the second half of the seventeenth century, owing to the efforts of Isaac Newton (1642–1727) and Gottfried Wilhelm Leibniz (1646–1716), that a unified range of processes and an effective notation were invented that came to be known as the calculus. Both men owed much to their predecessors, and recent researches show that they had relatively little influence on each other during the periods of their crucial and independent inventions.

Newton learnt mathematics by studying books, including the works of Euclid, and by receiving instruction from some of the finest mathematicians of the period. By the end of 1664, he was at the frontiers of mathematical knowledge and began his own investigations. Most of his major discoveries were made in the period 1665–1666, while the plague raged in England; these included the general binomial theorem, the calculus, the law of gravitation and the nature of colours. The history of Newton's discovery of the general binomial series is a strange one; one that led him to develop his 'method of infinite series', an indispensable tool in his approach to problems of quadrature. At that time the quadrature of curves with equation $y = (1 - x^2)^n$ was known for positive integral values of n: for $n = 0$, 1, 2, and 3 these were x, $x - \frac{1}{3}x^3$, $x - \frac{2}{3}x^3 + \frac{1}{5}x^5$, $x - \frac{3}{3}x^3 + \frac{3}{5}x^5 - \frac{1}{7}x^7$ respectively. In order to find the quadrature of a (semi)circle with equation $y = (1 - x^2)^{1/2}$, and, more generally, the quadrature of curves with equation $y = (1 - x^2)^n$ when n was any fraction, positive or negative, Newton looked for patterns in the coefficients of x, x^3, x^5, x^7 and so on in the quadrature formulae then known. In so doing, he arrived at the infinite expression

$$x - \frac{\frac{1}{2}x^3}{3} - \frac{\frac{1}{8}x^5}{5} - \frac{\frac{1}{16}x^7}{7} + \ldots$$

for the quadrature of $y = (1 - x^2)^{1/2}$. He then noted that this could be obtained directly if

$$(1 - x^2)^{1/2} = 1 - \tfrac{1}{2}x^2 - \tfrac{1}{8}x^4 - \tfrac{1}{16}x^8 + \ldots$$

Similar series could be generated for $(1 - x^2)^n$ for any value of n. It soon became clear to Newton that he could operate with infinite series in much the same way that the algebraists of the day dealt with finite polynomial expressions. In fact he wrote in his work, *De Analysi per Aequationes numero terminorum infinitus* (completed in 1669 and published in 1711)

> And whatever the common Analysis [that is, algebra] performs by Means of Equations of a finite number of Terms ... this new method can always perform the same by Means of infinite Equations. So that I have not made any Question of giving this the Name of *Analysis* likewise.

It was from this time onwards that infinite processes became accepted as legitimate in mathematics. The *De Analysi* also contains the first account of Newton's principal mathematical discovery – the calculus.

Newton regarded a curve as generated by the motion of a point whose coordinates x and y were changing quantities that he called **fluents**. The rate at which these fluents change were called the **fluxions** and were denoted by $\dot{x}$ and $\dot{y}$ (dx/dt and dy/dt in modern notation, where t is time). If o denotes an infinitely small amount of time then, during that time, a fluent such as x changes by an infinitely small amount, namely $\dot{x}o$, which Newton called the **moment** of the fluent. The ratio $\dot{y}/\dot{x}$ gives the slope of the tangent to the curve at any point. For example, if $y^2 = x^3$ then

$$(y + \dot{y}o)^2 = (x + \dot{x}o)^3$$

Expanding both sides, removing common terms, dividing through by o, and disregarding terms still containing o gives

$$2y\dot{y} = 3x^2\dot{x}$$

Hence

$$\frac{\dot{y}}{\dot{x}} = \frac{3x^2}{2y} = \frac{3x^2}{2x^{3/2}} = \tfrac{3}{2}x^{1/2}$$

So, given a relation amongst fluents (that is, an equation relating variables), Newton could find a corresponding relation between fluxions. This is of course just differentiation. Conversely, given a relation amongst fluxions, the process of finding the original relation amongst the fluents corresponds to integration, the reverse process to differentiation. The *De Analysi* also contains, for the first time, the use of integration to solve problems of quadrature.

Newton twice rewrote his original account of calculus in an attempt to circumvent its dependence on the use of infinitesimals, and he very nearly hit upon the use of limits. However, the mysterious moments remained, and, although Newton's calculus was extensively applied, its theoretical foundations remained a subject of intense debate throughout the eighteenth century.

Leibniz' early interests in mathematics centred around a desire to generate a general symbolic language into which all processes of reasoning and argument could be translated. He was also fascinated by calculating aids and invented, in 1671, a machine that could carry out multiplication. After visiting London in 1673 and meeting several eminent English scientists, he began studying mathematics in earnest. His early investigations were concerned with the summation of infinite sequences whose terms could be expressed as differences. For example, to find the sum of the reciprocals of the triangular numbers 1, 3, 6, 10, ..., $\frac{1}{2}n(n+1)$, ..., Leibniz observed that

$$\frac{2}{n(n+1)} = \frac{2}{n} - \frac{2}{n+1}$$

Hence

$$\frac{1}{1} + \frac{1}{3} + \frac{1}{6} + \frac{1}{10} + \ldots + \frac{2}{n(n+1)}$$

$$= \left(\frac{2}{1} + \frac{2}{2}\right) + \left(\frac{2}{2} - \frac{2}{3}\right) + \left(\frac{2}{3} - \frac{2}{4}\right) + \left(\frac{2}{4} - \frac{2}{5}\right) + \ldots + \left(\frac{2}{n} - \frac{2}{n+1}\right)$$

$$= 2 - \frac{2}{n+1}$$

and so the sum to infinity is 2.

Leibniz gained great facility in summing infinite series, and came to appreciate that the processes of summing and differencing were inversely related. He observed that the determination of the tangent to a curve depended on the ratio of the differences in the coordinates of nearby points, and that quadratures depended on the sums of the heights of infinitely thin rectangles. Leibniz then saw the possibility of a **calculus** of infinitely small differences and sums of sequences of heights that could solve the problems of tangent and quadrature for a given curve. Taking differences would correspond to finding tangents, and summation of sequences would correspond to the determination of quadratures, and these operations would be inverse to each other. His initial work in this area centred on generalizing the work of previous mathematicians whereby curves whose quadrature was required were transformed into curves of identical quadrature whose quadrature was known. This 'transmutation' of areas involved the use of an infinitesimal triangle, a device that was to loom large in his development of differential calculus. Leibnizian calculus took shape in the period 1673–1676, although publication was delayed until 1684 since Leibniz was aware of the furore his indiscriminate use of infinitesimals would cause.

Leibniz regarded the **differential** dx of a variable x to be an infinitely small difference between two successive values of x. The ratio of the differentials dy and dx of two variables y and x was taken to be finite and to represent the slope of the tangent of the curve whose Cartesian equation related x and y. See Figure P.7(a). However, relative to finite quantities, differentials were negligible and were ignored (in other words, $x + dx = x$). Similarly, products of differentials were negligible relative to differentials (hence $a\,dx + dy\,dx = a\,dx$). If y and x are related by an equation then the ratio of dy to dx can be obtained by taking the differential of the equation. In so doing, one has to follow the **rules of calculus**, examples of which are $da = 0$ for a constant a, $d(u + v) = du + dv$ and $d(uv) = u\,dv + v\,du$. Leibniz derived these and other rules using the general properties of differentials enunciated above; for example, to find the differential of uv, proceed as follows:

$$d(uv) = (u + du)(v + dv) - uv$$
$$= u\,dv + v\,du + du\,dv$$
$$= u\,dv + v\,du$$

It is a credit to the succinct notation introduced by Leibniz that his calculus is so easy to apply. For instance, given the hyperbola with equation $xy - k = 0$, take differentials to obtain

$$d(xy - k) = d0 = 0$$

Apply the rules to get

$$0 = d(xy - k) = d(xy) - dk = x\,dy + y\,dx - 0$$

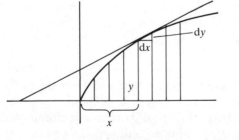

(a) The differential triangle

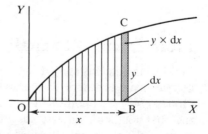

(b) Definite integration

Figure P.7

Hence

$$x\,dy + y\,dx = 0$$

leading to

$$\frac{dy}{dx} = \frac{-y}{x} = \frac{-k}{x^2}$$

as the formula for determining the slope of the tangent. Leibniz' notation for integration, $\int$, has also stood the test of time. From Figure P.7(b), the quadrature of the curve $y(x)$, written as $\int y\,dx$, is the sum of infinitely many, infinitesimally small rectangular areas $y\,dx$. Now the differential of the area OCB is the difference of two successive values of area, and this is none other than the area $y\,dx$ of the furthest right-hand rectangle. Hence $d(\int y\,dx) = y\,dx$, establishing the inverse relation of d and $\int$.

Leibniz was well aware that he had not precisely defined what differentials were, nor furnished any proof that they could be discarded relative to finite quantities. As did Newton, he made strenuous efforts to remedy these foundational flaws but failed to do so. The situation was not helped by the bitter controversy that surrounded the claims and counter-claims of the followers of Newton and Leibniz about who first discovered the calculus. The contributions made by both men were, and still are, impressive. Both had developed methods with wide applicability, and both, Leibniz more so than Newton, had developed notation and symbols by which their methods could be applied analytically – that is, by means of formulae instead of by verbal descriptions and geometrical arguments. Also they had established clearly the inverse linkage between differentiation and integration. Both calculi were to undergo fundamental changes before they came to resemble modern calculus; variables and relations between them were to be superseded by the function concept, moments and differentials were to be swept aside, and, through the concept of a limit, be replaced by derivatives of functions.

From calculus to analysis

Leibnizian calculus spread rapidly in Europe owing initially to the efforts of two Swiss brothers, Jaques Bernoulli (1654–1705) and Jean Bernoulli (1667–1748). A pupil of the latter, the Marquis de L'Hôpital, published the first textbook on calculus in 1696; this well-written text dominated most of the eighteenth-century developments in, and applications of, calculus. Leibniz, L'Hôpital and the Bernoullis took it as self-evident that two quantities differing by an infinitely small amount were equal and simultaneously that infinitely small quantitites were not zero: these infinitely small quantities were the differentials of Leibniz' calculus and the moments of Newton's calculus. This inevitably gave rise to foundational questions such as 'Do infinitely small quantities exist?' and 'Is the use of such quantities in calculus reliable?' One of the most famous attacks on the foundations of calculus came from Bishop George Berkeley, who in 1734 published a vitriolic tract aimed at 'infidel mathematicians'. In this tract infinitesimals were discredited and rejected as 'ghosts of departed quantities'. Despite all the criticisms, the second half of the eighteenth century witnessed an explosive growth in the use of infinite processes, justified largely on the grounds that they worked.

Calculus, initially a geometric field of study, developed into the subject we now call analysis owing to the epoch-making work of Leonard Euler (1707–1783). Euler's seminal work, *Introductio in analysin infinitorum* (1748) gives central place to the concept of a function. A function of a variable quantity is defined as an analytic expression, composed in whatever way, of that variable quantity and of numbers and of constant quantities. What was important about this concept of a function was that for the first time the idea of one quantity changing relative to another was established. This notion, allied with the concept of a limit, which emerged later in the eighteenth century, was to revolutionize the calculus and lead to the establishment of firm foundations for the subject. Euler contributed prolifically to virtually every branch of pure and applied mathematics, and is responsible for much of the notation in use today. His collected works run to 75 volumes and, although he warned against the risks involved in using infinite series, in the *Introductio* he uses infinite series in a remarkably unrestrained manner, and absurdities do arise. For example, the binomial series

$$(1 - x)^{-1} = 1 + x + x^2 + x^3 + \ldots$$

is used for values of $x \geq 1$, and the series

$$\frac{x}{1 - x} = x + x^2 + x^3 + \ldots$$

and

$$\frac{x}{x - 1} = 1 + \frac{1}{x} + \frac{1}{x^2} + \ldots$$

are combined to conclude (erroneously!) that

$$\ldots + \frac{1}{x^2} + \frac{1}{x} + 1 + x + x^2 + x^3 + \ldots = 0$$

One person who regarded Euler's infinite series manipulations with suspicion was Jean le Rond d'Alembert (1717–1783); he also objected to the use of infinitesimals in calculus. He was one of several mathematicians who responded to Berkeley's attack on calculus by advocating the use of limits. D'Alembert stressed that in differential calculus one was interested in the limiting value of ratios of finite differences of inter-related variable quantities rather than Leibniz' ratio of differentials. He regarded an infinitesimal as a non-zero quantity whose 'limiting value' was zero, and he also defined indefinitely large quantities in terms of limits. But his limit concept lacked clarity, and continental mathematicians continued to use the language and methods of Leibniz and Euler. Even so, the increasing number of absurd deductions that could be made from the indiscriminate use of infinite processes gave renewed impetus to efforts to resolve the foundational crisis in calculus. In England attempts to introduce rigour into Newton's calculus sought to tie the calculus more strongly to Euclid's geometry or to the physical notion of velocity. The first really serious effort to give rigour to calculus was published in 1797 by Joseph Louis Lagrange (1736–1813). His influential work, *Théorie des fonctions analytiques contenant les principes du calculi différentiel*, developed calculus as a 'theory of functions of a real variable' in which functions possessed derivatives as opposed to curves possessing tangents. Unfortunately, as was the case with the users of the new idea of functions, the work paid scant attention to matters of convergence and divergence. In particular, functions were considered to be 'well behaved' in that the variables involved were assumed to vary continuously and smoothly throughout their domains.

The successful merging of the idea of limit and function as the concepts required to provide a sound foundation for the calculus, and as the tools necessary to handle problems of convergence and divergence adequately, occurred in 1821. The publication in that year of the *Cours D'Analyse de l'École Polytechnique* by Augustin Louis Cauchy (1789–1857) brought about a fundamental change in attitudes in that rigour was applied to defining the basic concepts, and also to the manner in which proofs were presented. The Cauchy approach to calculus is essentially the modern approach in which differential calculus involves the definition, through a limit, of the derivative of a function, and integral calculus involved the limit of successive approximations to the area under the graph of a function. The fundamental theorem of calculus appears as a theorem in Cauchy's treatment. In addition, Cauchy uses his limit concept to define, for the first time, the idea of a continuous function as one whose values $f(x)$ vary continuously with x. Cauchy took great care in defining what he meant by a limit, and, although the use of limits allowed him to dispense with infinitesimals and geometric notions in developing the calculus, there still remained some residual geometric notions as regards the behaviour of variable quantities.

Conclusion

The nineteenth century was very much a Golden Age in mathematics, a century in which more was added to the sum of mathematical knowledge than was known at its beginning. The improved understanding of infinite processes was but one achievement in a period of intense activity that saw the introduction of new areas of mathematics such as non-Euclidean geometries, multidimensional spaces and non-commutative algebras. Analytic methods were applied to good effect in these areas, which in turn led to further developments in analysis itself. However, it became apparent that Cauchy had not reached the true foundations of analysis. In 1874, Karl Weierstrass (1815–1897) produced, counter to intuitive expectations, a function that was continuous but had no derivative anywhere (that is, a continuous curve with no tangent at any point). George Bernhard Riemann (1826–1866) produced a function discontinuous at infinitely many points and yet possessing a continuous integral. It was realized that the theory of limits upon which analysis had been built relied on simple intuitive notions concerning the real numbers. A programme was advocated to give rigour to the system of real numbers itself. This **arithmetization of analysis** was a difficult and intricate task ultimately completed by Weierstrass and his followers, thus completing a process begun in 1700 to separate analysis from its geometric basis. Today, as in the presentation of mathematical analysis given in this book, all of analysis can be logically derived from a set of postulates (or axioms) characterizing the real number system.

The process of introducing rigour into first calculus and then analysis was a process mirrored in all branches of mathematics. This careful attention to detail as regards basic concepts has in turn led to intricate generalizations of concepts such as space, dimension, convergence and integrability; topics covered in more advanced texts on analysis. One of the striking features of twentieth-century mathematics has been the move to greater generalization and abstraction, and this, in turn, has thrown up the deeper foundational problems that occupy today's generation of research mathematicians.

The reader interested in the history of mathematics, whose appetite has been whetted by this brief survey of the history of calculus and the early development of analysis, is recommended to dip into the following texts:

Boyer and Merzbach (1989). *A History of Mathematics* 2nd edn. New York: Wiley

Eves H. (1987). *An Introduction to the History of Mathematics* 5th edn. New York: Holt Rinehart Winston

These survey the historical development of mathematics from ancient times up to the twentieth century, and both contain extensive bibliographies directing the reader to more detailed expositions of particular topics.

1

Preliminaries

1.1 Logic
1.2 Sets
1.3 Functions

The main substance of analysis begins in Chapter 2. The purpose of the present chapter is to give a brief survey of prerequisite material, which the reader can safely skim through and refer back to as necessary.

In Section 1.1 there is a relatively formal treatment of logic, intended to alert the reader to the basic logical ideas required in the remainder of this book. In mathematics one is concerned with statements about the mathematical objects being studied. Examples of mathematical objects are whole numbers, real numbers, sets, functions, and so on, and examples of mathematical statements are as follows:

(i)   There are infinitely many prime numbers.

(ii) The equation $x^2 + 1$ has a real root.

(iii) If n is a positive whole number and n^2 is odd then n is odd.

Interest in such statements is focused on whether they are true or false; the truth or falsity of a given statement is demonstrated by a proof. Statements (i) and (iii) above are true, whereas (ii) is false. The latter is easily proved by noting that the square of a real number cannot be negative. Statement (i) can be proved as follows: since there are prime numbers such as 2, 3, 5 and so on, list, in increasing order, the first n primes $p_1, p_2, p_3, \ldots, p_n$. Now consider the number

$$p = (p_1 p_2 p_3 \cdots p_n) + 1$$

Either p is a prime, or there exists a prime $q < p$ such that q is a divisor of p. In the latter case, since p divided by any of p_1, p_2, p_3, ..., p_n leaves a remainder of 1, q must be a prime number greater than p_n. Hence there is always a prime number greater than p_n, whatever the size of n. It follows that there are infinitely many primes. Various alternative proofs of statement (iii) are given later. Clearly the proof (either of the truth or of the falsity) of a mathematical statement needs to be a precise and carefully reasoned argument based on a collection of basic mathematical truths concerning the objects being studied. This creates a difficulty in that, although the mathematical objects used may have been carefully defined, the English language used in mathematical statements also needs to be precise; but the use of the English language in real life is often ambiguous. However, as statement (iii) above demonstrates, it is only necessary to make precise what is meant by a few key words, such as **if, and** and **then**.

The aim then of Section 1.1 is to make precise the meaning of various key words, or connectives, a task achieved by the use of truth tables. In addition, truth tables are used to check whether two statements built up from the same collection of basic statements, but using different connectives, are logically equivalent or not. Practice is provided in working with the statements of the form 'if P then Q', which can be interpreted as saying that if (the statement) P is true then (the statement) Q is also true; when P is false, no conclusion may be drawn. A great deal of mathematicians' effort is devoted to proving the truth or falsity of statements of this type.

In Section 1.2 the concept of a set is introduced – a concept that pervades the whole of present-day mathematics. Any well-defined collection of objects is a set, and in Section 1.2 various ways of combining sets to produce new sets are discussed. A link is established between the equality of different combinations of sets and the logical equivalence of related logical statements. Readers familiar with the language of sets need only browse through this section to familiarize themselves with the notation.

Analysis is, in part, concerned with the study of functions from a very precise point of view, in order to establish, in a logically sound manner, the properties of such functions. The concept of a function is carefully discussed in Section 1.3, where numerous methods of combining functions to produce new functions are described. The history of the term **function** provides a good example of the mathematician's tendency to abstract and generalize concepts. The actual word first appears, in Latin, in 1694 in Leibniz' work on calculus, where it denotes any quantity connected with a curve, such as its slope or the coordinates of a point on the curve. By 1718 a function was regarded as any expression made up of a variable and some constants. This evolved into the idea of any equation involving variables and constants, and in Euler's *Introductio in analysin infinitorum* in 1748 the functional notation $f(x)$ first appears. By the early nineteenth century, a more general notion of function emerged, and the following formulation is due to Lejeune Dirichlet (1805–1859):

> A **variable** is a symbol that represents any one of a set of numbers; if two variables x and y are so related that whenever a value is assigned to x there is automatically assigned, by some rule or correspondence, a value to y, then we say that y is a (single-valued) **function** of x.

This definition is a very broad one, in that it does not imply that there is necessarily any way of expressing the relationship between x and y by some kind of analytic expression, only that a function sets up a particular sort of relationship between two sets of numbers. The notion of a function plays a central and unifying role in contemporary mathematics, and, as the reader will see, is of particular relevance in analysis.

1.1 Logic

A sentence that can be labelled T (for **true**) or F (for **false**) is called a **statement**. Several statements can be combined to produce a larger composite statement whose truth value depends not only on the truth values of the constituent statements but also on the manner in which they are connected. This means that precise meanings have to be attached to those words that link, or join, simpler statements together. Such words are called **connectives**, examples of which are 'and', 'or' and 'if ... then'. Shortly, precise meanings will be given to these connectives, but first consideration is given to the process of negating a given statement.

If P denotes a statement, its negation (not P) is obtained by inserting the word 'not' in the appropriate place. For example, if P denotes the (false) statement 'Glasgow is the capital of Scotland', (not P) denotes the statement 'Glasgow is not the capital of Scotland'. The statement (not P) has the opposite truth value to P. This fact can be summarized neatly in the **truth table** in Figure 1.1.

If P and Q are two statements, the composite statement (P and Q) is deemed to be true provided both P and Q are true, and false otherwise.

P	(not P)
T	F
F	T

Figure 1.1

The composite statement $(P \text{ or } Q)$ is true provided *at least* one of P and Q is true, and false otherwise. This information appears in the truth table in Figure 1.2, where four rows are required since there are four possible combinations of truth values when two simple statements are involved.

Notice that the connective 'and' is used in the ordinary English language sense but that the connective 'or' is used in the inclusive sense. In other words, $(P \text{ or } Q)$ is true when either P is true, or Q is true, or both P and Q are true.

It is possible to combine a collection of statements in different ways and end up with composite statements which have the same truth value for any specified truth values of the constituent statements. In this situation the composite statements are said to be **logically equivalent**.

■■ EXAMPLE 1

Show that the composite statements $R = (\text{not} \, (P \text{ and } (\text{not} \, Q)))$ and $S = ((\text{not} \, P) \text{ or } Q)$ are logically equivalent.

Solution

The truth tables for R and S appear in Figure 1.3. Notice the use of intermediary columns in building up the statements R and S from the statements P and Q. Inspection of these tables shows that the truth values of R and S, as given in the last column of each table, are identical. Hence R and S are logically equivalent. ■

Suppose that P and Q are two statements. Then the statement $(P \Rightarrow Q)$ is defined by the truth table in Figure 1.4. There are numerous English language equivalents of the statement $(P \Rightarrow Q)$, such as 'P implies Q', 'if P then Q' and 'P is a sufficient condition for Q'. Whichever interpretation is used, the final two lines in the truth table in Figure 1.4 often cause confusion. They allow sentences such as '$x > 3 \Rightarrow x > 0$' to be counted as true for all values of x. In practice, the mathematician is

P	Q	$(P \text{ and } Q)$	$(P \text{ or } Q)$
T	T	T	T
T	F	F	T
F	T	F	T
F	F	F	F

Figure 1.2

P	Q	$(\text{not } Q)$	$(P \text{ and } (\text{not } Q))$	R
T	T	F	F	T
T	F	T	T	F
F	T	F	F	T
F	F	T	F	T

P	Q	$(\text{not } P)$	S
T	T	F	T
T	F	F	F
F	T	T	T
F	F	T	T

Figure 1.3

P	Q	$(P \Rightarrow Q)$
T	T	T
T	F	F
F	T	T
F	F	T

Figure 1.4

interested in situations where $(P \Rightarrow Q)$ is true. From the truth table, this means that

(i) if P is true then Q must be true (lines 1 and 2);

(ii) if P is false then Q may be either true or false, and so no conclusion can be drawn in this case (lines 3 and 4).

■■   EXAMPLE 2

Show that $(P \Rightarrow Q)$ is logically equivalent to $((\text{not } Q) \Rightarrow (\text{not } P))$.

Solution

The truth table in Figure 1.5 demonstrates the required logical equivalence. The statement $((\text{not } Q) \Rightarrow (\text{not } P))$ is called the **contrapositive** of $(P \Rightarrow Q)$.   ■

P	Q	(not P)	(not Q)	$(P \Rightarrow Q)$	$((\text{not } Q) \Rightarrow (\text{not } P))$
T	T	F	F	T	T
T	F	F	T	F	F
F	T	T	F	T	T
F	F	T	T	T	T

Figure 1.5

The statement $(P \Rightarrow Q)$ does not mean that the statements P and Q are in any sense equal or equivalent; indeed, the statement $(Q \Rightarrow P)$ is a different statement, called the **converse** of $(P \Rightarrow Q)$. However, the composite statement $((P \Rightarrow Q) \text{ and } (Q \Rightarrow P))$, abbreviated as $(P \Leftrightarrow Q)$, does embody the notion that P and Q are logically equivalent. The statement $(P \Leftrightarrow Q)$ is true either when P and Q are both true, or when they are both false. See Figure 1.6. The statement $(P \Leftrightarrow Q)$ can be read as 'P (is true) if and only if Q (is true)', or as 'P is a necessary and sufficient condition for Q'.

Many theorems in mathematics take the form $(P \Rightarrow Q)$. To show that P implies Q, one usually adopts one of the following schemas:

(1) The first method, which is a **direct** method of proof, assumes that P is true and endeavours, by some process, to deduce that Q is true. Since $(P \Rightarrow Q)$ is true whenever P is false, there is no need to consider the case where P is false.

(2) The second method is **indirect**. First write down the contrapositive, $((\text{not } Q) \Rightarrow (\text{not } P))$, and try to prove this equivalent statement directly. In other words, assume that Q is false (that is, (not Q) is true) and deduce that P is false (that is, (not P) is true).

(3) The third commonly used method is to employ a **proof by contradiction** (also known as *reductio ad absurdum*). For this argument, assume that P is true and Q is false (that is, (not Q) is true) and deduce an obviously false statement. This shows that the original hypothesis $(P \text{ and } (\text{not } Q))$ must be false. In other words, the statement (not

P	Q	$(P \Rightarrow Q)$	$(Q \Rightarrow P)$	$(P \Leftrightarrow Q)$
T	T	T	T	T
T	F	F	T	F
F	T	T	F	F
F	F	T	T	T

Figure 1.6

(P and (not Q))) is true. But this is logically equivalent to $(P \Rightarrow Q)$. See Question 1(b) of Exercises 1.1.

These three methods of proof are illustrated in the next example, where the unique factorization of a positive whole number into a product of its prime factors is assumed.

■■   EXAMPLE 3

Assume that n is a positive whole number. Prove that n^2 odd $\Rightarrow n$ odd.

Solution

1   The direct method of proof

Write $n = p_1 p_2 \cdots p_r$, where $p_1, p_2, \ldots, p_r$ are the (not necessarily distinct) prime factors of n (for instance, if $n = 12$ then $n = 2 \cdot 2 \cdot 3$). Then $n^2 = p_1^2 p_2^2 \cdots p_r^2$. Assume that n^2 is odd. Then none of the p_i equals 2. Hence 2 is not a prime factor of n, and therefore n is also odd.

2   The indirect method of proof

The contrapositive of the given statement is n even $\Rightarrow n^2$ even. Assume that n is even and write $n = 2m$, where m is another positive whole number. Then $n^2 = (2m)^2 = 4m^2 = 2(2m^2)$; another even number. Therefore n^2 is even, as required.

3   Proof by contradiction

Assume that n^2 is odd and that n is even. Then $n^2 + n$ is the sum of an odd and an even number. Hence $n^2 + n$ is odd. But $n^2 + n = n(n + 1)$ is a product of consecutive positive whole numbers, one of which must be even. Hence $n^2 + n$ is also an even number. This conclusion that $n^2 + n$ is both even and odd is the desired contradiction. Therefore if n^2 is odd then n must necessarily be odd.   ■

It is easy to construct a direct proof that the converse of the result in Example 3 is also true, and hence for positive whole numbers n, n^2 odd $\Leftrightarrow n$ odd. In general, to prove theorems of the form $(P \Leftrightarrow Q)$ requires that both $(P \Rightarrow Q)$ and $(Q \Rightarrow P)$ be established. See, for instance, Example 1 in Section 1.2.

Exercises 1.1

1. Use truth tables to determine which of the following pairs of composite statements are logically equivalent.

    (a) $(\text{not}\,(P\,\text{and}\,(\text{not}\,P)))$; $(P\,\text{or}\,(\text{not}\,P))$
    (b) $(P\Rightarrow Q)$; $(\text{not}\,(P\,\text{and}\,(\text{not}\,Q)))$
    (c) $((P\Rightarrow Q)\,\text{and}\,R)$; $(P\Rightarrow(Q\,\text{and}\,R))$

2. What conclusion, if any, can be drawn from

    (a) the truth of $((\text{not}\,P)\Rightarrow P)$
    (b) the truth of P and the truth of $(P\Rightarrow Q)$
    (c) the truth of Q and the truth of $(P\Rightarrow Q)$
    (d) the truth of $(\text{not}\,Q)$ and the truth of $(P\Rightarrow Q)$

3. A **tautology** is a statement that is true no matter what the truth values of its constituent statements are. Decide which of the following are tautologies.

    (a) $(P\,\text{or}\,(\text{not}\,P))$
    (b) $(P\,\text{and}\,(\text{not}\,P))$
    (c) $(P\Rightarrow(\text{not}\,P))$
    (d) $(((P\Rightarrow Q)\,\text{or}\,(Q\Rightarrow P))\,\text{and}\,(\text{not}\,Q))$

4. Let n be a positive whole number. Which of the following conditions imply that the n is divisible by 6?

    (a) n is divisible by 3
    (b) n is divisible by 9
    (c) n is divisible by 12
    (d) n^2 is divisible by 12
    (e) $n = 24$
    (f) n is even and divisible by 3
    (g) $n = m^3 - m$ for some positive whole number m

    Which of (a)–(g) are logically equivalent to the statement 'n is divisible by 6'?

5. Let n be a positive whole number. Find three different proofs (as illustrated in Example 3) of the fact that n^2 even $\Rightarrow n$ even.

6.   Let m and n be positive whole numbers. Prove that mn^2 even $\Rightarrow$ at least one of m and n is even.

1.2   Sets

The statements of many theorems often involve variables drawn from some collection of objects. The technical term for a collection of objects is a **set**, and the objects in a set are called its **elements**. Sets are usually denoted by upper-case letters; if x belongs to some set S, this is written as $x \in S$, and if x does not belong to S, this is written as $x \notin S$. If, in a given context, *all* the objects under consideration lie in a particular set, this set is called the **universal** set for the problem; the universal set is denoted by $\mathcal{U}$.

A sentence involving a variable x cannot be counted as a statement, since it may be true for some values of x and false for others. If a sentence containing a variable x is true for certain values of $x \in \mathcal{U}$ and false for all others (in $\mathcal{U}$), the proposition is called a **predicate**. Much of the discussion of the logic of statements in Section 1.1 can be extended to cope with predicates in place of statements.

■■   EXAMPLE 1

Let the universal set $\mathcal{U}$ be the set of real numbers (these are discussed in detail in Chapter 2). The sentence $P(x)$ given by '$2x^2 = x$' is a predicate since, for any value of x, either $2x^2 = x$ or else $2x^2 \neq x$. The sentence $Q(x)$ given by '$x = 0$ or $x = 0.5$' is also a predicate. It is straightforward now to construct a proof that $P(x) \Leftrightarrow Q(x)$. First,

$$2x^2 = x \Rightarrow 2x^2 - x = 0$$
$$\Rightarrow x(2x - 1) = 0$$
$$\Rightarrow x = 0 \text{ or } 2x - 1 = 0$$
$$\Rightarrow x = 0 \text{ or } x = 0.5$$

Conversely,

$$x = 0 \text{ or } x = 0.5 \Rightarrow x = 0 \text{ or } 2x - 1 = 0$$
$$\Rightarrow x(2x - 1) = 0$$
$$\Rightarrow 2x^2 - x = 0$$
$$\Rightarrow 2x^2 = x$$ ■

Predicates are useful for specifying sets. The simplest method of specifying a set is to list its elements between braces. So, for example, $A = \{0, 0.5\}$ denotes the set A containing the numbers 0 and 0.5 and no others. This device is of course impracticable for sets containing an infinite number of elements. This is where predicates can be used to determine which elements of the universal set in question lie in the desired set, and which do not. For example,

$$B = \{x : 1 < x \leqslant 2\}$$

where x is a real number, uses the predicate $P(x)$ given by '$1 < x \leqslant 2$' to select those real numbers greater than 1 and less than or equal to 2 for membership of B. It can occur that the predicate used is universally false, as with the set $C = \{x : x^2 + 1 = 0\}$. Since there are no real numbers for which $x^2 = -1$, the set C consists of no elements at all. It is called the **empty set**, and is conventionally denoted by the symbol $\varnothing$.

Certain sets of numbers are denoted by special symbols; these sets are discussed in more detail in Chapter 2. The standard symbols denoting them are as follows:

$\mathbb{N} = \{n : n \text{ is a positive whole number}\}$, the set of **natural numbers**

$\mathbb{Z} = \{n : n \text{ is a whole number}\}$, the set of **integers**

$\mathbb{Q} = \left\{ \dfrac{m}{n} : m \text{ and } n \text{ are integers and } n \neq 0 \right\}$, the set of **rationals**

$\mathbb{R} = \{x : x \text{ is a real number}\}$, the set of **real numbers**

Hence $\mathbb{N} = \{1, 2, 3, \ldots\}$, $\mathbb{Z} = \{\ldots, -2, -1, 0, 1, 2, \ldots\}$, $\mathbb{Q}$ contains all fractions and $\mathbb{R}$ contains all decimals.

A **subset** of a set is a set all of whose elements belong to the original set. Thus A is a subset of B, written as $A \subseteq B$, if and only if $x \in A \Rightarrow x \in B$. If, as this definition allows, B contains elements that do not belong to A then A is called a **proper** subset of B. This is denoted by $A \subset B$. For the number sets listed above, $\mathbb{N} \subset \mathbb{Z} \subset \mathbb{Q} \subset \mathbb{R}$. To see that the subsets are proper, note that $-1 \in \mathbb{Z}$ but $-1 \notin \mathbb{N}$, $\frac{1}{2} \in \mathbb{Q}$ but $\frac{1}{2} \notin \mathbb{Z}$, and $\sqrt{2} \in \mathbb{R}$ but $\sqrt{2} \notin \mathbb{Q}$ (for a proof of this latter statement see 2.1.1). Sets can be manipulated using standard operations. For sets A and B the **union, intersection** and **difference**, denoted respectively by $A \cup B$, $A \cap B$ and $A - B$, are defined by

$$A \cup B = \{x : x \in A \text{ or } x \in B\}$$
$$A \cap B = \{x : x \in A \text{ and } x \in B\}$$
$$A - B = \{x : x \in A \text{ and } x \notin B\}$$

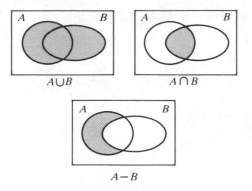

Figure 1.7

These new sets are illustrated by the Venn diagrams in Figure 1.7, where the bounding rectangle denotes the universal set $\mathcal{U}$, of which A and B are subsets. In each case the shaded area indicates the new set formed from A and B by the relevant set operation. If the universal set $\mathcal{U}$ is clear from the context, the difference $\mathcal{U} - A$ is called the **complement** of A and is alternatively denoted by $\mathcal{C}A$ or A^c.

As with composite logical statements, seemingly different combinations of sets can produce the same final set. Two sets A and B are **equal**, written as $A = B$, if they contain the same elements. Equivalently, $A = B$ if and only if $A \subseteq B$ and $B \subseteq A$.

■■   **EXAMPLE 2**

Prove that for any sets A and B

$$\mathcal{C}A \cup B = \mathcal{C}(A \cap \mathcal{C}B)$$

Solution

$$\mathcal{C}A \cup B = \{x : x \in \mathcal{C}A \text{ or } x \in B\}$$
$$= \{x : (\text{not } (x \in A)) \text{ or } x \in B\}$$
$$= \{x : \text{not} ((x \in A) \text{ and } (\text{not } (x \in B)))\}$$

using the logical equivalence given in Example 1 of Section 1.1, where P and Q are replaced by the predicates $x \in A$ and $x \in B$ respectively. Hence $\mathcal{C}A \cup B = \mathcal{C}(A \cap \mathcal{C}B)$.   ■

In order to facilitate the simplification of set-theoretic expressions, the following laws can be established. Each law can be established in a similar manner to Example 2 above.

1.2.1 Laws of the algebra of sets

Associative laws

$$A \cup (B \cup C) = (A \cup B) \cup C \qquad A \cap (B \cap C) = (A \cap B) \cap C$$

Commutative laws

$$A \cup B = B \cup A \qquad A \cap B = B \cap A$$

Identity laws

$$A \cup \varnothing = A \qquad A \cap \mathcal{U} = A$$
$$A \cup \mathcal{U} = \mathcal{U} \qquad A \cap \varnothing = \varnothing$$

Idempotent laws

$$A \cup A = A \qquad A \cap A = A$$

Distributive laws

$$A \cap (B \cup C) = (A \cap B) \cup (A \cap C) \qquad A \cup (B \cap C) = (A \cup B) \cap (A \cup C)$$

Complement laws

$$A \cup \mathcal{C}A = \mathcal{U} \qquad A \cap \mathcal{C}A = \varnothing$$
$$\mathcal{C}\mathcal{U} = \varnothing \qquad \mathcal{C}\varnothing = \mathcal{U}$$
$$\mathcal{C}(\mathcal{C}A) = A \qquad \mathcal{C}(\mathcal{C}A) = A$$

De Morgan's laws

$$\mathcal{C}(A \cup B) = \mathcal{C}A \cap \mathcal{C}B \qquad \mathcal{C}(A \cap B) = \mathcal{C}A \cup \mathcal{C}B$$

The laws in the right-hand column are called the **duals** of those in the left-hand column. As a consequence, every identity involving sets, deducable from 1.2.1 yields another **dual identity** if we interchange the symbols $\cap$ and $\cup$ and the symbols $\varnothing$ and $\mathcal{U}$. The third complement law is self-dual, and so the same statement has been written twice.

The **Cartesian product** of two sets A and B is defined by

$$A \times B = \{(x, y) : x \in A \text{ and } y \in B\}$$

The Cartesian product of the set $\mathbb{R}$ of real numbers with itself contains all ordered pairs of real numbers. The set $\mathbb{R} \times \mathbb{R}$ is usually denoted by $\mathbb{R}^2$ and is called the **Cartesian plane**. Diagrammatically, $\mathbb{R}^2$ is represented by the familiar picture in Figure 1.8.

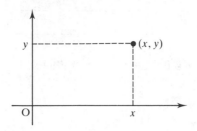

Figure 1.8

A predicate may be converted into a statement by substituting particular values for the variables occurring in the predicate. So, for example, the predicate '$x < 1$' yields a true statement for $x = 0$, but a false one for $x = 2$. Another valuable way of producing statements from predicates, which will occur frequently in this text, is to use the **quantifiers** 'for any' and 'there exists'. Although not adopted in the sequel, there are abbreviations for these quantifiers; $\forall$ stands for 'for any', and $\exists$ stands for 'there exists'. The symbol $\forall$ may be expressed verbally in several alternative ways, such as 'for all' or 'for every'; the reader should be aware that in subsequent definitions the symbols $\forall$ and $\exists$ are suppressed by using suitable English language equivalents.

■■   EXAMPLE 3

Determine which of the following statements are true and which are false. The variables x and y are assumed to be drawn from the set of real numbers.

(a) $\forall x\,(x^3 > x)$

(b) $\forall x\,(\cos^2 x + \sin^2 x = 1)$

(c) $\exists x\,(\exists y\,(x + y = y))$

Solution

(a) This is false, as can be seen by substituting $x = \frac{1}{2}$. This value of x provides a **counterexample** to the claim that $x^3 > x$ for all possible real values of x.

(b) This is true, and is proved in Example 3 of Section 4.3.

(c) This is true, since all that is required is to produce a value of x and a value of y for which $x + y = y$. Hence, for example, $x = 0$ and $y = 2$ will suffice.   ■

Exercises 1.2

---

1. Indicate on a diagram the following subsets of $\mathbb{R}^2$:

   $$A = \{(x, y) : x^2 + y^2 \leqslant 2\}$$
   $$B = \{(x, y) : x = 1\}$$
   $$C = \{(x, y) : y < 1\}$$

   Hence sketch the sets $A \cap B$, $A \cap C$ and $\mathscr{C}A \cup C$

2. Prove the following laws of the algebra of sets:

   (a) $\mathscr{C}(\mathscr{C}A) = A$
   (b) $A \cap (B \cup C) = (A \cap B) \cup (A \cap C)$
   (c) $\mathscr{C}(A \cap B) = \mathscr{C}A \cup \mathscr{C}B$

3. The **symmetric sum**, $A \oplus B$, of sets A and B is defined by

   $$A \oplus B = (A \cup B) \cap \mathscr{C}(A \cap B)$$

   Use the algebra of sets to establish the following:

   (a) $A \oplus B = B \oplus A$
   (b) $A \oplus A = \varnothing$
   (c) $A \cap (B \oplus C) = (A \cap B) \oplus (A \cap C)$

   (For (c), first establish that $X \oplus Y = (\mathscr{C}X \cap Y) \cup (X \cap \mathscr{C}Y)$.)

4. Determine which of the following statements about whole numbers are true and which are false.

   (a) $\forall n \, (2n^2 - 8n + 7 > 0)$
   (b) $\exists n \, (11n - n^2 > 29)$
   (c) $\forall n \, (n^2 - n \text{ is even})$
   (d) $\exists n \, (n^3 - n \text{ is odd})$

1.3 Functions

Let A and B be sets. A **function** f from A to B may be regarded as a rule that associates with each element of A a unique element of B. This

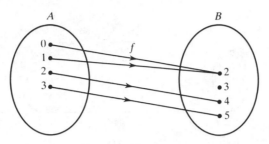

Figure 1.9

definition begs the question 'what is a rule?' and indeed it is possible to give a more precise definition of a function in terms of subsets of $A \times B$. However, for the purposes of this text, the informal specification above will suffice. The set A is called the **domain** of f and the set B is called the **codomain** of f. It is customary to write $f: A \to B$ for a function with domain A and codomain B, thus emphasizing the fact that f maps (or transforms) elements of A to uniquely determined elements of B. For each $x \in A$ the unique value in B corresponding to x is called the **image of x under** f, and is written as $f(x)$. For example, the process of squaring real numbers may be described by the function $f: \mathbb{R} \to \mathbb{R}$ given by $f(x) = x^2$. One of the consequences of a course in mathematical analysis is that it is then possible to give precise definitions of the rule(s) that determine the values of certain elementary functions. If, in a particular context, the domain of a function is understood, the shorthand notation $x \mapsto f(x)$ may be employed, provided that care is exercised not to confuse the function f with its value $f(x)$ at x.

In simple cases a schematic diagram may be drawn to represent a particular function. The diagram in Figure 1.9 represents the function $f: A \to B$ where $A = \{0, 1, 2, 3\}$, $B = \{2, 3, 4, 5\}$ and $f(0) = f(1) = 2$, $f(2) = 4$ and $f(3) = 5$.

For a function $f: A \to B$ the set of images of the elements of the domain need not give the whole of the codomain B. The set of images of elements of A is called the **image** of f, and is denoted by $f(A)$. Formally,

$$f(A) = \{f(x) : x \in A\}$$

If C is a subset of A then the image of C under f is the set $f(C) = \{f(x) : x \in C\}$. See Figure 1.10.

■■   EXAMPLE 1

Suppose that $f: \mathbb{R} \to \mathbb{R}$ is given by $f(x) = 1/(x^2 + 1)$. Show that the image of f is the set $\{x : x \in \mathbb{R} \text{ and } 0 < x \leq 1\}$.

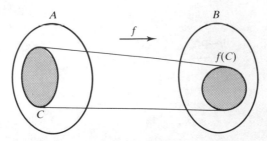

Figure 1.10

Solution

The image of f is

$$f(\mathbb{R}) = \left\{ \frac{1}{x^2 + 1} : x \in \mathbb{R} \right\}$$

Now, for all $x \in \mathbb{R}$, $x^2 \geqslant 0$, and so $x^2 + 1 \geqslant 1$. But then, $0 < 1/(x^2 + 1) \leqslant 1$ for all $x \in \mathbb{R}$. Therefore

$$f(\mathbb{R}) \subseteq \{x : x \in \mathbb{R} \text{ and } 0 < x \leqslant 1\}$$

Conversely, if $0 < y \leqslant 1$, the equation $y = 1/(x^2 + 1)$ admits two solutions, namely

$$x = \pm \sqrt{\frac{1 - y}{y}}$$

In other words, there is a real value of x with $f(x) = y$. Hence

$$\{x : x \in \mathbb{R} \text{ and } 0 < x \leqslant 1\} \subseteq f(\mathbb{R})$$

Therefore the image of f is the set $\{x : x \in \mathbb{R} \text{ and } 0 < x \leqslant 1\}$. ■

The set $\{x : x \in \mathbb{R} \text{ and } 0 < x \leqslant 1\}$ is an example of an **interval** of real numbers. Given real numbers a and b, with $a \leqslant b$, the **closed interval** $[a, b]$ denotes the set $\{x : x \in \mathbb{R} \text{ and } a \leqslant x \leqslant b\}$. If one of the inequalities in the defining predicate is strict (and so one or other of a or b is omitted), the relevant square bracket is replaced by a round bracket. For example the set $f(\mathbb{R})$ in Example 1 may be written as $(0, 1]$. When both endpoints of an interval are omitted, the **open interval** (a, b) denotes the set $\{x : x \in \mathbb{R} \text{ and } a < x < b\}$.

This book is concerned with functions $f: A \to \mathbb{R}$ where the domain A is a subset of $\mathbb{R}$. Two functions $f: A \to \mathbb{R}$ and $g: A \to \mathbb{R}$ are deemed to be equal if they have the same domain A and if $f(x) = g(x)$ for all x in their common domain.

If $f: A \rightarrow \mathbb{R}$ and $g: A \rightarrow \mathbb{R}$ are two functions, it is possible to form the **sum** function, $f + g$, and the **product** function, fg; obtained respectively by adding and multiplying function values. Hence $(f + g)(x) = f(x) + g(x)$ and $(fg)(x) = f(x)g(x)$ for all $x \in A$. Similarly, $f - g$ may be formed, where $(f - g)(x) = f(x) - g(x)$ for all $x \in A$. Also, if $\lambda \in \mathbb{R}$, the function λf may be formed, where $(\lambda f)(x) = \lambda f(x)$ for all $x \in A$. It now follows that $f - g$ is the same function as $f + (-1)g$. Finally, if $g: A \rightarrow \mathbb{R}$ is a function such that $g(x)$ is never zero, the **reciprocal** function $1/g$ may be formed by defining $(1/g)(x) = 1/g(x)$ for all $x \in A$. The product of this reciprocal function with any other function $f: A \rightarrow \mathbb{R}$ produces the quotient f/g where $(f/g)(x) = f(x)/g(x)$ for all $x \in A$. More subtle ways exist of combining functions.

Suppose that $f: A \rightarrow B$ and $g: B \rightarrow C$ are functions such that the codomain of f coincides with the domain of g. For any $x \in A$, $f(x)$ lies in the domain of g, and so $g(f(x))$ may be calculated. In fact, $g(f(x))$ can be calculated under the weaker condition that the image of f lies in the domain of g. In either case the formula $g(f(x))$ is used to define the **composite** function $g \circ f: A \rightarrow C$ by the rule $x \mapsto g(f(x))$. See Figure 1.11. Composition of functions is an extremely useful way of constructing new functions from more basic ones. Properties of functions that are preserved under sums, products, quotients, composites and so on are a key feature of courses in mathematical analysis.

■■   **EXAMPLE 2**

Find the composites $g \circ f$ and $f \circ g$ for the functions $f: \mathbb{R} \rightarrow \mathbb{R}$ and $g: \mathbb{R} \rightarrow \mathbb{R}$ given by $f(x) = x^3$ and $g(x) = x + 2$.

Solution

Since f and g have domains and codomains equal to $\mathbb{R}$, both of the composites $g \circ f$ and $f \circ g$ are defined. The composite $g \circ f: \mathbb{R} \rightarrow \mathbb{R}$ is given by $(g \circ f)(x) = g(f(x)) = x^3 + 2$, and the composite

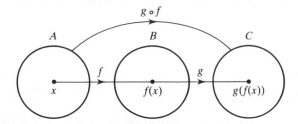

Figure 1.11

$f \circ g: \mathbb{R} \to \mathbb{R}$ is given by $(f \circ g)(x) = f(g(x)) = (x+2)^3$. Notice that $g \circ f \neq f \circ g$. ∎

The last way of producing new functions from old functions is now considered; the basic idea is one of *reversing* the effect of the original function, and it only applies to functions with particular properties. Consider the function $f: \mathbb{R} \to \mathbb{R}$ given by $f(x) = 3x + 2$. To compute a value of this simple function, take $x \in \mathbb{R}$, multiply it by three and then add two. To reverse this process, first subtract two and then divide by three. See Figure 1.12.

The reason that f can be reversed is that for *any* $y \in \mathbb{R}$ the equation $y = 3x + 2$ admits a solution for x, and this solution is unique; in fact $x = \frac{1}{3}(y-2)$. It is thus possible to define an **inverse function**, denoted by $f^{-1}: \mathbb{R} \to \mathbb{R}$, by the rule $x \mapsto \frac{1}{3}(x-2)$. Notice the use of x as the domain variable for both f and f^{-1}. The reader should be careful not to confuse the inverse function f^{-1} with the reciprocal function $(1/f)$, which in this case is not defined! In general, inverse functions do not exist; the general criterion for an inverse function to exist is now developed.

A function $f: A \to B$ is **injective** if distinct elements of A have distinct images in B. In other words, $x_1 \neq x_2 \Rightarrow f(x_1) \neq f(x_2)$. The contrapositive of this condition is more useful in practice; namely, f is injective when $f(x_1) = f(x_2) \Rightarrow x_1 = x_2$. A function $f: A \to B$ is **surjective** if every element of the codomain B lies in the image of f (that is, $f(A) = B$). If f is both injective and surjective, it is called a **bijective** function. A bijective function $f: A \to B$ is thus characterized by the property that for *every* element y in the codomain of f there exists a *unique* element x in the domain of f such that $f(x) = y$. This is precisely what is required for one to be able to define the inverse function, since the effect of f can be reversed simply by sending each element of B back to the unique element of A from whence it came. This **inverse function** is denoted by $f^{-1}: B \to A$ and it is itself bijective; moreover, $(f^{-1})^{-1} = f$. The reader has the opportunity to calculate inverse functions in Exercises 1.3.

■■　　EXAMPLE 3

Determine which of the following functions is bijective:

(a) $f: \mathbb{R} \to \mathbb{R}$ given by $f(x) = 3x + 2$
(b) $g: \mathbb{R} \to \mathbb{R}$ given by $g(x) = 1/(x^2 + 1)$

Solution

(a) First, f is injective, since

$$f(x_1) = f(x_2) \Rightarrow 3x_1 + 2 = 3x_2 + 2$$
$$\Rightarrow x_1 = x_2$$

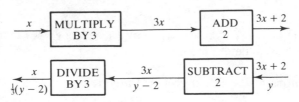

Figure 1.12

Secondly, if $y \in \mathbb{R}$ then $y = 3x + 2$ if and only if $x = \frac{1}{3}(y - 2)$. Hence for each real number y there exists a real number x with $f(x) = y$. Therefore the image of f coincides with the codomain of f. Thus f is bijective.

(b) The function g fails to be bijective on two counts (only one is required of course). As seen in Example 1, $g(\mathbb{R}) = (0, 1]$, and so the image of g is a proper subset of the codomain of g. In one sense this deficiency can be overcome by defining a (technically new) function $G: \mathbb{R} \to (0, 1]$ by the rule $G(x) = 1/(x^2 + 1)$. The codomain of G and the image of G now coincide by design. However, G (and g) fails to be injective. To see this, observe that $G(1) = G(-1)$.

At this point it is worth noting that the function $h: A \to B$, where $A = [0, \infty)$, $B = (0, 1]$ and $h(x) = 1/(x^2 + 1)$, is injective and has image B coinciding with its codomain. Hence h is a bijective function, which, it must be emphasized, is a different function from the original function g. This is a good illustration of the fact that there is more to a function than the formula defining it! ∎

As will be seen in later chapters, functions that can be built up from a set of basic functions by adding, subtracting, multiplying, dividing (when possible), scaling, composing (when possible) and inverting (for bijective functions only) inherit many important analytic properties of the original basic functions. It is for this reason that attention is focused both on the properties possessed by certain elementary functions such as polynomial functions, trigonometric functions, and the exponential function, and on those properties inherited by the various ways of combining these basic functions.

Exercises 1.3

1. By determining the set of real numbers for which each of the following formulae are valid, specify a suitable domain A for the function $f: A \to \mathbb{R}$:

(a) $f(x) = \sqrt{-x^2}$ (b) $f(x) = \sqrt{x-3}$
(c) $f(x) = 1/(x^2 - 1)$ (d) $f(x) = \log_e(\sin x)$
In each case, determine the image of f.

2.   Functions $f: \mathbb{R} \to \mathbb{R}$ and $g: \mathbb{R} \to \mathbb{R}$ are given by

$$f(x) = x^2 \quad \text{and} \quad g(x) = \begin{cases} x - 1 & \text{if } x \geq 0 \\ -x & \text{if } x < 0 \end{cases}$$

Determine $f \circ f$, $g \circ g$, $f \circ g$ and $g \circ f$.

3.   Determine which of the following functions is bijective.

(a) $f: \mathbb{R} \to \mathbb{R}$, $f(x) \quad = x^4$

(b) $g: \mathbb{R} \to \mathbb{R}$, $g(x) \quad = \begin{cases} 2x + 1 & \text{if } x \geq 0 \\ x - 1 & \text{if } x < 0 \end{cases}$

(c) $h: \mathbb{R} \to \mathbb{R}$, $h(x) \quad = \begin{cases} x^2 & \text{if } x \geq 0 \\ 2x & \text{if } x < 0 \end{cases}$

(d) $k: \mathbb{R} \to \mathbb{R}$, $k(x) \quad = \begin{cases} 2x & \text{if } x \in \mathbb{Z} \\ x & \text{if } x \notin \mathbb{Z} \end{cases}$

4.   Find the inverse of each of the following bijective functions.

(a) $f: \mathbb{R} - \{\frac{1}{2}\} \to \mathbb{R} - \{0\}$, $f(x) = \dfrac{1}{2x - 1}$

(b) $g: [0, \infty) \to (0, 1]$, $g(x) = \dfrac{1}{x^2 + 1}$

(c) $h: \mathbb{R} \to \mathbb{R}$, $h(x) = \begin{cases} \dfrac{3x + 2}{x - 1} & \text{if } x \neq 1 \\ 3 & \text{if } x = 1 \end{cases}$

5.   Let A be any set. The **identity** function $\text{id}_A: A \to A$ is given by $\text{id}_A(x) = x$ for all $x \in A$. Suppose that $f: A \to A$ is a bijective function. Prove that $f^{-1} \circ f = f \circ f^{-1} = \text{id}_A$.

Problems 1

1.   The logical connective $*$ is defined by the truth table shown in Figure 1.13. Show that the following pairs of statements are logically equivalent.

P	Q	$(P*Q)$
T	T	F
T	F	T
F	T	T
F	F	T

Figure 1.13

(a) $(\text{not } P)$; $(P*P)$

(b) $(P \text{ and } Q)$; $(P*Q)*(P*Q)$

(c) $(P \text{ or } Q)$; $(P*P)*(Q*Q)$

(d) $(P \Rightarrow Q)$; $P*(Q*Q)$

Deduce that any composite statement may be written solely in terms of the connective $*$.

2. Write down the converse of each of the following statements, where m, n and p denote whole numbers (positive, negative or zero). In each case where the statement or its converse is false, give a counterexample.

(a) $m + n \geqslant p \Rightarrow m^2 + n^2 \geqslant p^2$

(b) $m^2 \geqslant n^2 \Rightarrow m \geqslant n$

(c) $(m + n)(m - n) = m - n \Rightarrow m + n = 1$

Would your answers be different if m, n and p were restricted to be positive whole numbers?

3. Use proof by contradiction to establish the following:

(a) If a positive whole number n can be expressed as $n_1 n_2$, where $n_1 \geqslant 2$ and $n_2 \geqslant 2$, then at least one of n_1 and n_2 is less than $\sqrt{n}$.

(b) If the product of three positive whole numbers exceeds 1000 then at least one of the numbers exceeds 10.

4. The statement

$$(\forall n)(\exists m)(m > 3n)$$

which refers to positive whole numbers m and n, asserts that for each number n there exists a number m such that m exceeds $3n$. Is this statement true?

Express each of the following statements in words:

$(\exists m)(\forall n)(m > 3n)$

$(\exists n)(\forall m)(m > 3n)$

Are they true?

5. The **difference** $A - B$ of two sets A and B is given by

$A - B = A \cap \mathscr{C}B$

Use the laws of the algebra of sets to establish the following:

(a) $(A - B) \cup (A - C) = A - (B \cap C)$
(b) $(A - B) - C = A - (B \cup C)$
(c) $(A \oplus B) - C \subseteq A \oplus (B - C)$

Show by example that equality need not occur in part (c).

6. Prove that for any sets A, B and C

$A \times (B \cap C) = (A \times B) \cap (A \times C)$

7. The **power set** of a set A is the set $\mathscr{P}(A)$ consisting of all subsets of A. For example, $\mathscr{P}(\{0, 1\}) = \{\varnothing, \{0\}, \{1\}, \{0, 1\}\}$. Prove that

(a) $\mathscr{P}(A) \cap \mathscr{P}(B) = \mathscr{P}(A \cap B)$
(b) $\mathscr{P}(A) \cup \mathscr{P}(B) \subseteq \mathscr{P}(A \cup B)$

Show by example that equality need not hold in part (b).

8. Functions $f: \mathbb{R} \to \mathbb{R}$ and $g: \mathbb{R} \to \mathbb{R}$ are given by

$$f(x) = \begin{cases} x & \text{if } x \geqslant 0 \\ x - 1 & \text{if } x < 0 \end{cases}$$

$$g(x) = \begin{cases} x^2 & \text{if } x \geqslant 0 \\ x + 1 & \text{if } x < 0 \end{cases}$$

Find formulae for $f \circ g$ and $g \circ f$. Explain briefly why none of f, g and $f \circ g$ possess an inverse. Find the inverse of $g \circ f$.

9.   Let A be any set and let $f: A \to A$ be a function satisfying $f^3 = \mathrm{id}_A$ (that is, $f \circ f \circ f$ is the identity function on A). Prove that f is bijective.

Suppose now that $A = \{x \in \mathbb{R} : x \neq 0, 1\}$ and that $f: A \to A$ is given by $f(x) = 1 - 1/x$. Calculate f^3 and hence deduce that f is bijective. Determine f^{-1}.

10.   (a) Suppose that the functions $f: A \to B$, $g: A \to B$ and $h: B \to C$ satisfy $h \circ g = h \circ f$. If h is injective then prove that $g = f$.

(b) Suppose that the functions $f: A \to B$, $g: B \to C$ and $h: B \to C$ satisfy $g \circ f = h \circ f$. If f is surjective then prove that $g = h$.

CHAPTER TWO

2

The Real Numbers

2.1 Numbers
2.2 Axioms for the real numbers
2.3 The completeness axiom

During the nineteenth century, mathematicians involved in the process of producing a rigorous foundation for analysis found that their work was based upon simple intuitive notions concerning the system of real numbers. Since the analysis being developed involved the study of real functions, that is, functions whose domains and images are subsets of the real line, there was a clear need to develop firm foundations for the real number system itself. This feat was accomplished by the turn of the nineteenth century, and is commonly referred to as the **arithmetization of analysis**.

In Section 2.1 there is heavy reliance on readers' intuitive notions regarding numbers. In particular it is assumed that numbers can be represented geometrically as points on an infinite straight line where, after choosing a unit of length, the whole numbers are used to label a set of evenly spaced points on this line. This number line is illustrated in Figure 2.1.

Fractions (or, as they will be called from now on, **rationals**) can also be marked on this number line; for example, $\frac{1}{2}$ is placed exactly halfway between the points labelled 0 and 1. This means that there is a natural ordering of the rationals on the number line. For example $\frac{7}{22}$ lies to the

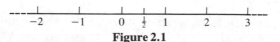

Figure 2.1

left of $\frac{3}{8}$ since $\frac{7}{22} = \frac{28}{88}$ and $\frac{3}{8} = \frac{33}{88}$. This fact is expressed as $\frac{7}{22} < \frac{3}{8}$, which is read as '$\frac{7}{22}$ is less than $\frac{3}{8}$'. It now seems reasonable, though sadly wrong, to suppose that *every* point on the number line can be labelled with a rational. To see why this intuition is wrong, consider a point to the right of the point labelled 0 whose distance L from 0 satisfies $0 < L < 1$. Since there are infinitely many, infinitely closely packed rationals between 0 and 1, surely one of them must coincide with L? But the rationals $0, 1, \frac{1}{2},$ $\frac{1}{4}, \frac{3}{4}, \frac{1}{8}, \frac{3}{8}, \frac{5}{8}, \frac{7}{8}, \frac{1}{16}, \ldots$ and so on, obtained by successive bisection, are infinitely closely packed, and they do not even account for the point labelled by the fraction $\frac{1}{3}$. What is surprising is that even if rationals between 0 and 1 with any denominator are included, there will be un-labelled points. This is proved in theorem 2.1.1, where it is shown that the distance usually denoted by $\sqrt{2}$ does not correspond to any rational. The remainder of Section 2.1 uses familiar notions such as the decimal expansion of numbers to show that between any two points on the number line there are infinitely many points that do not correspond to rationals as well as infinitely many that do.

The existence, at least in geometric terms, of non-rational (or **irra-tional**) numbers was a source of great confusion for early mathematicians, and it was only in the nineteenth century that the entire system of real numbers was successfully developed from the set $\mathbb{N}$ of natural numbers. Since this theory is fairly complicated, it is not appropriate to present it here. However, the material presented in Section 2.1 does use various properties of the real number system, and, although these are applied mainly to rationals, there is the tacit assumption that they hold for the entire set of real numbers. Since the real numbers are to form the starting point for the development of analysis, it is necessary to spell out precisely what properties of real numbers are being assumed. Then, as far as analysis is concerned, the whole theory can be deduced from these prop-erties, or **axioms**. The axioms for the real numbers form the substance of Section 2.2, and the results of Section 2.1 can all be deduced from these axioms. Of particular importance in analysis is the last of the axioms, the completeness axiom. This axiom is treated separately in Section 2.3, where it is used to prove that between any two distinct real numbers, however close, there is a rational number. It is also used to prove that $\sqrt{2}$ exists as a real number and hence that the set of rational numbers satisfies all of the axioms for the real numbers with the exception of the completeness axiom. Thus it is the completeness axiom that distinguishes between the set $\mathbb{Q}$ of rational numbers and the set $\mathbb{R}$ of real numbers.

2.1 Numbers

Our first encounter with numbers is with the set $\mathbb{N}$ of counting numbers or **natural numbers**. These are $1, 2, 3, \ldots$ and so on. Young children learn them by rote, and contact with adults leads to an awareness of their significance in phrases such as 'two eyes' and 'four marbles'. The concept of 'no marbles' and hence the existence of the number zero comes later and is a little harder to assimilate. Next we learn basic arithmetic where certain truths such as $a + b = b + a$ and $(a + b) + c = a + (b + c)$ may or may not appear obvious, depending on the methods used. For example, if addition is performed by 'counting on', a sum like $2 + 19$ is more difficult than $19 + 2$, whereas the use of the coloured rods of the appropriate lengths laid end to end makes $2 + 19 = 19 + 2$ a triviality.

When we learn subtraction, we find that $5 - 2 = 3$ is analogous to the problem of removing two marbles from five marbles to leave three marbles. This is fairly easy to comprehend. However, $3 - 5$ leads to negative numbers, which have to be interpreted using a model involving debts or the descent of ladders. We extend our horizons and end up with the set $\mathbb{Z}$ of **integers**, consisting of zero and all positive and negative whole numbers. Arithmetic in $\mathbb{Z}$ is more complicated, and, when we come to multiply integers, facts like -1 times -1 equals 1 seem far from obvious.

Fractions are introduced to facilitate division. To share six marbles between three people to give them two apiece is written symbolically as

$$6 \div 3 = 2 \quad \text{or} \quad \frac{6}{3} = 2$$

Although we cannot divide 11 marbles equally between three people, we can still attach a meaning to the fraction $\frac{11}{3}$, thus arriving at the set $\mathbb{Q}$ of **rationals**. Formally,

$$\mathbb{Q} = \left\{ \frac{m}{n} : m \text{ and } n \text{ are integers and } n \neq 0 \right\}$$

Addition and multiplication of rationals are given by

$$\frac{m}{n} + \frac{p}{q} = \frac{(mq + np)}{nq} \quad \text{and} \quad \frac{m}{n} \times \frac{p}{q} = \frac{mp}{nq}$$

One quirk of this notation is that apparently different fractions define the same rational. For example, $\frac{1}{2} - \frac{1}{3} = \frac{1}{6}$ is the same as $\frac{1}{4} \times \frac{2}{3} = \frac{2}{12}$. However, every positive rational can be expressed in the form m/n where m and n are positive integers with no common factors greater than one. The rational in question is then said to be in its 'lowest form'.

The rationals can be placed on the number line as in Figure 2.1, and the question can then be posed as to whether or not every point on this number line corresponds to some rational. Put another way, is every length a rational length? The answer is no! The following result is attributed to Pythagoras.

2.1.1 Theorem

There is no rational number whose square is 2.

Proof

First note that it is possible to construct a line segment of length x with $x^2 = 2$. See Figure 2.2. Suppose that x is rational. Then, as observed above, x can be expressed as a rational in its lowest form. Hence $x = m/n$, where m and n are integers, $n \neq 0$ and m and n have no common factor greater than one. Now $m^2 = 2n^2$, and so m^2 is an even integer. Hence m is also even (see Question 5 of Exercises 1.1) and so $m = 2m_1$ for some other integer m_1. Hence $m^2 = 4m_1^2 = 2n^2$, leading to $n^2 = 2m_1^2$. Thus n^2 is even and so n is also even. But then m and n have a common factor of 2. This contradiction can only be resolved by concluding that x is not a rational number. $\qquad\square$

The proof of Theorem 2.1.1 is a classic example of a proof by contradiction (see Section 1.1). In such a proof the required result is assumed to be false, and it is shown by some means that this assumption is inconsistent with the given information. Since it is deemed impossible to deduce any inconsistencies from true statements, the original assumption must be wrong. In other words, the required result is in fact true. This method of proof is particularly useful when dealing with results that are phrased in a negative manner, such as Theorem 2.1.1.

The theorem itself demonstrates that there are simple equations, such as $x^2 = 2$, that have no rational solutions. In order to handle such equations, it is necessary to introduce new numbers that are not rationals. So let $\sqrt{2}$ denote the solution of $x^2 = 2$, which corresponds to the length of the hypotenuse of the triangle in Figure 2.2. A number, such as $\sqrt{2}$, that is not rational is called **irrational**. Other examples of irrational numbers are $\sqrt{3}$, $\sqrt{6}$, $2\sqrt{5}$, π, e, and $\log_e 2$. The question now arises as to how irrational numbers can be incorporated into arithmetic calculations. One way forward is to introduce a representation for rational numbers that can be extended to represent irrational numbers as well.

$$x^2 = 1^2 + 1^2 = 2$$

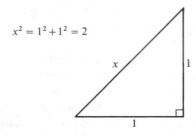

Figure 2.2

CHESTER COLLEGE LIBRARY

Pythagoras (born 572 BC) was one of the outstanding mathematicians of Ancient Greece. The advent of the Greek civilization in the second millennium BC saw a growth in rationalization in all areas of human activity. Modern mathematics was born in this atmosphere – a mathematics that not only asked 'how?' but also asked the scientific question 'why?'. Pythagoras founded the famous Pythagorean school, an academy for the study of philosophy, mathematics and the natural sciences. Their philosophy rested on the assumption that whole numbers were the cause of the various qualities of man and matter. Daily life required the measurement of quantities such as length, weight and time, and the rational numbers were at first deemed sufficient for such practical purposes. However, the Pythagoreans showed that there is no rational number corresponding to the point P on the number line where the distance OP is equal to the diagonal of a square having unit side (see Figure 2.3).

The discovery of irrational magnitudes such as $\sqrt{2}$ upset the underlying philosophy of the Pythagoreans that everything depended on whole numbers. For a time, efforts were made to keep the whole matter secret, though a treatment of irrational numbers is recorded in the fifth book of Euclid's *Elements*. This account accords well with nineteenth-century European expositions.

A **decimal** is an expression of the form $\pm a_0.a_1a_2a_3\ldots$, where a_0 is a non-negative integer and a_1, a_2, a_3, ..., are digits. If only a finite number of the digits a_1, a_2, $a_3 \ldots$ are non-zero then the decimal is called a **terminating** or **finite** decimal, and the infinite tail of zeros is usually omitted. Such decimals represent rationals; for example, $1.63 = 1 + \frac{6}{10} + \frac{3}{100} = \frac{163}{100}$. It can be shown that any rational whose denominator contains only powers of 2 and/or 5 can be represented by a terminating decimal, which can be found by long division. However, there are rationals for which the process of long division yields a **non-terminating** or **infinite** decimal; for example, $\frac{3103}{9990} = 0.310610610610\ldots$ These infinite decimals have the property that they are **recurring**. In other words, they have a recurring block of digits, and so may be written in shorthand form as follows: $0.310610610661\ldots = 0.31\overline{06}$.

To see why the decimal expansion of a rational number is either a terminating or recurring decimal, consider the way in which the decimal is

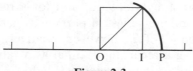

Figure 2.3

generated. Consider a rational p/q lying in the interval $[0, 1)$, so that $0 \le p < q$, and apply the method of long division. Then

$$10p = a_1 q + r_1$$

where a_1 is a digit and r_1 is an integer satisfying $0 \le r_1 < q$. In other words 'q goes into $10p$, a_1 times with remainder r_1'. Now consider the rational r_1/q, which also lies in the interval $[0, 1)$. Then

$$10r_1 = a_2 q + r_2$$

where a_2 is a digit and r_2 is an integer satisfying $0 \le r_2 < q$. Continue this process for n steps, arriving at

$$10r_{n-1} = a_n q + r_n$$

where a_n is a digit and r_n is an integer satisfying $0 \le r_n < q$. Since

$$0 \le \frac{p}{q} - \left(\frac{a_1}{10} + \frac{a_2}{10^2} + \ldots + \frac{a_n}{10^n} \right)$$

$$= \left(\frac{r_n}{q} \right) 10^{-n}$$

$$< 10^{-n}$$

this process generates the first n decimal digits in the decimal expansion $0.a_1 a_2 a_3 \ldots a_n \ldots$ of p/q. At each step, the remainder r_i is one of $0, 1, 2, \ldots, q - 1$, and so after $q + 1$ steps the same remainder must have occurred more than once. This means that (provided zero is not one of the remainders) a recurring pattern will be established in which the block of digits which repeat is no longer than $q - 1$ digits in length. If zero occurs as a remainder then all successive decimal digits are zero and the decimal terminates. Since any rational can be written as $\pm a_0 + p/q$ where a_0 is a non-negative integer and p/q is a rational in the interval $[0, 1)$, the decimal expansion of any rational number is either a terminating or a recurring decimal.

Conversely, a terminating or recurring decimal represents a rational number. This is not difficult to establish when the decimal is finite. The infinite case is not so straightforward, and for a rigorous treatment requires knowledge of infinite series, in particular, geometric series, which are covered in Section 4.1. It is, however, relatively easy to see how the process of representing a recurring decimal as a rational works. For instance, consider the recurring decimal $0.3\overline{12}$. Let $x = 0.\overline{12}$ and multiply both sides by 10^2 (because the recurring block has length 2). Then

$$100x = 12.\overline{12} = 12 + x$$

Hence

$$99x = 12$$

and so

$$x = \frac{12}{99} = \frac{4}{33}$$

Therefore

$$0.3\overline{12} = \frac{3}{10} + \frac{x}{10} = \frac{3}{10} + \frac{4}{330} = \frac{103}{330}$$

It is now natural to ask if irrational numbers can be represented by decimals, especially since it is possible to find finite decimals very close to numbers like $\sqrt{2}$; the reader may like to check that $(1.41421356)^2$ does give an answer very close to 2. The answer is yes, but this is a consequence of the axiomatic description of the set of real numbers presented in the next section, and not a consequence of the material presented so far in this chapter. Under the assumption that irrational numbers can be represented by decimals, then, these decimals are neither finite nor recurring. It is also the case that infinite non-recurring decimals, such as

0.101 001 000 100 001 . . .

and

0.123 456 789 101 112 . . .

represent irrational numbers, and so the set $\mathbb{R}$ of **real numbers** can, naively, be considered as the set of all decimals. Each point of the number line in Figure 2.1 then corresponds to some decimal. The following results show that irrationals are far from rare and that they are intimately mixed up with rationals on the number line.

2.1.2 Theorem

If m/n and p/q are rationals, $p \neq 0$, then $m/n + \sqrt{2}(p/q)$ is irrational.

Proof

The proof is by contradiction. Suppose that $m/n + \sqrt{2}(p/q)$ is a rational number. Then the following can be written:

$$\frac{m}{n} + \sqrt{2}\left(\frac{p}{q}\right) = \frac{r}{s}$$

where m, n, p, q, r and s are all integers, and n, q and s are non-zero. Rearranging the equation gives

$$\sqrt{2} = \frac{q(rn - ms)}{pns}$$

which is a rational. This contradicts 2.1.1, and so the hypothesis that $m/n + \sqrt{2}(p/q)$ was rational is wrong. Hence $m/n + \sqrt{2}(p/q)$ is irrational. □

2.1.3 Theorem

Between any two distinct rationals there is an irrational.

Proof

Suppose that $m/n < p/q$. This gives $p/q - m/n > 0$. Hence

$$\frac{m}{n} < \frac{m}{n} + \frac{\sqrt{2}}{2}\left(\frac{p}{q} - \frac{m}{n}\right)$$

and, since $\sqrt{2}/2 < 1$,

$$\frac{m}{n} + \frac{\sqrt{2}}{2}\left(\frac{p}{q} - \frac{m}{n}\right) < \frac{m}{n} + \left(\frac{p}{q} - \frac{m}{n}\right) = \frac{p}{q}$$

Thus the irrational

$$\frac{m}{n} + \frac{\sqrt{2}}{2}\left(\frac{p}{q} - \frac{m}{n}\right)$$

lies between the rationals m/n and p/q. □

2.1.4 Theorem

Between any two distinct irrationals there is a rational.

Proof

Suppose that a and b are irrationals with $a < b$. Since $a \neq b$, there is a first decimal digit where they differ. Thus

$$a = a_0.a_1a_2 \ldots a_{n-1}a_n \ldots \quad \text{and} \quad b = a_0.a_1a_2 \ldots a_{n-1}b_n \ldots$$

where $a_n < b_n$. Now let

$$x = a_0.a_1a_2 \ldots a_{n-1}b_n 00000 \ldots$$

Hence $x = (a_0a_1a_2 \ldots b_n)/10^n$ is rational and $a < x < b$.   □

These results can be further refined to show that between any two distinct real numbers there are infinitely many rationals and infinitely many irrationals.

The above results rely, of course, on the many assumptions about real numbers that have been made in this section. Questions such as whether irrational numbers exist have not been satisfactorily resolved. As mentioned in the introduction to this chapter, the system of real numbers can be constructed from the set $\mathbb{N}$ of natural numbers. This is a complicated process that, once achieved, enables certain properties possessed by the real numbers to be proved. With care, a selection of these properties can be listed that completely characterize the set $\mathbb{R}$ of real numbers. In Section 2.2 this list of properties of $\mathbb{R}$ is presented as a set of axioms, and further properties which are required in the formal development of analysis are deduced. Material in this section will only be used when it is clear that it follows from the axioms.

Exercises 2.1

1.   Prove that $\sqrt{5}$ is irrational and hence prove that $a + b\sqrt{5}$ is irrational for all rationals a and b, $b \neq 0$. Deduce that the **golden ratio** r, defined by $r = 1 + 1/r$, $r > 0$, is irrational.

2.   Which of the following statements are true?

   (a) x rational, y irrational $\Rightarrow x + y$ irrational.
   (b) x rational, y rational $\Rightarrow x + y$ rational.
   (c) x irrational, y irrational $\Rightarrow x + y$ irrational.

   Prove the true ones and give a counterexample for each of the false ones.

3.   Show that between any two distinct real numbers there are infinitely many rationals and infinitely many irrationals.

4. Prove that there is no rational number x such that $10^x = 2$. Deduce that $\log_{10} 2$ is irrational.

5. Let $x = \sqrt{3 + 2\sqrt{2}} - \sqrt{3 - 2\sqrt{2}}$ and calculate x^2. Is x irrational?

2.2 Axioms for the real numbers

From now on it is assumed that there is a non-empty set $\mathbb{R}$ of real numbers with a certain structure satisfying a number of properties, or axioms. The axioms fall under three headings:

(1) the axioms of arithmetic

(2) the axioms of order

(3) the completeness axiom

The first two groups of axioms are well-known properties from which all the usual rules for manipulating real numbers may be derived. In more advanced texts $\mathbb{R}$ is constructed from something more basic, for example from the set of integers, or from set theory, and it is then necessary to prove that the numbers constructed satisfy these axioms. In this text we start from the assumption that $\mathbb{R}$ exists. We assume then that there are two functions from $\mathbb{R} \times \mathbb{R}$ to $\mathbb{R}$. These are called **addition** and **multiplication**, and the images of (a, b) under these functions are denoted by $a + b$ and $a \cdot b$ respectively.

2.2.1 The axioms of arithmetic

A1 $a + (b + c) = (a + b) + c$

A2 $a + b = b + a$

A3 There exists a unique element 0 in $\mathbb{R}$ satisfying $a + 0 = a$

A4 For any a in $\mathbb{R}$, there exists a unique element x in $\mathbb{R}$ satisfying $a + x = 0$

A5 $a \cdot (b \cdot c) = (a \cdot b) \cdot c$

A6 $a \cdot b = b \cdot a$

A7 There exists a unique element 1 in $\mathbb{R}$, $1 \neq 0$, satisfying $a \cdot 1 = a$

A8 For any a in $\mathbb{R}$, $a \neq 0$, there exists a unique element y in $\mathbb{R}$ satisfying $a \cdot y = 1$

A9 $a \cdot (b + c) = a \cdot b + a \cdot c$

Notes

(1) Each axiom holds for all $a, b, c \in \mathbb{R}$.

(2) Any set satisfying A1–A9 is called a **field**.

(3) The x in A4 is called the **negative** of a and is usually denoted by $(-a)$.

(4) The y in A8 is called the **reciprocal** of a and is written as $1/a$ or a^{-1}.

(5) Axioms A1 and A5 allow us to omit brackets in expressions such as $a + b + c$ or $a \cdot b \cdot c \cdot d$.

(6) The axioms give names to only two particular elements of $\mathbb{R}$, namely 0 and 1, whose roles are defined in A3 and A7.

(7) 0^{-1} is not defined – see A8. In fact, no such element exists by Example 1(a) below.

(8) **Subtraction** can be defined by $a - b = a + (-b)$.

(9) **Division** can be defined by $a \div b = a \cdot (b^{-1})$, $b \neq 0$.

It is assumed that the reader is quite familiar with axioms A1–A9. From these axioms, many further algebraic properties can be deduced. Since these axioms are so basic, several elementary consequences are needed first.

■■ EXAMPLE 1

Deduce that

(a) for all $x \in \mathbb{R}$, $x \cdot 0 = 0$

(b) for all $x, y \in \mathbb{R}$, $x \cdot (-y) = -(x \cdot y)$

Solution

(a) Now $0 + 0 = 0$ by A3, and so $x \cdot (0 + 0) = x \cdot 0$. Thus, by A9, $x \cdot 0 + x \cdot 0 = x \cdot 0$. Since $(x \cdot 0) \in \mathbb{R}$, it possesses a negative $-(x \cdot 0)$ by A4. Now

$$((x \cdot 0) + (x \cdot 0)) + (-(x \cdot 0)) = (x \cdot 0) + (-(x \cdot 0))$$

and hence

$$(x \cdot 0) + ((x \cdot 0) + (-(x \cdot 0))) = (x \cdot 0) + (-(x \cdot 0)) \quad \text{by A1}$$

So $(x \cdot 0) + 0 = 0$ by A4, and finally $x \cdot 0 = 0$ by A3.

(b) Now

$$(x \cdot y) + x \cdot (-y) = x \cdot (y + (-y)) \quad \text{by A9}$$
$$= x \cdot 0 \qquad\qquad \text{by A4}$$
$$= 0 \qquad\qquad \text{by (a)}$$

By A4, the negative of $(x \cdot y)$ is the unique real number $-(x \cdot y)$ satisfying $(x \cdot y) + (-(x \cdot y)) = 0$. Hence $x \cdot (-y) = -(x \cdot y)$. ∎

The reader is invited to derive similar properties in Exercises 2.2. Of course, in practice, one would not write out detailed proofs of results like these. What is important is that the algebraic manipulations are firmly based on those properties of numbers (the axioms of arithmetic) that form our logical starting point. In principle, all the elementary algebraic operations that are carried out routinely can be justified in detail by deducing them from the axioms. This is quite a lengthy business, as the following example shows.

■■ EXAMPLE 2

Prove the algebraic identity $(a - b) \cdot (a + b) = a^2 - b^2$.

Solution

$$(a - b) \cdot (a + b) = (a + (-b)) \cdot (a + b) \qquad \text{using Note (8)}$$
$$= (a + (-b)) \cdot a + (a + (-b)) \cdot b$$
$$\text{by A9}$$
$$= a \cdot (a + (-b)) + b \cdot (a + (-b))$$
$$\text{by A6}$$
$$= a \cdot a + a \cdot (-b) + b \cdot a + b \cdot (-b)$$
$$\text{by A9 and A1}$$
$$= a \cdot a + (-(a \cdot b)) + b \cdot a + (-(b \cdot b))$$
$$\text{using Example 1(b)}$$
$$= a \cdot a + (a \cdot b + (-(a \cdot b)) + (-(b \cdot b)))$$
$$\text{by A6, A2 and A1}$$
$$= (a \cdot a + 0) + (-(b \cdot b)) \quad \text{by A4 and A1}$$
$$= a \cdot a + (-(b \cdot b)) \qquad \text{by A3}$$
$$= a^2 - b^2 \qquad\qquad \text{using Note (8)} \quad ■$$

The axioms of arithmetic alone are not sufficient to describe $\mathbb{R}$ completely, since many proper subsets of $\mathbb{R}$ also satisfy A1–A9, for example $\mathbb{Q}$.

A relation $\leq$ can be defined on $\mathbb{R}$ such that, for certain elements, a, $b \in \mathbb{R}$, $a \leq b$. In terms of the heuristic ideas in Section 2.1, $a \leq b$, read as 'a is less than or equal to b', occurs when a corresponds to a point on the real number line (Figure 2.1) to the left of the point corresponding to b. It is required that this **order** relation $\leq$ satisfy the following axioms.

2.2.2 The axioms of order

A10	$a \leq b$ or $b \leq a$
A11	$a \leq b$ and $b \leq a \Rightarrow a = b$
A12	$a \leq b$ and $b \leq c \Rightarrow a \leq c$
A13	$a \leq b \Rightarrow a + c \leq b + c$
A14	$a \leq b$ and $0 \leq c \Rightarrow a \cdot c \leq b \cdot c$

Notes

(1) Each axiom is satisfied by any a, b, $c \in \mathbb{R}$.

(2) Any set satisfying axioms A1–A14 is called an **ordered** field.

(3) A relation $<$, read as 'less than', can be defined on $\mathbb{R}$ by

$$a < b \Leftrightarrow a \leq b \text{ and } a \neq b$$

The axioms of order are not as straightforward as the axioms of arithmetic, but, just as with the latter, some elementary consequences can be deduced before moving on to more sophisticated results. First, the set $\mathbb{R}^+$ of **positive** real numbers and the set $\mathbb{R}^-$ of **negative** real numbers are introduced by defining

$$\mathbb{R}^+ = \{x : x \in \mathbb{R} \text{ and } 0 < x\}$$

and

$$\mathbb{R}^- = \{x : x \in \mathbb{R} \text{ and } x < 0\}$$

By A10 and A11, every real number is either positive, negative or zero.

■■   EXAMPLE 3

Prove that

(a) $x \in \mathbb{R}^+ \Leftrightarrow (-x) \in \mathbb{R}^-$

(b) $x^2 \in \mathbb{R}^+$ for all $x \neq 0$

Solution

(a) First,

$$0 \leqslant x \quad \Rightarrow 0 + (-x) \leqslant x + (-x) \quad \text{by A4 and A13}$$
$$\Rightarrow (-x) \leqslant 0 \qquad\qquad \text{by A2–A4}$$

Conversely,

$$(-x) \leqslant 0 \Rightarrow (-x) + x \leqslant 0 + x \quad \text{by A13}$$
$$\Rightarrow 0 \leqslant x \qquad\qquad \text{by A2–A4}$$

Since $x \neq 0$, the result now follows.

(b) If $x \in \mathbb{R}^+$ then $0 \leqslant x$, and so $0 \leqslant x^2$ using A14 (with $a = 0$ and $b = c = x$). Since $x \neq 0$, $x^2 \in \mathbb{R}^+$. If $x \in \mathbb{R}^-$ then $(-x) \in \mathbb{R}^+$ by (a), and so $0 \leqslant (-x)$. From A14, $0 \leqslant (-x) \cdot (-x) = x^2$ (see Question 1(a) of Exercises 2.2). Since $x \neq 0$, $x^2 \in \mathbb{R}^+$. ■

The axioms of order, in particular A14, do not tell us directly the effect of multiplying both sides of an inequality by a negative real number. The answer is provided in the next example.

■■ EXAMPLE 4

Prove that $a \leqslant b$ and $c \leqslant 0 \Rightarrow b \cdot c \leqslant a \cdot c$. In other words, multiplying both sides of an inequality by a negative real number *reverses* the inequality sign.

Solution

Now

$$a \leqslant b \text{ and } c \leqslant 0 \Rightarrow a \leqslant b \text{ and } 0 \leqslant (-c) \quad \text{by Example 3}$$
$$\Rightarrow a \cdot (-c) \leqslant b \cdot (-c) \qquad \text{by A14}$$
$$\Rightarrow -(a \cdot c) \leqslant -(b \cdot c) \qquad \text{by Example 1(b)}$$

Adding $a \cdot c + b \cdot c$ to both sides of the last inequality and invoking A13 and the axioms of arithmetic as required, it soon follows that $b \cdot c \leqslant a \cdot c$. ■

■■　EXAMPLE 5

Establish the inequality

$$a \cdot b + b \cdot c + c \cdot a \leqslant a^2 + b^2 + c^2$$

for all $a, b, c \in \mathbb{R}$.

Solution

By Example 3(b), observe that $0 \leqslant (x - y)^2$ for any real numbers x and y. Thus, from A12 and A13, for any a, b and c,

$$
\begin{aligned}
0 \leqslant (a - b)^2 &+ (b - c)^2 + (c - a)^2 \\
&= (a^2 - 2a \cdot b + b^2) + (b^2 - 2b \cdot c + c^2) \\
&\quad + (c^2 - 2c \cdot a + a^2)
\end{aligned}
$$

using the axioms of order and arithmetic (2.2.1 and 2.2.2) as required. Again using the axioms, it is soon deduced that

$$a \cdot b + b \cdot c + c \cdot a \leqslant a^2 + b^2 + c^2 \qquad\qquad ■$$

■■　EXAMPLE 6

Solve the inequality $x^2 + x - 2 < 0$.

Solution

By factorization, $x^2 + x - 2 = (x - 1)(x + 2)$. Hence

$$x^2 + x - 2 < 0 \Leftrightarrow (x - 1)(x + 2) < 0$$

This last inequality holds only when $x - 1$ and $x + 2$ have opposite signs. This observation follows from Question 2 of Exercises 2.2. Hence

$$
\begin{aligned}
x^2 + x - 2 < 0 &\Leftrightarrow (x - 1) < 0 \text{ and } (x + 2) > 0 \\
&\Leftrightarrow x < 1 \text{ and } x > -2 \\
&\Leftrightarrow -2 < x < 1
\end{aligned}
$$

The solution to the problem is $-2 < x < 1$. ■

■■   EXAMPLE 7

Solve the inequality

$$\frac{5x + 6}{x + 2} > \frac{2x - 3}{x - 1}$$

Solution

Now

$$\frac{5x + 6}{x + 2} > \frac{2x - 3}{x - 1} \Leftrightarrow \frac{5x + 6}{x + 2} - \frac{2x - 3}{x - 1} > 0$$

$$\Leftrightarrow \frac{(5x + 6)(x - 1) - (2x - 3)(x + 2)}{(x - 1)(x + 2)} > 0$$

$$\Leftrightarrow \frac{3x^2}{(x - 1)(x + 2)} > 0$$

Since $x^2 \geq 0$ for all x, the inequality is equivalent to

$$(x - 1)(x + 2) > 0.$$

But the inequality $(x - 1)(x + 2) \geq 0$ describes those values of x that do not satisfy the inequality in Example 6. Hence the solution set is $x < -2$ or $x > 1$.   ■

The next example establishes the **triangle inequality**, which finds much application in what follows. First it is necessary to introduce $|x|$, the **modulus** (or **absolute value**) of a real number x:

$$|x| = \begin{cases} x & \text{if } 0 \leq x \\ -x & \text{if } x < 0 \end{cases}$$

By considering the signs of the numbers involved, it is easy to establish the following properties of the modulus:

$$x \leq |x|$$
$$|x|^2 = x^2$$
$$|x \cdot y| = |x| \cdot |y|$$

■■   EXAMPLE 8

For all real numbers x and y, $|x + y| \leq |x| + |y|$.

Solution

If a, $b \in \mathbb{R}$ and $a > b \geqslant 0$ then $a^2 > b^2$. This shows by a contradiction argument that for a, $b \geqslant 0$, $a^2 \leqslant b^2$ implies that $a \leqslant b$. It is thus sufficient to prove that $|x + y|^2 \leqslant (|x| + |y|)^2$. Now

$$
\begin{aligned}
|x + y|^2 &= (x + y)^2 \\
&= x \cdot x + 2x \cdot y + y \cdot y \\
&\leqslant x \cdot x + 2|x| \cdot |y| + y \cdot y \\
&= |x|^2 + 2|x| \cdot |y| + |y|^2 \\
&= (|x| + |y|)^2 \qquad\qquad \blacksquare
\end{aligned}
$$

At this stage it is instructive to check that $\mathbb{Q}$ satisfies the axioms of arithmetic and order. First, addition and multiplication of rationals produce rationals. In other words, *addition* and *multiplication* restricted to $\mathbb{Q} \times \mathbb{Q}$ define functions from $\mathbb{Q} \times \mathbb{Q}$ to $\mathbb{Q}$. Secondly, axioms A1, A2, A5, A6 and A10–A14 hold for any subset of $\mathbb{R}$. Finally, 0 and 1 are rationals and the negatives and reciprocals of rationals are rational. Hence the remaining axioms hold. So $\mathbb{Q}$ is an ordered field. Hence, since it has been shown that there is no rational number whose square is two, the axioms of an ordered field cannot be sufficient to ensure the existence of the real number $\sqrt{2}$. This comes from the final axiom, which may be unfamiliar to the reader.

2.2.3   The completeness axiom

> A15   Every non-empty set of real numbers that is bounded above has a least upper bound.
> Every non-empty set of real numbers that is bounded below has a greatest lower bound.

Notes

(1) A subset S of $\mathbb{R}$ is said to be **bounded above** if there is some real number M such that $x \leqslant M$ for all $x \in S$. This M is called an **upper bound** for S.
  A subset S of $\mathbb{R}$ is said to be **bounded below** if there is some real number m such that $m \leqslant x$ for all $x \in S$. This m is called a **lower bound** for S.

(2) Any set satisfying A1–A15 is called a **complete** ordered field. Essentially, $\mathbb{R}$ is the only complete ordered field, and so axioms A1–A15 characterize $\mathbb{R}$.

(3) The least upper bound is called the **supremum** of S, or sup S. The greatest lower bound is called the **infimum** of S, or inf S.

(4) One half of A15 is redundant (see Question 5 of Exercises 2.2).

This final axiom is a subtle one, which will be discussed in more detail in the next section. To claim that an upper bound M of a bounded set S is the least upper bound or supremum of S is saying that $M \leq M_1$ for any other upper bound M_1. Put another way, if $M' < M$ then there must exist at least one element $x \in S$ with $x > M'$, thus preventing M' from being an upper bound and so making M the least upper bound. Similar remarks apply to the infimum of a bounded set. As will be seen, the completeness axiom is the one that guarantees the existence of irrational numbers such as $\sqrt{2}$.

Exercises 2.2

1. Use the axioms of arithmetic (see 2.2.1) and Example 1 to prove the following.

    (a) $(-x) \cdot (-y) = x \cdot y$ for all $x, y \in \mathbb{R}$
    (b) $-(x^{-1}) = (-x)^{-1}$ for all $x \in \mathbb{R}, x \neq 0$
    (c) $(x^{-1})^{-1} = x$ for all $x \in \mathbb{R}, x \neq 0$

2. Use the axioms of arithmetic and order (see 2.2.1 and 2.2.2) to establish the following.

    (a) $0 \leq x$ and $0 \leq y \Rightarrow 0 \leq x \cdot y$
    (b) $0 \leq x$ and $y \leq 0 \Rightarrow x \cdot y \leq 0$
    (c) $x \leq 0$ and $y \leq 0 \Rightarrow 0 \leq x \cdot y$

3. Solve the following inequalities:

    (a) $\dfrac{1}{4} < \dfrac{1}{x + 3}$ \qquad (b) $\dfrac{x - 1}{x + 1} \leq \dfrac{x + 1}{x - 1}$

4. Use the triangle inequality (Example 8 of Section 2.2) to prove that for all $x, y \in \mathbb{R}$

    $$\left| |x| - |y| \right| \leq |x - y|$$

5. Let S be a subset of $\mathbb{R}$ that is bounded above and let

    $$T = \{-x : x \in S\}.$$

    Show that T is bounded below and that

    $$\inf T = -\sup S.$$

    Explain why one half of A15 is redundant.

2.3   The completeness axiom

This section will prove that $\sqrt{2}$ exists as a real number. It will then prove that $\mathbb{Q}$ is not complete by showing that A15 fails for $\mathbb{Q}$. In other words, there are sets of rational numbers, bounded above, that do not possess a rational least upper bound. Consequently, the completeness axiom is the one that distinguishes $\mathbb{Q}$ from $\mathbb{R}$. This section begins with a theorem that is a direct consequence of the completeness axiom.

2.3.1   The Archimedean postulate

$\mathbb{N}$ is not bounded above.

Proof

Suppose that $\mathbb{N}$ is bounded above. In other words, there exists a real number M such that $n \leqslant M$ for all $n \in \mathbb{N}$. By the completeness axiom, M can be chosen to be $\sup \mathbb{N}$. Now $M - 1$ cannot be an upper bound for $\mathbb{N}$, and so there is a natural number m with $M - 1 < m$. Now $M < m + 1$ and $m + 1 \in \mathbb{N}$, which contradicts the choice of M. Hence $\mathbb{N}$ is not bounded above.   □

Archimedes (287–212 BC) was a native of Sicily and the son of an astronomer. His writings are masterpieces of mathematical exposition and are written in a manner that resembles modern journal articles. His most remarkable work on a method for finding areas (and volumes) by dividing the area into a large number of thin parallel strips is closely connected with the modern idea of definite integration. A copy of this work, called *Method*, was not discovered until 1906. It had been written in the tenth century on parchment, washed off in the thirteenth century and reused for a religious text. Luckily, the underlying text could be restored. The *Method* is a particularly significant work, since, although Archimedes' other treatises are logically precise, there is little hint of the preliminary analysis required in their formulation. In the *Method*, however, Archimedes describes, in detail, the 'mechanical' processes underlying most of his discoveries concerning areas and volumes. The principal reason why this is the only work where this vital background information is provided is that Archimedes himself had profound reservations about the rigour of his 'method', in particular the assumption that an area is a sum of line segments.

The Archimedean postulate that bears his name has been variously credited, and appears in the treatment of irrational numbers in the fifth book of Euclid's *Elements*. Little is known of the life of

Euclid, but his seminal work, the *Elements*, represent the first full mathematical text preserved from Greek antiquity. There have been over 1000 editions of the 13 books of the *Elements* since the invention of printing. The books are based on a strict logical deduction of theorems from a set of definitions, postulates and axioms, and have influenced scientific thinking more than any other books.

Another way of phrasing the Archimedean postulate is to say that, given any real number x, there exists an integer n with $n \geqslant x$.

■■ EXAMPLE 1

Prove that the supremum of $S = \{(n - 1)/2n : n \in \mathbb{N}\}$ is $\frac{1}{2}$.

Solution

Since $(n - 1)/2n = \frac{1}{2} - 1/2n$ and n is positive, S is bounded above by $\frac{1}{2}$. Suppose that $M < \frac{1}{2}$ is an upper bound for S. Now

$$(n - 1)/2n \leqslant M \quad \text{for all } n \Leftrightarrow n - 1 \leqslant 2nM \quad \text{for all } n$$
$$\Leftrightarrow (1 - 2M)n \leqslant 1 \quad \text{for all } n$$
$$\Leftrightarrow n \leqslant 1/(1 - 2M) \quad \text{for all } n$$

The latter step is valid since $0 < 1 - 2M$. Thus $\mathbb{N}$ is bounded above, which contradicts the Archimedean postulate. Hence no such M exists. In other words, $\frac{1}{2}$ is the least upper bound of S. ■

■■ EXAMPLE 2

Identify the supremum and infimum of the following subsets of $\mathbb{R}$:

$E_1 = \{1, 2, 3\}$

$E_2 = \{x : |x - 1| < 2\}$

$E_3 = \mathbb{R}^+$

$E_4 = \{x : 0 < x < 1$ and the decimal expansion of x contains no nines$\}$

Solution

Clearly $\sup E_1 = 3$ and $\inf E_1 = 1$. In fact, for any finite set the supremum is merely the maximum element and the infimum is the minimum element in the set.

Now $|x - 1| < 2$ gives $-1 < x < 3$, and so $\sup E_2 = 3$ and $\inf E_2 = -1$. Note that neither the supremum nor the infimum is an element of E_2, since E_2 contains neither a maximum nor a minimum element.

$\mathbb{R}^+$ is not bounded above and so $\sup E_3$ does not exist; $\inf E_3 = 0$, which is not an element of $\mathbb{R}^+$. So $\mathbb{R}^+$ does not contain a least element.

E_4 contains arbitrarily small positive real numbers; however, the largest number it contains is $0.888\,88\ldots$ which equals $\frac{8}{9}$. Thus $\sup E_4 = \frac{8}{9}$ and $\inf E_4 = 0$.   ∎

■■   EXAMPLE 3

Prove that if $C = \{x + y : x \in A \text{ and } y \in B\}$, where A and B are non-empty sets of $\mathbb{R}$ bounded above, then C is bounded above and $\sup C = \sup A + \sup B$.

Solution

Since A and B are non-empty and bounded above, $\sup A$ and $\sup B$ exist. Then $x + y \leqslant \sup A + \sup B$ for all $x \in A$ and for all $y \in B$, and hence C is bounded above. Clearly C is non-empty, and so $\sup C$ exists. Now $\sup C \leqslant \sup A + \sup B$ because, as has been demonstrated, $\sup A + \sup B$ is an upper bound for C. Suppose, by way of contradiction, that $\sup C < \sup A + \sup B$. Then $\sup C - \sup B < \sup A$, and hence there exists an $x' \in A$ with $\sup C - \sup B < x'$ (otherwise $\sup C - \sup B$ would be an upper bound for A smaller than $\sup A$; this contradicts the definition of a least upper bound). But now $\sup C - x' < \sup B$, and so there exists a $y' \in B$ with $\sup C - x' < y'$ (otherwise $\sup C - x'$ would be an upper bound for B smaller than $\sup B$). Finally then, $\sup C < x' + y'$, contradicting the fact that $\sup C$ is an upper bound for C. The only conclusion now is that $\sup C = \sup A + \sup B$.   ∎

It can now be proved that $\mathbb{Q}$ is a **dense** subset of $\mathbb{R}$; in other words, given any two real numbers x and y with $x < y$, there is a rational number r such that $x < r < y$. Although this may appear to be a consequence of the results obtained in Section 2.1, some of the arguments there depend on the unproved assertion that elements of $\mathbb{R}$, as axiomatized in Section 2.2, can be represented by infinite decimals.

2.3.2 Density of the rationals

Let $x, y \in \mathbb{R}$ and $x < y$. There exists a rational number r, $x < r < y$.

Proof

By the Archimedean postulate an integer n can be chosen with $n > 1/(y - x) > 0$. Hence $1/n < y - x$. Since nx and $-nx$ are real numbers, there exist integers m and m' satisfying $m > nx$ and $m' > -nx$ by the Archimedean postulate again. In other words,

$$-\frac{m'}{n} < x < \frac{m}{n}$$

Now there are a *finite* number of integers from $-m'$ to m, so the smallest one m'' can be selected with $x < m''/n$. The element m'' exists, since every finite set has a minimal element. See Question 6 of Exercises 2.3. Therefore $(m'' - 1)/n \leqslant x$ and

$$x < \frac{m''}{n} = \frac{m'' - 1}{n} + \frac{1}{n} \leqslant x + \frac{1}{n} < x + (y - x) = y$$

Hence $r = m''/n$ is a rational between x and y. $\qquad\qquad\square$

It can now be proved that $\sqrt{2}$, the positive square root of 2, exists. In other words, there is a real number M satisfying $M^2 = 2$. To this end, let $S = \{x : x \in \mathbb{R}^+ \text{ and } x^2 \leqslant 2\}$, a non-empty subset of $\mathbb{R}$. The set S is bounded above, since $x^2 \leqslant 2 \Rightarrow x^2 \leqslant 4 \Rightarrow x \leqslant 2$. By the completeness axiom, S possesses a least upper bound. Call this least upper bound M. A proof by contradiction is now employed to prove that $M^2 = 2$. Suppose that $M^2 < 2$ and consider $M' = M + 1/n$, $n \in \mathbb{N}$. Then

$$(M')^2 = M^2 + \frac{2M}{n} + \frac{1}{n^2} \leqslant M^2 + \frac{2M + 1}{n}$$

since $n^2 \geqslant n$ for all $n \in \mathbb{N}$. Now $(M')^2 < 2$ provided that

$$M^2 + \frac{2M + 1}{n} < 2$$

which is equivalent to

$$n > \frac{2M + 1}{2 - M^2}$$

By 2.3.1, there exists an integer n satisfying this last inequality. Hence $M' \in S$ for such an n, and this contradicts the fact that M is an upper bound for S.

Suppose now that $M^2 > 2$, and let $M' = M - 1/n$, $n \in \mathbb{N}$. This time,

$$(M')^2 = M^2 - \frac{2M}{n} + \frac{1}{n^2} \geqslant M^2 - \frac{2M}{n}$$

since $1/n^2 \geqslant 0$ for all $n \in \mathbb{N}$. Now $(M')^2 > 2$ provided that $M^2 - 2M/n > 2$, which holds when $n > 2M/(M^2 - 2)$. By 2.3.1 again, such an n exists. This contradicts the fact that M is the least upper bound of S. Therefore $M^2 = 2$ as desired.

Now that the existence of $\sqrt{2}$ has been established and it is known that $\sqrt{2}$ is not a rational number, it is possible to prove that the set $\mathbb{Q}$ of rational numbers is not complete. In other words, there exist bounded sets of rational numbers that do not possess a *rational* upper bound.

Consider $T = \{x : x \in \mathbb{Q}, x > 0 \text{ and } x^2 \leqslant 2\}$. Suppose that T has a rational least upper bound r. Since T is a subset of $S = \{x : x \in \mathbb{R}^+ \text{ and } x^2 \leqslant 2\}$, $\sqrt{2}$ is an (irrational) upper bound for T. Hence $r < \sqrt{2}$. By 2.3.2, there is a rational r' such that $r < r' < \sqrt{2}$. But then $(r')^2 < 2$, and so $r' \in T$. This contradicts the assertion that r is the supremum of T. Hence T does not possess a rational least upper bound. Thus $\mathbb{Q}$ is not complete.

This section concludes by stating the principle of mathematical induction. This makes it possible to establish that certain results hold for all natural numbers n.

2.3.3   The principle of mathematical induction

Let $P(n)$ be a statement for each natural number n. If

(a) $P(1)$ is true, and
(b) $P(k)$ true $\Rightarrow P(k + 1)$ true for every $k \in \mathbb{N}$

then $P(n)$ is true for all $n \in \mathbb{N}$.

This result, which can be deduced from the axioms for $\mathbb{R}$, is really a statement about natural numbers. If (a) holds then $P(1)$ is true. Hence, by (b), $P(2)$ holds, and, by (b) again, $P(3)$ holds, and so on. To conclude that $P(n)$ holds for all $n \in \mathbb{N}$ amounts to the belief that $\mathbb{N}$ is the set $\{1, 1 + 1, 1 + 1 + 1, \ldots\}$ or, as it is more commonly denoted, $\{1, 2, 3, \ldots\}$.

■■   EXAMPLE 4

Prove that $\displaystyle\sum_{r=1}^{n} r = \frac{1}{2}n(n + 1)$ for all $n \in \mathbb{N}$.

Solution

Let $P(n)$ be the statement

$$\sum_{r=1}^{n} r = \tfrac{1}{2}n(n + 1)$$

Since $1 = \tfrac{1}{2}1(1 + 1)$, $P(1)$ is true. Now suppose that $P(k)$ is true for a particular $k \in \mathbb{N}$. Then

$$1 + 2 + 3 + \ldots + (k + 1) = 1 + 2 + 3 + \ldots + k + (k + 1)$$
$$= \tfrac{1}{2}k(k + 1) + (k + 1)$$

By factorization,

$$\tfrac{1}{2}k(k + 1) + (k + 1) = \tfrac{1}{2}(k + 1)(k + 2)$$

Hence

$$1 + 2 + 3 + \ldots + (k + 1) = \tfrac{1}{2}(k + 1)(k + 2)$$

and so $P(k + 1)$ is also true. By the principle of mathematical induction, $P(n)$ is true for all $n \in \mathbb{N}$. ∎

Exercises 2.3

1. Show that $S = \{a + b\sqrt{2} : a \text{ and } b \text{ are rational}\}$ satisfies axioms A1–A9. Show further that S is an ordered field. Use a proof by contradiction to prove that $\sqrt{3} \notin S$. Is S a complete field?

2. Determine $\sup S$ and $\inf S$, where appropriate, for the following subsets of $\mathbb{R}$:

   (a) $S = \{x : x \in \mathbb{R} \text{ and } |2x - 1| < 11\}$
   (b) $S = \{x + |x - 1| : x \in \mathbb{R}\}$
   (c) $S = \{1 - 1/n : n \text{ is a non-zero integer}\}$
   (d) $S = \{2^{-m} + 3^{-n} + 5^{-p} : m, n \text{ and } p \in \mathbb{N}\}$
   (e) $S = \left\{ \dfrac{(-1)^n n}{2n + 1} : n \in \mathbb{N} \right\}$

3. Prove your assertions for the set S in Question 2(e).

4.   Let A and B be non-empty bounded subsets of $\mathbb{R}^+$. Prove that the set $C = \{x \cdot y : x \in A \text{ and } y \in B\}$ is bounded and that $\sup C = \sup A \cdot \sup B$, and $\inf C = \inf A \cdot \inf B$.

5.   Prove by induction on n that

(a) $\sum_{r=1}^{n} r^2 = \frac{1}{6}n(n+1)(2n+1)$ for all $n \in \mathbb{N}$

(b) $1 + nx < (1+x)^n$ for $n \geq 2$, $x \in \mathbb{R}$, $x > -1$ and $x \neq 0$

(c) $2^n < n!$ for all $n \in \mathbb{N}$, $n \geq 4$

6.   Prove by induction on the size of the set that every finite set has a minimum element.

7.   Prove the following generalization of the triangle inequality (see Section 2.2, Example 8): if $a_1, a_2, \ldots, a_n \in \mathbb{R}$ then

$$|a_1 + a_2 + \ldots + a_n| \leq |a_1| + |a_2| + \ldots + |a_n|$$

Problems 2

1.   (a) Show that $\sqrt[3]{2}$ is irrational.

(b) Given that $\sqrt{6}$ is irrational, show that $\sqrt{2} + \sqrt{3}$ is also irrational. (*Hint*: first show that if x^2 is irrational then x is irrational.)

2.   The rational numbers x and y satisfy

$$x^2 - 2xy - 2y^2 = 0$$

Given that $\sqrt{3}$ is irrational, prove that $x = y = 0$.

3.   Which of axioms A1–A14 of Section 2.2 are possessed by

(a) $\mathbb{N}$, the natural numbers,

(b) $\mathbb{Z}$, the integers,

(c) $\mathbb{Q}^+$, the positive rationals?

4.   Solve the following inequalities

(a) $x - 1 > \dfrac{6}{(x-2)}$

(b) $|x+1| \geq |x+2| + |x+3|$

5.   Prove that for all a, b, c and $d \in \mathbb{R}$

$$(a^2 + b^2)(c^2 + d^2) \geqslant (ac + bd)^2$$

6.   Let x, $y \in \mathbb{R}$, and suppose that $y \leqslant x + \varepsilon$ for all $\varepsilon \in \mathbb{R}^+$. Prove by contradiction that $y \leqslant x$.

7.   Determine $\sup S$ and $\inf S$, provided they exist, for the following subsets of $\mathbb{R}$:

(a) $S = \{x : |x + 1| + |2 - x| = 3\}$

(b) $S = \left\{ \left(1 - \dfrac{1}{n} \right)^2 : n \text{ is a non-zero integer} \right\}$

(c) $S = \left\{ x + \dfrac{1}{x} : x \text{ is a non-zero real number} \right\}$

(d) $S = \{x : 0 < x < 1 \text{ and the decimal expansion of } x \text{ contains no even digits}\}$

(e) $S = \{x : 0 < x < 1 \text{ and the decimal expansion of } x \text{ contains no odd digits}\}$

8.   Let A and B be non-empty bounded subsets of $\mathbb{R}$ such that $x \leqslant y$ for all $x \in A$ and for all $y \in B$. Prove that $\sup A \leqslant \inf B$. (*Hint*: first show that $\sup A \leqslant y$ for all $y \in B$.)

9.   Which of the following subsets of $\mathbb{R}$ are dense subsets of $\mathbb{R}$:

(a) a finite set,

(b) the set of integers,

(c) those rational numbers that in their lowest form have an even denominator?

10.   Prove by induction on n that

(a) $\displaystyle\sum_{r=1}^{n} (-1)^{r-1} r^2 = \tfrac{1}{2}(-1)^{n-1} n(n + 1)$

(b) $3^n \geqslant 2n^2 + 1$

(c) $\displaystyle\sum_{r=0}^{n} a^r = \dfrac{1 - a^{n+1}}{1 - a} \quad (a \neq 1)$

Sequences

This chapter deals with sequences of real numbers, such as

$1, \frac{1}{2}, \frac{1}{3}, \frac{1}{4}, \frac{1}{5}, \frac{1}{6}, \ldots$

$1, 2, 3, \ldots$

$-1, 1, -1, \ldots$

Various properties of such sequences are described, including the central concept of a convergent sequence. Crudely speaking, a sequence converges if the terms of the sequence approach arbitrarily close to a unique real number, called the limit of the sequence.

As a concrete example, consider the problem of determining a decimal corresponding to the irrational number $x = \sqrt{2}$. Since $x^2 = 2$, some simple algebra gives the rather strange equation

$$x = \frac{x^2 + 2}{2x}$$

This is of the form $x = f(x)$, and such equations may often be solved by the method of iteration. This technique involves the generation of a sequence $x_1, x_2, x_3, \ldots$ of real numbers, where the iteration formula

$x_{n+1} = f(x_n)$ is used to generate successive terms of the sequence, given a starting value for x_1. The hope is that successive terms provide better and better approximations to a solution of the equation $x = f(x)$. The iteration formula

$$x_{n+1} = \frac{x_n^2 + 2}{2x_n}$$

does indeed generate a sequence of better and better approximations to $\sqrt{2}$, a fact proved in Example 1 of Section 3.4. With $x_1 = 1$, the formula $x_{n+1} = (x_n^2 + 2)/2x_n$ gives

$$x_2 = \frac{1 + 2}{2} = 1.5$$

$$x_3 = \frac{(1.5)^2 + 2}{2(1.5)} = 1.417$$

$$x_4 = \frac{(1.417)^2 + 2}{2(1.417)} = 1.4142$$

and so on. It can be shown that each successive term of this sequence is a better approximation to $\sqrt{2}$ than previous terms by at least two decimal places, and so the decimal expansion of $\sqrt{2}$ can be found to any degree of accuracy required. It should be noted, however, that every term of the sequence of approximations to $\sqrt{2}$ is a rational number, whereas the limit $\sqrt{2}$ is not.

The definition of a convergent sequence (3.1.1) appears early in Section 3.1; at first sight, the definition may seem rather obscure, and time will be needed for the reader to master the logic involved. Historically, it was not until the late nineteenth century that a satisfactory definition of convergence emerged as part of the efforts to place analysis on a rigorous non-intuitive footing. What the definition does is to make precise what is meant by the phrase 'approach arbitrarily close to'. The definition is then used to establish various rules that can be used to determine the limits of many convergent sequences. These rules require knowledge of the limits of certain basic sequences, namely convergent sequences with limit zero. These basic null sequences are presented in Section 3.2, and a general strategy for finding the limit of an arbitrary convergent sequence is described. Sequences that do not converge are called divergent sequences, and Section 3.3 is devoted to a study of those divergent sequences that diverge to plus or minus infinity. Finally, in Section 3.4 the principle of monotone sequences is proved, namely that any increasing (or decreasing) sequences whose terms are bounded must converge. The requirement in this result that the sequence be monotone is essential, since there are bounded sequences, such as $1, -1, 1, -1, 1, \ldots$, that are not convergent. What can be established, though, is the Bolzano–Weierstrass theorem, which states that every bounded sequence contains a convergent subsequence. This theorem is then used to develop a characterization of convergent sequences that, unlike Definition 3.1.1, does not need to refer to the limit of the sequence. The proofs of the theorems and of the numerous rules established in this chapter make extensive use of the completeness axiom for the real numbers.

3.1   Convergent sequences

An **infinite sequence** $a_1, a_2, a_3, \ldots$ of real numbers may be specified either by giving a formula for the nth term, a_n, or by prescribing a method of generating the terms of the sequence. Such a sequence will be denoted by (a_n).

■■   EXAMPLE 1

Calculate the first five terms of the sequence (a_n) in the following cases:

(a) $a_n = n$ (b) $a_n = 1 - 1/n$

(c) $a_n = (-1)^n/n$ (d) $a_n = (-1)^n$

Solutions

In each case it is simply a matter of substituting $n = 1$, $n = 2$ up to $n = 5$ in the formula for a_n. This gives the following:

(a) $1, 2, 3, 4, 5, \ldots$

(b) $0, \frac{1}{2}, \frac{2}{3}, \frac{3}{4}, \frac{4}{5}, \ldots$

(c) $-1, \frac{1}{2}, -\frac{1}{3}, \frac{1}{4}, -\frac{1}{5}, \ldots$

(d) $-1, 1, -1, 1, -1, \ldots$ ■

The above example illustrates some general facts about sequences. Either the terms of a sequence are unbounded, or else there exists a real number M such that $|a_n| \leqslant M$ for all $n \in \mathbb{N}$. In the latter case the sequence (a_n) is said to be a **bounded sequence**. With a bounded sequence, the terms either oscillate without 'settling down', or else they tend to get closer to some fixed real number L. In the latter case (a_n) is said to be **converging to** L, and this is denoted by $a_n \to L$ as $n \to \infty$. The number L is called the **limit** of the sequence (a_n). A sequence that does not converge is said to **diverge**; divergent sequences are discussed in more detail in Section 3.3.

Intuitively, it seems plausible that certain sequences converge while others do not. In Example 1 sequences (b) and (c) both converge (to the limits 1 and 0 respectively), but sequences (a) and (d) diverge, though in markedly different ways. However, in order to work seriously with convergent sequences, deduce their properties and decide whether or not a given sequence does converge, a rigorous definition of the concept of convergence is required. Such a definition emerged only in the late nineteenth century and at first sight seems daunting because of the subtleties of the logic involved. What the definition seeks to capture, in a very precise manner, is the idea that for a sequence (a_n) to converge to a limit L it is necessary that the terms of the sequence, from a certain point onwards, all lie within some previously specified (small) distance of L.

3.1.1 Definition

> A sequence (a_n) **converges** to a limit L if and only if for every $\varepsilon > 0$ there exists a natural number N such that
>
> $$n > N \Rightarrow |a_n - L| < \varepsilon.$$

This is denoted by $a_n \to L$ as $n \to \infty$, or, alternatively, $\lim_{n\to\infty} a_n = L$, when (a_n) is a convergent sequence with limit L. Either statement is read as 'a_n tends to L as n tends to infinity'. The statement $|a_n - L| < \varepsilon$ says that $a_n - L$ lies between $-\varepsilon$ and $+\varepsilon$. Hence $-\varepsilon < a_n - L < \varepsilon$, leading to $L - \varepsilon < a_n < L + \varepsilon$. Therefore Definition 3.1.1 says that (a_n) converges to L provided that for any specified $\varepsilon > 0$, there is a term a_N of the sequence such that *all* subsequent terms lies between $L - \varepsilon$ and $L + \varepsilon$. This is illustrated in Figure 3.1, where, for the sequence depicted and the value of ε given, a suitable value for N is 7; for $n > 7$ the terms a_n all lie inside the horizontal strip from $L - \varepsilon$ to $L + \varepsilon$. Another way of viewing a convergent sequence is to regard successive terms of the sequence as

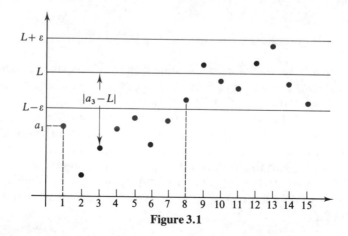

Figure 3.1

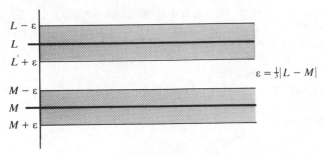

Figure 3.2

successive approximations to the limiting value of the sequence. The ε in Definition 3.1.1 then corresponds to a prescription of maximum error, and the N provided is such that *all* the terms corresponding to $n > N$ approximate the limit to the degree of accuracy specified by ε. The power of Definition 3.1.1 lies in the fact that a sequence (a_n) will converge to a limit L provided that for every $\varepsilon > 0$ it is possible to find a suitable N.

As it stands, Definition 3.1.1 defines precisely what a limit of a convergent sequence is, but does not immediately imply that there is only one limit. However, a convergent sequence (a_n) has a unique limit. To see this, suppose that L and M were different limits of the convergent sequence (a_n). Then, for any positive number ε and all sufficiently large n, the terms of the sequence would approximate both L and M with a margin of error no more than ε. But for small enough ε this would mean that the terms of the sequence eventually lay in two disjoint strips as illustrated in Figure 3.2 (the value of ε shown is one-third of the distance between L and M). This is impossible, and so a convergent sequence has a uniquely determined limit.

■■   **EXAMPLE 2**

Prove that $\dfrac{1}{n} \to 0$ as $n \to \infty$.

Solution

It seems intuitively obvious that for large values of n, the reciprocal $1/n$ is small and hence close to zero. However, in order to implement Definition 3.1.1, it has to be demonstrated that for every $\varepsilon > 0$ there is a suitable value for N.

Now $|a_n - L| = |1/n - 0| < \varepsilon$ provided that $1/n < \varepsilon$. By 2.3.1, there exists an integer N with $N > 1/\varepsilon$. For such an N,

$$n > N \Rightarrow n > \frac{1}{\varepsilon} \Rightarrow \frac{1}{n} < \varepsilon \Rightarrow \left| \frac{1}{n} - 0 \right| < \varepsilon$$

Thus for every $\varepsilon > 0$ a suitable N can be found. In this case N is any integer exceeding $1/\varepsilon$ (not necessarily the least such integer). Therefore $1/n \to 0$ as $n \to \infty$ ∎

■■ **EXAMPLE 3**

Prove that

$$\frac{4n + 1}{2n - 1} \to 2 \quad \text{as } n \to \infty$$

Solution

In this example

$$|a_n - L| = \left| \frac{4n + 1}{2n - 1} - 2 \right| = \frac{3}{2n - 1} \leqslant \frac{3}{n} < \varepsilon$$

provided that $n > 3/\varepsilon$. By 2.3.1, there exists an integer N exceeding $3/\varepsilon$. For such an N,

$$n > N \Rightarrow n > 3/\varepsilon$$

$$\Rightarrow \frac{3}{2n - 1} \leqslant \frac{3}{n} < \varepsilon$$

$$\Rightarrow \left| \frac{4n + 1}{2n - 1} - 2 \right| < \varepsilon$$

In other words,

$$\frac{4n + 1}{2n - 1} \to 2 \quad \text{as } n \to \infty$$

Incidentally, the inequality $3/(2n - 1) \leqslant 3/n$ used above was only employed in order to simplify the algebra before choosing the integer N. The reason that such tricks can be used is that the N required in Definition 3.1.1 need not be the least possible such N. ∎

These last two examples illustrate that the value of N depends very much on the value of ε. In general, the smaller ε is, the larger the value of N needs to be. Figure 3.3 details some possible values of ε and a corresponding value of N when $a_n = (4n + 1)/(2n - 1)$.

It is clearly tedious to use the definition of convergence to prove that a

ε	N
0·1	31
0·01	301
0·002	1501

Figure 3.3

given sequence has a prescribed limit L. In fact, it may not be known what L is! Once the following rules have been established, they can be used to evaluate the limits of quite complicated sequences from knowledge of the limits of simple sequences, like those in Examples 2 and 3.

3.1.2   Rules

Suppose that (a_n) and (b_n) are convergent sequences with limits A and B respectively; then the following rules apply:

Sum rule

$(a_n + b_n)$ converges to $A + B$.

Product rule

$(a_n b_n)$ converges to AB.

Quotient rule

(a_n/b_n) converges to A/B, provided that $b_n \neq 0$ for each n and $B \neq 0$.

Scalar product rule

(ka_n) converges to kA for every real number k.

Proof of the sum rule

Consider

$$|(a_n + b_n) - (A + B)| = |(a_n - A) + (b_n - B)|$$
$$\leqslant |a_n - A| + |b_n - B|$$

using the triangle inequality (see Section 2.2, Example 8). Given $\varepsilon > 0$, let $\varepsilon' = \frac{1}{2}\varepsilon$. Then $\varepsilon' > 0$ and, since $a_n \to A$ and $b_n \to B$ as $n \to \infty$, there exist natural numbers N_1 and N_2 such that

$$n > N_1 \Rightarrow |a_n - A| < \varepsilon'$$

and

$$n > N_2 \Rightarrow |b_n - B| < \varepsilon'$$

Let N be the maximum of N_1 and N_2, and so

$$n > N \Rightarrow |a_n - A| + |b_n - B| < \varepsilon' + \varepsilon' = 2\varepsilon' = \varepsilon$$

In other words, $a_n + b_n \rightarrow A + B$ as $n \rightarrow \infty$. $\qquad \square$

Proof of the product rule

First, it is shown that a convergent sequence is bounded. Since $b_n \rightarrow B$ as $n \rightarrow \infty$, if $\varepsilon = 1$ in the definition of convergence then there exists a natural number N such that $|b_n - B| < 1$ for $n > N$. Now $|b_n| = |b_n - B + B|$, so

$$|b_n| \leq |b_n - B| + |B| < 1 + |B| \quad \text{for } n > N$$

It is thus assumed that there is a real number M such that $|b_n| \leq M$ for all n. (M is the maximum of $|b_1|, |b_2|, \ldots, |b_N|$ and $1 + |B|$.) Now

$$\begin{aligned}
|a_n b_n - AB| &= |a_n b_n - Ab_n + Ab_n - AB| \\
&= |b_n(a_n - A) + A(b_n - B)| \\
&\leq |b_n||a_n - A| + |A||b_n - B|
\end{aligned}$$

using the triangle inequality and properties of the modulus. Given $\varepsilon > 0$, let $\varepsilon_1 = \varepsilon/2M$ and $\varepsilon_2 = \varepsilon/2(|A| + 1)$. Clearly, $\varepsilon_1 > 0$ and $\varepsilon_2 > 0$. Since $a_n \rightarrow A$ and $b_n \rightarrow B$ as $n \rightarrow \infty$, there exist natural numbers N_1 and N_2 such that

$$n > N_1 \Rightarrow |a_n - A| < \varepsilon_1$$

and

$$n > N_2 \Rightarrow |b_n - B| < \varepsilon_2$$

Let N_3 be the maximum of N_1 and N_2, and so conclude that if $n > N_3$ then

$$\begin{aligned}
|a_n b_n - AB| &\leq |b_n||a_n - A| + |A||b_n - B| \\
&< M\varepsilon/2M + |A|\varepsilon/2(|A| + 1) < \varepsilon
\end{aligned}$$

In other words, $a_n b_n \rightarrow AB$ as $n \rightarrow \infty$. $\qquad \square$

Proof of the quotient rule

First, it is shown that if $b_n \to B \neq 0$ as $n \to \infty$, and $b_n \neq 0$ for all n, then

$$\frac{1}{b_n} \to \frac{1}{B} \quad \text{as } n \to \infty$$

Now

$$\left| \frac{1}{b_n} - \frac{1}{B} \right| = \frac{|b_n - B|}{|b_n||B|}$$

If $\varepsilon = \frac{1}{2}|B|$ then $\varepsilon > 0$, and so there exists an integer N_1 such that $|b_n - B| < \varepsilon$ for all $n > N_1$. Hence

$$\tfrac{1}{2}|B| < |b_n| < \tfrac{3}{2}|B| \quad \text{for all } n > N_1$$

using Question 4 of Exercises 2.2. Let M be the maximum of

$$\frac{2}{|B|}, \frac{1}{|b_1|}, \ldots, \frac{1}{|b_{N_1}|}$$

Then $|1/b_n| \leq M$ for $n > N_1$. So, given any $\varepsilon > 0$, let $\varepsilon' = \varepsilon|B|/M$. Then $\varepsilon' > 0$, and so there exists an integer N_2 such that $|b_n - B| < \varepsilon'$ for all $n > N_2$. Hence

$$\left| \frac{1}{b_n} - \frac{1}{B} \right| = \frac{|b_n - B|}{|b_n||B|} < \frac{\varepsilon' M}{|B|} = \varepsilon \quad \text{for all } n > N$$

where N is the maximum of N_1 and N_2. In other words, $1/b_n \to 1/B$ as $n \to \infty$. Then, by the product rule,

$$\frac{a_n}{b_n} \to \frac{A}{B} \quad \text{as } n \to \infty \qquad\qquad \square$$

The scalar product rule is a special case of the product rule.

■■   EXAMPLE 4

Evaluate

$$\lim_{n \to \infty} \frac{n^2 + 2n + 3}{4n^2 + 5n + 6}$$

Solution

From Example 1, it can be taken that $1/n \to 0$ as $n \to \infty$, and it is easily shown that the constant sequence $k, k, k, \ldots$ converges to k (for every $\varepsilon > 0$ set $N = 0$ to fulfil Definition 3.1.1). Now the quotient rule cannot be applied directly, since neither the numerator nor denominator of

$$\frac{n^2 + 2n + 3}{4n^2 + 5n + 6}$$

converges to a finite limit. However, if the numerator and denominator are divided by the dominant term n^2, the following is obtained:

$$a_n = \frac{1 + 2/n + 3/n^2}{4 + 5/n + 6/n^2}$$

Hence

$$a_n = \frac{n^2 + 2n + 3}{4n^2 + 5n + 6} = \frac{1 + 2/n + 3/n^2}{4 + 5/n + 6/n^2} \to \frac{1 + 2 \cdot 0 + 3 \cdot 0^2}{4 + 5 \cdot 0 + 6 \cdot 0^2}$$

$$= \tfrac{1}{4} \quad \text{as } n \to \infty$$

freely using the sum, product, scalar product and quotient rules (see 3.1.2). ∎

3.1.3 Sandwich rule

Let (a_n), (b_n) and (c_n) be sequences satisfying $a_n \leqslant b_n \leqslant c_n$ for all $n \in \mathbb{N}$. If (a_n) and (c_n) both converge to the same limit L then (b_n) also converges to L.

Proof of the sandwich rule

First note that for any real number x

$$-|x| \leqslant x \leqslant |x|$$

Hence if x, y and z are real numbers satisfying $x \leqslant y \leqslant z$ then

$$-|x| \leqslant x \leqslant y \leqslant z \leqslant |z|$$

Thus

$$|y| \leqslant \max(|x|, |z|)$$

Therefore if $a_n \leqslant b_n \leqslant c_n$ then

$$a_n - L \leqslant b_n - L \leqslant c_n - L$$

and so $|b_n - L|$ is less than or equal to the maximum of $|a_n - L|$ and $|c_n - L|$. Given $\varepsilon > 0$, there exist natural numbers N_1 and N_2 such that $n > N_1 \Rightarrow |a_n - L| < \varepsilon$ and $n > N_2 \Rightarrow |c_n - L| < \varepsilon$. Let N be the maximum of N_1 and N_2. Then for $n > N$ it follows that $|b_n - L| < \varepsilon$. In other words, $b_n \to L$ as $n \to \infty$. □

■■　**EXAMPLE 5**

Show that $\lim_{n \to \infty} (-1)^n / n^2 = 0$.

Solution

Note that $|(-1)^n / n^2| \leqslant 1/n^2$. Now let $a_n = -1/n^2$, $b_n = (-1)^n / n^2$ and $c_n = 1/n^2$. By the product and scalar product rules (3.1.2), both (a_n) and (c_n) converge to 0. By the sandwich rule (3.1.3), (b_n) also converges to 0. ■

Remark

The condition in the sandwich rule that $a_n \leqslant b_n \leqslant c_n$ for all $n \in \mathbb{N}$ can be relaxed to $a_n \leqslant b_n \leqslant c_n$ for all $n > k$ (for some fixed $k \in \mathbb{N}$). This follows since the deletion of a finite number of terms from a sequence, or the addition of a finite number of new terms to a sequence, affects neither the convergence nor the limit.

The final rule involves the concept of a continuous function. Intuitively, a function $f: A \to \mathbb{R}$ whose domain A is a subset of $\mathbb{R}$ is continuous if its graph is a continuous curve. The rigorous definition of continuity is given in Section 5.2. It turns out that elementary functions such as polynomials, sine, cosine and exponential functions are all continuous on their domains. The proof of 3.1.4 is delayed until Section 5.2.

3.1.4　Composite rule

> Let (a_n) be a convergent sequence with limit L and let f be a continuous function whose domain contains $\{a_n : n \in \mathbb{N}\}$ and L. Then the sequence $(f(a_n))$ converges to $f(L)$.

■■ EXAMPLE 6

Show that

$$\lim_{n\to\infty} \cos\left(\frac{1}{n}\right) = 1$$

Solution

Knowing that $1/n \to 0$ as $n \to \infty$, and assuming that cos is a continuous function, then $\cos(1/n) \to \cos 0 = 1$ as $n \to \infty$. ■

In order to apply the rules 3.1.2–3.1.4 to good effect, knowledge of the convergence or divergence of certain basic sequences is required. Generating such a collection of basic sequences forms the substance of Sections 3.2 and 3.3. First note that if $a_n \to L$ as $n \to \infty$ then the sequence (b_n) given by $b_n = a_n - L$ converges to zero. The aim of the next section is to generate a collection of sequences that converge to zero. Such sequences are called null sequences.

Exercises 3.1

1. Use the definition of a convergent sequence to prove the following:

    (a) $\dfrac{n-1}{2n} \to \frac{1}{2}$ as $n \to \infty$

    (b) $\dfrac{(-1)^n}{n^2} \to 0$ as $n \to \infty$

2. Use the rules for convergent sequences to establish that if (a_n) converges to A and (b_n) converges to B then $(\alpha a_n + \beta b_n)$ converges to $\alpha A + \beta B$.

3. Use the rules for convergent sequences to evaluate the following limits:

    (a) $\lim\limits_{n\to\infty} \dfrac{4n^3 + 6n - 7}{n^3 - 2n^2 + 1}$

    (b) $\lim\limits_{n\to\infty} \dfrac{6 - n^2}{n^2 + 5n}$

    (c) $\lim\limits_{n\to\infty} [\log_e(n+1) - \log_e n]$

4.　　Use the sandwich rule to prove that each of the following sequences (a_n) converges to zero:

　　(a) $a_n = \dfrac{(-1)^n n}{\sqrt{n^3 + 1}}$　　　(b) $a_n = \dfrac{\cos n}{n}$

3.2　Null sequences

A sequence (a_n) is called a **null sequence** if (a_n) converges to zero. Application of the rules 3.1.2 gives immediately that if (a_n) and (b_n) are null sequences then so are the sequences $(a_n + b_n)$, $(a_n b_n)$ and (ka_n), $k \in \mathbb{R}$. In addition, the following rule will prove useful.

3.2.1　Power rule

> If (a_n) is a null sequence where $a_n \geq 0$ for all $n \in \mathbb{N}$, and if $p \in \mathbb{R}$, $p > 0$, then (a_n^p) is a null sequence.

Proof

For any $\varepsilon > 0$ set $\varepsilon' = \varepsilon^{1/p}$. Since (a_n) is a null sequence and $\varepsilon' > 0$, there exists a natural number N such that

$$n > N \Rightarrow |a_n| < \varepsilon'$$

But $a_n \geq 0$ and $p > 0$, so $|a_n^p| < (\varepsilon')^p = \varepsilon$ for $n > N$. In other words, (a_n^p) is a null sequence.　　　　　　　　　　　　　　　　　□

Proving that certain basic sequences are null sequences will require the use of the binomial expansion; for $x \in \mathbb{R}$ and $n \in \mathbb{N}$

$$(1 + x)^n = \sum_{k=0}^{n} \binom{n}{k} x^k$$

$$= 1 + nx + \tfrac{1}{2}n(n - 1)x^2 + \ldots + x^n$$

where

$$\binom{n}{k} = \frac{n!}{k!(n - k)!}$$

This formula can be established by induction on n.

3.2.2 Basic null sequences

The following are null sequences:

(a) $(1/n^p)$, for $p > 0$
(b) (c^n), for $|c| < 1$
(c) $(n^p c^n)$, for $p > 0$ and $|c| < 1$
(d) $(c^n/n!)$, for $c \in \mathbb{R}$
(e) $(n^p/n!)$, for $p > 0$

Proof

(a) By Example 2 of Section 3.1, $(1/n)$ is a null sequence. Hence, by the power rule (3.2.1), $(1/n^p)$ is a null sequence for $p > 0$.

(b) First note that a sequence (a_n) is null if and only if the sequence $(|a_n|)$ is null (see Question 1 of Exercises 3.2). It thus suffices to consider (c^n) when $0 \leqslant c < 1$. If $c = 0$, the sequence is clearly null, so it may be assumed that $0 < c < 1$. Write $c = 1/(1 + x)$, where $x > 0$. From the binomial expansion,

$$(1 + x)^n \geqslant 1 + nx \geqslant nx \quad \text{for } n \in \mathbb{N}$$

and hence

$$0 \leqslant c^n = \frac{1}{(1 + x)^n} \leqslant \frac{1}{nx} \quad \text{for } n \in \mathbb{N}$$

Since $(1/n)$ is a null sequence, the scalar product rule (3.1.2) gives that $(1/nx)$ is a null sequence. By the sandwich rule (3.1.3), (c^n) is a null sequence for $0 < c < 1$. Therefore (c^n) is a null sequence for $|c| < 1$.

(c) As in (b), it is sufficient to consider values of c satisfying $0 < c < 1$. Write $c = 1/(1 + x)$, where $x > 0$, and first consider the case $p = 1$. From the binomial expansion,

$$(1 + x)^n \geqslant 1 + nx + \tfrac{1}{2}n(n - 1)x^2 \geqslant \tfrac{1}{2}n(n - 1)x^2$$

$$\text{for } n \geqslant 2$$

and hence

$$0 \leqslant nc^n = \frac{n}{(1 + x)^n} \leqslant \frac{2n}{n(n - 1)x^2} = \frac{2}{(n - 1)x^2}$$

$$\text{for } n \geqslant 2$$

Since $(1/n)$ is a null sequence, the rules (3.1.2) give that

$$\frac{2}{(n-1)x^2} \to 0 \quad \text{as } n \to \infty$$

By the sandwich rule (3.1.3), (nc^n) is a null sequence.
For the general case where $p > 0$ and $0 < c < 1$, note that

$$n^p c^n = (nd^n)^p, \quad \text{where } d = c^{1/p}, \quad \text{for } n \in \mathbb{N}$$

Since $0 < d < 1$ and (nd^n) is a null sequence, the power rule
(3.2.1) gives that $(n^p c^n)$ is null for $p > 0$ and $0 < c < 1$. Hence
$(n^p c^n)$ is a null sequence for $p > 0$ and $|c| < 1$.

(d) It is sufficient to consider the case $c > 0$. By the Archimedean
postulate (2.3.1), there exists an integer m with $m + 1 > c$.
Then for $n > m + 1$

$$\frac{c^n}{n!} = \left(\frac{c}{1}\right)\left(\frac{c}{2}\right) \cdots \left(\frac{c}{m}\right)\left(\frac{c}{m+1}\right) \cdots \left(\frac{c}{n-1}\right)\left(\frac{c}{n}\right)$$

$$\leq \left(\frac{c}{1}\right)\left(\frac{c}{2}\right) \cdots \left(\frac{c}{m}\right)\left(\frac{c}{n}\right) = \left(\frac{c^m}{m!}\right)\left(\frac{c}{n}\right) = \frac{Kc}{n}$$

where $K = c^m/m!$ is a constant. Now $(1/n)$ is a null sequence,
and so (Kc/n) is a null sequence by the scalar product rule
(3.1.2). Since $0 \leq c^n/n! \leq Kc/n$ for $n > m + 1$, the sandwich
rule (3.1.3) gives that $(c^n/n!)$ is null.

(e) For $p > 0$ write

$$\frac{n^p}{n!} = \left(\frac{n^p}{2^n}\right)\left(\frac{2^n}{n!}\right) \quad \text{for } n \in \mathbb{N}$$

By part (c) with $c = \frac{1}{2}$, $(n^p/2^n)$ is null, and by part (d) with
$c = 2$, $(2^n/n!)$ is null. Hence by the product rule (3.1.2),
$(n^p/n!)$ is null. $\qquad\square$

The following general strategy may now be applied to evaluate the
limit of a complicated quotient of sequences:

(1) Identify the dominant term, using knowledge of basic null sequences.

(2) Divide both the numerator and denominator by the dominant term.

(3) Apply the various rules (3.1.2–3.1.4 and 3.2.1).

■■   EXAMPLE 1

Evaluate

$$\lim_{n \to \infty} \frac{n^3 + 2^n}{n^2 + 3^n}$$

Solution

The dominant term is 3^n. Hence

$$\frac{n^3 + 2^n}{n^2 + 3^n} = \frac{n^3/3^n + 2^n/3^n}{n^2/3^n + 1}$$

$$= \frac{n^3(\frac{1}{3})^n + (\frac{2}{3})^n}{n^2(\frac{1}{3})^n + 1}$$

Now $(n^2(\frac{1}{3})^n)$ and $(n^3(\frac{1}{3})^n)$ are null sequences by 3.2.2(c) with $c = \frac{1}{3}$ and with $p = 2$ and $p = 3$ respectively. By 3.2.2(b), $((\frac{2}{3})^n)$ is also null. Now apply the rules in 3.1.2 to deduce that

$$\frac{n^3 + 2^n}{n^2 + 3^n} \to \frac{0 + 0}{0 + 1} = 0 \quad \text{as } n \to \infty$$ ■

■■   EXAMPLE 2

Evaluate

$$\lim_{n \to \infty} \frac{n! + (-1)^n}{2^n + 3n!}$$

Solution

The dominant term is $3n!$. Hence

$$\frac{n! + (-1)^n}{2^n + 3n!} = \frac{1/3 + (-1)^n/3n!}{2^n/3n! + 1}$$

Both $((-1)^n/n!)$ and $(2^n/n!)$ are null by 3.2.2(d), and so, by the rules (3.1.2),

$$\frac{n! + (-1)^n}{2^n + 3n!} \to \frac{1}{3} \quad \text{as } n \to \infty$$ ■

The techniques employed in the proof of 3.2.2 have wide applicability, as the following example illustrates.

■■ **EXAMPLE 3**

(a) Prove that $1 + x/n \geqslant (1 + x)^{1/n}$ for $x > 0$ and $n \in \mathbb{N}$.

(b) Deduce that $\lim_{n \to \infty} c^{1/n} = 1$ for any $c > 0$.

Solution

(a) From the binomial expansion,

$$\left(1 + \frac{x}{n}\right)^n \geqslant 1 + n\left(\frac{x}{n}\right) = 1 + x$$

Since $x > 0$,

$$1 + \frac{x}{n} \geqslant (1 + x)^{1/n}$$

(b) The result is clear if $c = 1$. Consider the case $c > 1$, and write $c = 1 + x$, where $x > 0$. By part (a),

$$1 \leqslant c^{1/n} = (1 + x)^{1/n} \leqslant 1 + \frac{x}{n}$$

Since $\lim_{n \to \infty} (1 + x/n) = 1$, the sandwich rule (3.1.3) gives that $\lim_{n \to \infty} c^{1/n} = 1$. In the case $0 < c < 1$, $1/c > 1$ and so $\lim_{n \to \infty} (1/c)^{1/n} = 1$. By the quotient rule (3.1.2), $\lim_{n \to \infty} c^{1/n} = 1$, as required. Therefore, for any $c > 0$, $\lim_{n \to \infty} c^{1/n} = 1$. ■

Exercises 3.2

---

1.   Prove that (a_n) is a null sequence if and only if $(|a_n|)$ is a null sequence. Is the statement true if the word 'null' is replaced by the word 'convergent'?

2.   Show that the following sequences are null:

(a) $\left(\dfrac{n^2 + 2^n}{n! + 3n^3}\right)$ (b) $\left(\dfrac{2n!}{(n + 1)! + (n - 1)!}\right)$

(c) $\left(\dfrac{n^4 4^n}{n!}\right)$

3. Prove that if (a_n) is a null sequence and (b_n) is a bounded sequence then the sequence $(a_n b_n)$ is null.

4. Evaluate the following limits:

(a) $\lim\limits_{n\to\infty} \dfrac{n^2 - 2^n}{2^n + n}$

(b) $\lim\limits_{n\to\infty} \dfrac{3n! + 3^n}{n! + n^3}$

(c) $\lim\limits_{n\to\infty} (2^n + 3^n)^{1/n}$

5. Use the inequality $(1+x)^n \geqslant \frac{1}{2}n(n-1)x^2$ for $n \geqslant 2$ and $x \geqslant 0$ to prove that

$$n^{1/n} \leqslant 1 + \sqrt{\frac{2}{n-1}} \quad \text{for } n \geqslant 2$$

Hence deduce that $\lim_{n\to\infty} n^{1/n} = 1$.

3.3 Divergent sequences

A sequence that is not convergent is **divergent**. For example, the sequences $((-1)^n)$ and (n) are both divergent; although how to prove this by applying the definition of convergence (3.1.1) is unclear. The aim of this section is to provide criteria for divergence that avoid the need to argue directly from 3.1.1. Incidentally, showing that (a_n) is not convergent amounts to proving that the sequence $(a_n - L)$ is *not* null for any value of L. The criteria will be developed by establishing properties that are possessed by convergent sequences; if a sequence does not have these properties then it must be divergent.

Recall that a sequence (a_n) is bounded if $|a_n| \leqslant M$ for some real number M, for all $n \in \mathbb{N}$. The proof of the following result appeared in the proof of the product rule (3.1.2).

3.3.1 Result

> If (a_n) is convergent then (a_n) is bounded.

An immediate consequence of this result is that an unbounded sequence is necessarily divergent.

■■   **EXAMPLE 1**

Classify the following sequences as bounded or unbounded, and as convergent or divergent.

(a) $(1 + (-1)^n)$ (b) $((-1)^n n)$

(c) $\left(\dfrac{2n}{n+1}\right)$ (d) $(2n)$

Solution

(a) Bounded, since $1 + (-1)^n = 0$ or 2 depending on whether n is odd or even. Intuitively, the sequence $(1 + (-1)^n)$ is divergent – a fact that will be demonstrated later in this section.

(b) Unbounded, since for any real number M there exists an integer n such that $2n > M$, and $(-1)^{2n}2n = 2n$ is a term of the given sequence. Hence $((-1)^n n)$ is divergent by the contrapositive of 3.3.1.

(c) Since $2n/(n+1) \to 2$ as $n \to \infty$, $(2n/(n+1))$ is a convergent sequence. Hence, by 3.3.1, $(2n/(n+1))$ is bounded.

(d) Unbounded, using a similar argument to that in (b) above. Hence, by the contrapositive of 3.3.1, $(2n)$ is divergent.   ■

Note that there are bounded divergent sequences, as the example of $(1 + (-1)^n)$ demonstrates. This shows that the converse of 3.3.1 is false; in Section 3.4 some partial converses will be presented. Also observe that, although both $((-1)^n n)$ and $(2n)$ are unbounded, and hence are divergent, they behave in different ways. The terms of both become arbitrarily large, but in the sequence $(2n)$ they also remain positive. This notion of the terms of an unbounded sequence becoming large and positive is formalized in the next definition.

3.3.2   **Definition**

A sequence (a_n) **tends to infinity** if and only if for every positive real number K there exists an integer N such that

$$n > N \Rightarrow a_n > K.$$

This is denoted by $a_n \to \infty$ as $n \to \infty$, or, alternatively, $\lim_{n\to\infty} a_n = \infty$. The latter notation is a little unfortunate, since it seems to suggest that ∞ is a real number. This is not the case; the statement $\lim_{n\to\infty} a_n = \infty$ only expresses the fact that a_n is arbitrarily large and positive for sufficiently large positive values of n. Note immediately that if (a_n) tends to infinity, it is unbounded and hence (a_n) is divergent.

■■   EXAMPLE 2

Show that

$$\lim_{n\to\infty} \frac{n^2}{2n-1} = \infty$$

Solution

Observe that

$$\frac{n^2}{2n-1} > \frac{n^2}{2n} \quad \text{for all } n \in \mathbb{N}$$

Also, $n^2/2n = \frac{1}{2}n > K$, provided that $n > 2K$. For $K \in \mathbb{R}$, choose N to be any integer exceeding $2K$. Hence

$$n > N \Rightarrow n > 2K \Rightarrow \tfrac{1}{2}n > K \Rightarrow \frac{n^2}{2n-1} > K$$

In other words, $n^2/(2n-1) \to \infty$ as $n \to \infty$ ■

The above example demonstrates that, as with convergent sequences, arguing directly from a formal definition can prove technically demanding. The following rules, whose proofs are left as an exercise, may be used to identify sequences that tend to infinity.

3.3.3   Rules

Suppose that (a_n) tends to infinity and (b_n) tends to infinity; then the following hold:

Sum rule

$(a_n + b_n)$ tends to infinity

Product rule

$(a_n b_n)$ tends to infinity

Scalar product rule

$(k a_n)$ tends to infinity for any positive real number k

Basic null sequences (3.2.2) may also be used in the identification of sequences that tend to infinity, as the following result shows.

3.3.4 Reciprocal rule

> Suppose that (a_n) is a sequence whose terms are eventually positive and $(1/a_n)$ is null. Then $a_n \to \infty$ as $n \to \infty$.

Proof

The phrase 'eventually positive' means that there exists an integer N_1 such that $a_n > 0$ for all $n > N_1$. Let K be a positive real number. Then $\varepsilon = 1/K > 0$. Since $(1/a_n)$ is null, there exists an integer N_2 such that $n > N_2 \Rightarrow |1/a_n| < \varepsilon$. For such values of n, $|a_n| > 1/\varepsilon = K$. Choose N to be the greater of N_1 and N_2. Then

$$n > N \Rightarrow a_n = |a_n| > K$$

In other words, $a_n \to \infty$ as $n \to \infty$. □

■■ EXAMPLE 3

Prove that $(n! - 2^n)$ tends to infinity.

Solution

The sequence $(n! - 2^n)$ is eventually positive since, by Question 5(c) of Exercises 2.3, $n! > 2^n$ for all $n \geqslant 4$. Also,

$$\frac{1}{n! - 2^n} = \frac{1/n!}{1 - 2^n/n!} \to \frac{0}{1 - 0} = 0 \quad \text{as } n \to \infty$$

using the basic null sequences (3.2.2) and the rules for convergent sequences (3.1.2). By the reciprocal rule, $n! - 2^n \to \infty$ as $n \to \infty$. ■

A sequence (a_n) **tends to minus infinity** if the sequence $(-a_n)$ tends to infinity. This is written as $a_n \to -\infty$ as $n \to \infty$, or $\lim_{n \to \infty} a_n = -\infty$. Sequences that tend to minus infinity are unbounded and hence are divergent. However, not all unbounded sequences tend to either infinity or minus infinity. For instance, consider the unbounded sequence (a_n) given by $a_n = (-1)^n n$, in which the terms alternate in sign. The sequence obtained by writing down only the even terms $a_2, a_4, a_6, \ldots$ is a sequence that tends to infinity and hence is divergent. As will be seen shortly, this is sufficient to prove that (a_n) is divergent. The sequence obtained by writing down the odd terms $a_1, a_3, a_5, \ldots$ is also a divergent sequence (which tends to minus infinity). In a similar vein, if the sequence (a_n) is given by $a_n = 1 + (-1)^n$, the sequence obtained by writing down the even terms is 0, 0, 0, . . . and the sequence obtained by writing down the odd terms is 2, 2, 2, . . . These both converge, but to different limits; again, as will be seen, this suffices to prove that (a_n) is divergent. Clearly it is of interest to consider sequences obtained by restricting the subscripts of the a_n to be some increasing sequence of natural numbers.

3.3.5 Definition

> The sequence (a_{n_r}) is a **subsequence** of the sequence (a_n) if (n_r) is a strictly increasing sequence of natural numbers.

Certain properties of sequences are inherited by their subsequences, as the following results shows.

3.3.6 Result

> Suppose that (a_{n_r}) is a subsequence of (a_n):
>
> (i) if $a_n \to L$ as $n \to \infty$ then $a_{n_r} \to L$ as $r \to \infty$
>
> (ii) if $a_n \to \infty$ as $n \to \infty$ then $a_{n_r} \to \infty$ as $r \to \infty$

Proof

(i) Since $a_n \to L$ as $n \to \infty$, for each $\varepsilon > 0$ there exists an integer N such that

$$n > N \Rightarrow |a_n - L| < \varepsilon$$

Choose R such that $n_R \geq N$. Then

$$r > R \Rightarrow n_r > n_R > N \Rightarrow |a_{n_r} - L| < \varepsilon$$

In other words, $a_{n_r} \to L$ as $r \to \infty$.

(ii) is left as an exercise. □

This result leads directly to two useful criteria for establishing the divergence of certain sequences. These strategies validate the claims made earlier concerning the (divergent) sequences $(1 + (-1)^n)$ and $((-1)^n n)$.

Strategy 1

The sequence (a_n) is divergent if (a_n) has two convergent subsequences with different limits.

Strategy 2

The sequence (a_n) is divergent if (a_n) has a subsequence that tends to infinity or minus infinity.

■■ EXAMPLE 4

Prove that

$$\left(\frac{(-1)^n n + 1}{2n - 1}\right)$$

is a divergent sequence.

Solution

Let

$$a_n = \frac{(-1)^n n + 1}{2n - 1}$$

The subsequence (a_{2k}) is given by

$$a_{2k} = \frac{(-1)^{2k} 2k + 1}{2(2k) - 1} = \frac{2k + 1}{4k - 1}$$

Since $(2k + 1)/(4k - 1) \to \frac{1}{2}$ as $k \to \infty$, the subsequence (a_{2k}) converges to $\frac{1}{2}$. The subsequence (a_{2k-1}) is given by

$$a_{2k-1} = \frac{(-1)^{2k-1}(2k - 1) + 1}{2(2k - 1) - 1} = \frac{-2k + 2}{4k - 3}$$

Since $(-2k + 2)/(4k - 3) \to -\frac{1}{2}$ as $k \to \infty$, the subsequence (a_{2k-1}) converges to $-\frac{1}{2}$. By strategy 1, (a_n) is divergent. ■

Exercises 3.3

1.   Use the rules 3.3.3 and 3.3.4 to prove that the following sequences tend to infinity:

(a) $\left(\dfrac{2^n}{n^3}\right)$ (b) $\left(\dfrac{n! - n^3}{3^n}\right)$ (c) $(n^{1+1/n})$

2. Prove that if $a_n \neq 0$ for all $n \in \mathbb{N}$ and $a_n \to \infty$ as $n \to \infty$ then $(1/a_n)$ is a null sequence.

3. By looking for suitable subsequences, prove that the following sequences are divergent:

   (a) $(2n + (-1)^n)$ (b) $\left(\dfrac{(-1)^n n}{2n + 1} \right)$ (c) $(\sin \frac{1}{3} n\pi)$

4. Let (a_n) be a sequence such that the subsequences (a_{2k}) and (a_{2k-1}) both converge to the same limit L.
   Prove that $\lim_{n \to \infty} a_n = L$.

3.4 Monotone sequences

In Sections 3.1 and 3.2 various techniques for finding the limit of a convergent sequence were given. In Section 3.3 techniques for showing that certain sequences are divergent were developed. But an arbitrary sequence (a_n) may not yield to any of these methods, and hence whether or not (a_n) is convergent may be left unanswered. However, it is sometimes possible to prove that a sequence is convergent, even though its limit is unknown. This situation arises for certain bounded sequences, as the first result in this section establishes. Some new terminology will be required first.

An **increasing sequence** (a_n) is one in which $a_n \leq a_{n+1}$ for all $n \in \mathbb{N}$. Similary, a **decreasing sequence** (a_n) is one in which $a_n \geq a_{n+1}$ for all $n \in \mathbb{N}$. A sequence that is either increasing or decreasing is called a **monotone sequence**. Note that not all sequences are monotone.

3.4.1 Principle of monotone sequences

A bounded monotone sequence is convergent.

Proof

The result for a bounded monotone increasing sequence is proved, the proof being similar for a decreasing sequence.

Let (a_n) be such that $a_1 \leq a_2 \leq \ldots$ and $|a_n| \leq M$ for all $n \in \mathbb{N}$. Let $M_0 = \sup \{a_n : n \in \mathbb{N}\}$, the least upper bound of the set of numbers appearing in the sequence; M_0 exists by the completeness axiom for the real numbers (see 2.2.3).

Given $\varepsilon > 0$, $M_0 - \varepsilon$ cannot be an upper bound for $\{a_n\}$. Hence there exists a value of N such that $a_N > M_0 - \varepsilon$. If $n > N$ then $a_n \geq a_N$, and so $a_n > M_0 - \varepsilon$. Furthermore, $a_n \leq M_0$ by

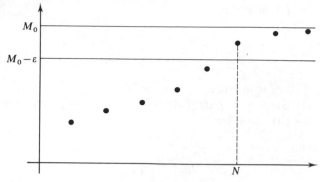

Figure 3.4

the definition of M_0, and hence, for $n > N$, $|a_n - M_0| < \varepsilon$. This proves that $a_n \to M_0$ as $n \to \infty$.   □

The proof is illustrated in Figure 3.4.

■■   **EXAMPLE 1**

A sequence (a_n) is defined by

$$a_1 = 1 \quad \text{and} \quad a_{n+1} = \frac{a_n^2 + 2}{2a_n} \quad \text{for } n \geq 1$$

Prove that (a_n) converges to $\sqrt{2}$.

Solution

First of all, observe that if the sequence does converge it must do so to $\sqrt{2}$. To see this, suppose that $a_n \to L$ as $n \to \infty$, where $L \neq 0$. Then $a_{n+1} \to L$ as $n \to \infty$ (why is this so?). By the product, sum and reciprocal rules,

$$\frac{a_n^2 + 2}{2a_n} \to \frac{L^2 + 2}{2L} \quad \text{as } n \to \infty$$

and so, by the uniqueness of limits, $L = (L^2 + 2)/2L$. Solving for L gives $L = \pm\sqrt{2}$.

As will be shown below, $a_n \geq \sqrt{2}$ for all $n > 1$, and so $L = \sqrt{2}$, as claimed.

To complete the argument, it is now necessary to prove that (a_n) is indeed a convergent sequence. Two facts will be established

(i) $a_n \geq \sqrt{2}$ for all $n > 1$

(ii) $a_{n+1} \leq a_n$ for all $n > 1$

For (i), note that

$$a_n \geqslant \sqrt{2} \Rightarrow a_{n+1} = \frac{a_n^2 + 2}{2a_n} \geqslant \frac{2\sqrt{2}a_n}{2a_n} = \sqrt{2}$$

using the inequality $x^2 + y^2 \geqslant 2xy$ (why does this inequality hold?). Since $a_2 = \frac{3}{2}$ and $(\frac{3}{2})^2 \geqslant 2$, (i) follows by induction on n.

For (ii), consider

$$a_n - a_{n+1} = a_n - \frac{a_n^2 + 2}{2a_n} = \frac{a_n^2 - 2}{2a_n}$$

Since $a_n \geqslant \sqrt{2}$ for all $n > 1$, it follows immediately that $a_n - a_{n+1} \geqslant 0$, and so (ii) is established.

By virtue of (i) and (ii), (a_n) is, after the first term, a decreasing sequence bounded below. Hence, by the principle of monotone sequences, (a_n) is convergent. Since $a_n \geqslant \sqrt{2}$ for all $n > 1$, the limit L of the sequence is non-zero, and hence the earlier argument shows that $a_n \to \sqrt{2}$ as $n \to \infty$. ∎

■■ EXAMPLE 2

A sequence (a_n) is defined by $a_1 = 1$ and $a_{n+1} = \sqrt{a_n + 1}$ for $n \geqslant 1$. Show that

$$\lim_{n \to \infty} a_n = \tfrac{1}{2}(1 + \sqrt{5})$$

Solution

First it is shown by induction on n that (a_n) is an increasing sequence. Since $a_1 = 1$ and $a_2 = \sqrt{2}$, $a_1 \leqslant a_2$. Now

$$a_{n+1} - a_n = \sqrt{a_n + 1} - \sqrt{a_{n-1} + 1}$$

$$= \frac{a_n - a_{n-1}}{\sqrt{a_n + 1} + \sqrt{a_{n-1} + 1}}$$

using the easily established fact that

$$\sqrt{x} - \sqrt{y} = \frac{x - y}{\sqrt{x} + \sqrt{y}}$$

This expression for $a_{n+1} - a_n$ has a positive denominator. If

$a_{n-1} \leqslant a_n$ then $a_n \leqslant a_{n+1}$, and so, by induction, (a_n) is an increasing sequence. Now

$$a_n^2 - a_{n+1}^2 = a_n^2 - a_n - 1$$
$$= (a_n - \tfrac{1}{2})^2 - \tfrac{5}{4}$$

and, since (a_n) is increasing, $(a_n - \tfrac{1}{2})^2 - \tfrac{5}{4} \leqslant 0$. This quickly leads to (a_n) being bounded above by $\tfrac{1}{2}(1 + \sqrt{5})$. By the principle of monotone sequences, (a_n) is convergent.

Let $a_n \to L$ as $n \to \infty$, so that $a_{n+1} \to L$ as $n \to \infty$. By the product rule, $a_{n+1}^2 \to L^2$ as $n \to \infty$. But $a_{n+1}^2 = a_n + 1 \to L + 1$ as $n \to \infty$, and so, by the uniqueness of limits, $L^2 = L + 1$. The quadratic equation $L^2 = L + 1$ has two roots, namely $\tfrac{1}{2}(1 \pm \sqrt{5})$. Since $a_n \geqslant 1$ for all $n \in \mathbb{N}$, the positive root is required. Hence $L = \tfrac{1}{2}(1 + \sqrt{5})$.   ■

■■   EXAMPLE 3

Show that the sequence $((1 + 1/n)^n)$ is convergent.

Solution

By the binomial expansion,

$$a_n = \left(1 + \frac{1}{n}\right)^n = 1 + n\left(\frac{1}{n}\right) + \frac{n(n-1)}{2!}\left(\frac{1}{n}\right)^2$$

$$+ \frac{n(n-1)(n-2)}{3!}\left(\frac{1}{n}\right)^3 + \ldots + \left(\frac{1}{n}\right)^n$$

$$= 1 + 1 + \frac{1}{2!}\left(1 - \frac{1}{n}\right) + \frac{1}{3!}\left(1 - \frac{1}{n}\right)\left(1 - \frac{2}{n}\right)$$

$$+ \ldots + \frac{1}{n!}\left(1 - \frac{1}{n}\right)\left(1 - \frac{2}{n}\right) \cdots \left(1 - \frac{n-1}{n}\right)$$

The analogous formula for a_{n+1} shows that, apart from the first two terms on the right-hand side of the expression for a_n, the terms increase in value, and an extra term appears. Hence $a_{n+1} > a_n$, and so (a_n) is increasing. Also

$$a_n < 1 + 1 + \frac{1}{2!} + \frac{1}{3!} + \ldots + \frac{1}{n!}$$

$$\leq 1 + 1 + \frac{1}{2} + \frac{1}{2^2} + \ldots + \frac{1}{2^{n-1}}$$

since $2^n \leq (n+1)!$ for all $n \in \mathbb{N}$ (see Question 5(c) of Exercises 2.3). Therefore $a_n < 1 + 2(1 - 1/2^n) < 3$. Hence (a_n) is bounded, and so, by the principle of monotone sequences, (a_n) is convergent. In fact, $\lim_{n \to \infty} (1 + 1/n)^n = e$ (see after Example 5 in Section 6.2). In elementary texts, e is defined to be the limit of the sequence $((1 + 1/n)^n)$, but in practice it is more convenient to define e via the exponential function, a topic discussed in Section 4.3. ∎

The principle of monotone sequences is clearly a partial converse to the result that every convergent sequence is bounded (3.3.1). The next result is also a partial converse to 3.3.1.

3.4.2 Bolzano–Weierstrass theorem

Any bounded sequence (a_n) of real numbers contains a convergent subsequence.

Proof

Let $S_N = \{a_n : n > N\}$. If every S_N has a maximum element then define a subsequence of (a_n) as follows: $b_1 = a_{n_1}$ is the maximum of S_1, $b_2 = a_{n_2}$ is the maximum of S_{n_1}, $b_3 = a_{n_3}$ is the maximum of S_{n_2}, and so on. Therefore (b_n) is a monotone decreasing subsequence of (a_n). Since (a_n) is bounded, so too is (b_n). By 3.4.1, (b_n) is a convergent subsequence of (a_n).

On the other hand, if, for some M, S_M does not have a maximum element then for any a_m with $m > M$ there exists an a_n following a_m with $a_n > a_m$ (otherwise the largest of $a_{M+1}, \ldots, a_m$ would be the maximum of S_M). Now let $c_1 = a_{M+1}$ and let c_2 be the first term of (a_n) following c_1 for which $c_2 > c_1$. Now let c_3 be the first term of (a_n) following c_2 for which $c_3 > c_2$, and so on. Therefore (c_n) is a monotone increasing subsequence of (a_n). Since (c_n) is bounded, 3.4.1 implies that (c_n) is a convergent subsequence of (a_n). □

Bernhard Bolzano (1781–1848) was a priest living in Prague whose mathematical discoveries were largely ignored in his lifetime. They were 'rediscovered' by subsequent mathematicians whose names adorn many of his results. The exception to this is the Bolzano–Weierstrass theorem. This reflects the fact that it was Weierstrass who independently (and some 80 years later) constructed functions originally discovered by Bolzano – functions whose behaviour conflicted with the Newtonian ideal that curves were generated by smooth and continuous motion, with only isolated abrupt changes. Bolzano found curves with no tangent at any point! His posthumous work, *Paradoxes of the Infinite*, appeared in 1850 and contained a version of the Bolzano–Weierstrass theorem later made famous to mathematicians by Weierstrass.

Karl Weierstrass (1815–1897) was renowned as a great and influential teacher. Originally he studied law and finance and taught in elementary school until he was 40. Weierstrass then became an instructor at the University of Berlin and obtained a full professorship in 1864. From then on, he devoted himself to advanced mathematics. His greatest contribution to mathematics was the founding of a theory of complex functions on power series, but it was his recognition of the need for the arithmetization of analysis that earned him the nickname of 'the father of modern analysis'. Weierstrass' great gifts as a teacher allied with his careful attention to logical reasoning in the field of analysis established an ideal for future generations of mathematicians. His death in 1897 was exactly 100 years after the first published attempt (by Lagrange) at a rigorous calculus.

An important consequence of the Bolzano–Weierstrass theorem is that it is now possible to formulate a definition of convergence for a sequence (a_n) that does not explicitly involve the limit L of that sequence. First note that if $a_n \to L$ as $n \to \infty$ then for any given $\varepsilon > 0$ there exists, by 3.1.1, a natural number N such that

$$n > N \Rightarrow |a_n - L| < \tfrac{1}{2}\varepsilon$$

Suppose now that $m, n > N$; then

$$\begin{aligned}
|a_m - a_n| &= |(a_m - L) + (L - a_n)| \\
&\leqslant |a_m - L| + |L - a_n| \\
&< \tfrac{1}{2}\varepsilon + \tfrac{1}{2}\varepsilon = \varepsilon
\end{aligned}$$

In other words, any two terms of the convergent sequence after an appropriate a_N differ by no more than the ε originally specified. This **Cauchy condition** can now be used to define the concept of a Cauchy sequence.

3.4.3 Definition

A sequence (a_n) is a **Cauchy sequence** if and only if for every $\varepsilon > 0$ there exists a natural number N such that $m, n > N \Rightarrow |a_m - a_n| < \varepsilon$.

What is surprising is that all Cauchy sequences are convergent, and hence that the Cauchy condition gives a characterization of convergence that involves no reference to the limit in question.

3.4.4 Theorem

A sequence (a_n) is convergent if and only if (a_n) is a Cauchy sequence.

Proof

It has already been demonstrated that a convergent sequence is a Cauchy sequence. For the converse, suppose that (a_n) is a Cauchy sequence. Then for $\varepsilon = 1$ there exists a natural number N such that

$$m, n > N \Rightarrow |a_m - a_n| < 1$$

In particular,

$$|a_m - a_{N+1}| < 1 \quad \text{for all} \quad m > N$$

Therefore the set

$$S = \{a_m : m > N\}$$

is a bounded set. Since S contains all but a finite number of terms of (a_n), the sequence (a_n) is itself bounded. By the Bolzano–Weierstrass theorem (3.4.2), (a_n) contains a convergent subsequence (a_{n_r}). Suppose that $a_{n_r} \to L$ as $r \to \infty$. Then, given $\varepsilon > 0$, there exists a natural number N_1 such that

$$n_r > N_1 \Rightarrow |a_{n_r} - L| < \tfrac{1}{2}\varepsilon$$

Since (a_n) is a Cauchy sequence, there exists a natural number N_2 such that

$$n, m > N_2 \Rightarrow |a_m - a_n| < \tfrac{1}{2}\varepsilon$$

Now let N be the maximum of N_1 and N_2. Then for $n, n_r > N$

$$
\begin{aligned}
|a_n - L| &= |(a_n - a_{n_r}) + (a_{n_r} - L)| \\
&\leq |a_n - a_{n_r}| + |a_{n_r} - L| \\
&< \tfrac{1}{2}\varepsilon + \tfrac{1}{2}\varepsilon = \varepsilon
\end{aligned}
$$

In other words, (a_n) converges to L, and so (a_n) is a convergent sequence as required. ∎

Augustin-Louis Cauchy (1789–1857) was a prolific French analyst whose contributions to research include work on the convergence and divergence of infinite series. He developed a theory of real and complex functions, and published papers on differential equations, determinants, probability theory and mathematical physics. In 1821 he successfully developed an acceptable theory of limits, and defined convergence, continuity, differentiability and integrability in terms of this limit concept. Our present definitions of these same concepts are essentially those given by Cauchy. Cauchy wrote books at all levels and took great pride in communicating his discoveries to others in relatively clear terms. For instance, his definition of limit reads:

When successive values attributed to a variable approach indefinitely a fixed value so as to end by differing from it by as little as one wishes, this last is called the limit of all the others.

(A not dissimilar definition is also attributed to Bolzano.)

Exercises 3.4

1. Determine which of the following sequences are monotone:

   (a) $\left(\dfrac{n+1}{n+2} \right)$ (b) $\left(n + \dfrac{8}{n} \right)$

   (c) $(n + (-1)^n)$ (d) $(2n + (-1)^n)$

2. Let

   $$
   a_n = 1 + \frac{1}{2^2} + \frac{1}{3^2} + \ldots + \frac{1}{n^2} \quad \text{for each } n \in \mathbb{N}
   $$

(a) Show that (a_n) is increasing.

(b) Prove by induction on n that $a_n \leqslant 2 - 1/n$ for $n \geqslant 1$.

(c) Deduce that (a_n) is convergent.

(The limit is $\frac{1}{6}\pi^2$, but this is harder to establish.)

3. By considering the product of the first n terms of the sequence $((1 + 1/n)^n)$, prove that

$$n! > \left(\frac{n + 1}{e}\right)^n$$

Problems 3

---

1. Use the definition of a convergent sequence (3.1.1) to prove the following.

(a) $\dfrac{2n + 5}{n + 3} \to 2$ as $n \to \infty$

(b) $\dfrac{3n^2 + 4}{n^2 + 1} \to 3$ as $n \to \infty$

2. The sequence (a_n) satisfies $a_n > 0$ and $a_{n+1} < ka_n$ for all n, where $0 < k < 1$. Prove that (a_n) is a null sequence.

3. Evaluate, where possible, $\lim_{n \to \infty} a_n$ for

(a) $a_n = \dfrac{n^2 - 2\sqrt{n} + 1}{1 - n - 3n^2}$ (b) $a_n = \sqrt{n^2 + n} - \sqrt{n^2 - 1}$

(c) $a_n = (-1)^n + \dfrac{1}{n}$ (d) $a_n = \left(\dfrac{n - 3}{n}\right)^n$

(e) $a_n = \dfrac{5n! + 5}{n^{100} + n!}$

4. Which of the following statements are true and which are false? Justify your assertions.

(a) If $\lim_{n \to \infty} a_n = \infty$ and $\lim_{n \to \infty} b_n = L > 0$ then $\lim_{n \to \infty} (a_n b_n) = \infty$.

(b) If $\lim_{n \to \infty} a_n = \infty$ and $\lim_{n \to \infty} b_n = -\infty$ then $\lim_{n \to \infty} (a_n + b_n) = 0$.

(c) If $\lim_{n\to\infty} a_n = \infty$ and $\lim_{n\to\infty} b_n = -\infty$ then
   $\lim_{n\to\infty} (a_n b_n) = -\infty$.

(d) If (a_n) and (b_n) are divergent then $(a_n b_n)$ is also divergent.

5.   Find unbounded sequences (a_n) and (b_n) that do not diverge to $\pm\infty$ such that

(a) $(a_n + b_n)$ is convergent.

(b) $(a_n + b_n)$ is bounded but divergent.

(c) $(a_n + b_n)$ is neither convergent, nor divergent to $\pm\infty$.

6.   For each of the following sequences, decide whether it is

(i)   bounded (above or below),

(ii) (eventually) positive or negative,

(iii) (eventually) increasing, decreasing or neither,

(iv) convergent, divergent or divergent to $\pm\infty$.

(a) $\left(\dfrac{n^2}{n^2 + 1} \right)$ (b) $(2n - |10 - n|)$

(c) $(n + (-1)^n n^2)$ (d) $\left((-1)^n + \dfrac{1}{n} \right)$

(e) $\left(1 - \dfrac{(-1)^n}{n} \right)$

7.   Suppose that (a_n) is an increasing sequence and let (a_{n_r}) denote a subsequence of (a_n). Prove that

(a) if $(a_{n_r}) \to L$ as $r \to \infty$ then $(a_n) \to L$ as $n \to \infty$,

(b) if $(a_{n_r}) \to \infty$ as $r \to \infty$ then $(a_n) \to \infty$ as $n \to \infty$.

(*Hint*: results 3.3.6 and 3.4.1 are useful.)

8.   A sequence (a_n) is defined by

$$a_1 = \alpha \quad \text{and} \quad a_{n+1} = \tfrac{1}{4}(1 + a_n) \quad \text{for } n \geqslant 1$$

(a) When $\alpha < \tfrac{1}{3}$, show that (a_n) is increasing and bounded above by 1.

(b) When $\alpha > \tfrac{1}{3}$, show that (a_n) is decreasing and bounded below by 0.

What happens in the case $\alpha = \tfrac{1}{3}$?

9.   A sequence (b_n) is defined by

$$b_1 = 0 \quad \text{and} \quad b_{n+1} = \tfrac{1}{2}(1 - b_n) \quad \text{for } n \geqslant 1$$

Calculate the first six terms of (b_n), and then, by considering the subsequences (b_{2n}) and (b_{2n-1}), discuss the convergence of (b_n). (*Hint*: Problem 8 above is useful.)

10.   A sequence (a_n) satisfies $|a_{n+1} - a_n| \leqslant \alpha^n$ for all $n \geqslant 1$, where $0 < \alpha < 1$. Show that for $m > n$

$$|a_m - a_n| \leqslant \alpha^n/(1 - \alpha)$$

(*Hint*: Problem 10(c) of Chapter 2 is useful.) Deduce that (a_n) is a convergent (Cauchy) sequence.

CHAPTER FOUR

4

Series

Absurdities arising from the indiscriminate use of infinite series were among the catalysts for attempts to introduce rigour into analysis. However, as pointed out in the Prologue, paradoxes of the infinite had been in existence since the time of Ancient Greece. For instance, the Greek philosopher Zeno claimed that it was impossible to travel any distance, since to do so required one to first travel half the distance, then half the remaining distance, and so on. Since some distance always remained, one could not complete the journey. The essence of this paradox was the feeling that it was impossible to 'add up' an infinite number of positive quantities and obtain a finite answer. However, the distance from 0 to 1 can be split up into an infinite sequence of distances $\frac{1}{2}, \frac{1}{4}, \frac{1}{8}, \ldots$ as shown in Figure 4.1, and, as will be shown in Example 1 of Section 4.1,

$$\tfrac{1}{2} + \tfrac{1}{4} + \tfrac{1}{8} + \ldots = 1$$

Nevertheless, the temptation to treat these 'infinite sums' as if they were exceedingly long finite sums has to be resisted. For, suppose that it were possible to add up the numbers 2, 4, 8, 16, ... and that the resulting

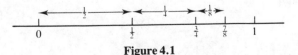

Figure 4.1

sum was S. Then

$$2 + 4 + 8 + 16 + \ldots = S$$

Multiplying throughout by $\frac{1}{2}$ gives

$$1 + 2 + 4 + 8 + \ldots = \frac{1}{2}S$$

Therefore $1 + S = \frac{1}{2}S$, leading to the nonsensical result that $S = -2$.

To avoid such absurdities, a careful definition of the sum of (convergent) infinite series is required. This is given in Section 4.1, where some elementary properties of convergent series are also established. In Section 4.2 several tests are developed for testing whether or not a given series is convergent. The important concept of absolute convergence is discussed; in crude terms, absolutely convergent series can be manipulated much like finite sums without any absurdities arising. The section finishes with a suggested strategy for testing an arbitrary series for convergence. In Section 4.3 infinite series consisting of ascending powers of a real number x are dealt with. Such series are used to give rigorous definitions of elementary functions such as the exponential and trigonometric functions.

4.1 Infinite series

If a sequence (a_n) is given, the finite sum $s_n = (a_1 + a_2 + \ldots + a_n)$ for each $n \in \mathbb{N}$ can be formed. If (s_n) converges to some limit s then s can justifiably be called the sum of the **infinite series**

$$\sum_{r=1}^{\infty} a_r = a_1 + a_2 + \ldots$$

To be precise, the series $\sum_{r=1}^{\infty} a_r$ is a **convergent series**, with **sum** s, if and only if the sequence (s_n) of **nth partial sums** converges to s. If (s_n) is a divergent sequence then, irrespective of its precise behaviour, $\sum_{r=1}^{\infty} a_r$ is called a **divergent series**. Rather regrettably, $\sum_{r=1}^{\infty} a_r$ is still used to denote a divergent series, even though it does not possess a sum.

■■ EXAMPLE 1

Show that $\frac{1}{2} + \frac{1}{4} + \frac{1}{8} + \ldots$ has sum 1.

Solution

The nth partial sum of $\sum_{r=1}^{\infty} \left(\frac{1}{2}\right)^r$ is

$$s_n = \frac{1}{2} + \frac{1}{4} + \ldots + \left(\frac{1}{2}\right)^n$$

Multiplying by 2 gives

$$2s_n = 1 + \tfrac{1}{2} + \tfrac{1}{4} + \ldots + (\tfrac{1}{2})^{n-1}$$

Subtracting these two expressions gives $s_n = 1 - (\tfrac{1}{2})^n$. Since $s_n \to 1$ as $n \to \infty$, it can be deduced that $\sum_{r=1}^{\infty} (\tfrac{1}{2})^r$ converges and has sum 1. ∎

■■ EXAMPLE 2

Show that $1 + 2 + 3 + \ldots$ is a divergent series.

Solution

The nth partial sum is $s_n = 1 + 2 + \ldots + n = \tfrac{1}{2}n(n+1)$. Since (s_n) is a divergent sequence, $\sum_{r=1}^{\infty} r$ is a divergent series. ∎

■■ EXAMPLE 3

Show that

$$\sum_{r=1}^{\infty} \frac{1}{r^2 + r} = 1$$

Solution

Since

$$\frac{1}{r^2 + r} = \frac{1}{r(r+1)} = \frac{1}{r} - \frac{1}{r+1}$$

the nth partial sum of $\sum_{r=1}^{\infty} 1/(r^2 + r)$ can be written as

$$s_n = \left(1 - \frac{1}{2}\right) + \left(\frac{1}{2} - \frac{1}{3}\right) + \ldots + \left(\frac{1}{n} - \frac{1}{n+1}\right) = 1 - \frac{1}{n+1}$$

Now $s_n \to 1$ as $n \to \infty$, as required. ∎

Example 1 is a special case of an important class of infinite series, namely the **geometric series** $\sum_{r=0}^{\infty} ax^r$, where x is a real number. Note that the summation here begins at $r = 0$ and not at $r = 1$. For this series the sum of the first n terms is

$$s_n = a + ax + ax^2 + \ldots + ax^{n-1}$$

so $xs_n = ax + ax^2 + \ldots + ax^n$, and, by subtraction,

$$s_n = \frac{a(1 - x^n)}{1 - x} \quad \text{for } x \neq 1$$

If $|x| < 1$, (x^n) is a basic null sequence (3.2.2), and hence

$$s_n \to \frac{a}{1 - x} \quad \text{as } n \to \infty$$

If $x > 1$, (x^n) tends to infinity by application of the reciprocal rule (3.3.4) to the null sequence $(1/x^n)$.

If $x < -1$, (x^n) contains the subsequence (x^{2n}), which tends to infinity, and therefore, by Strategy 2 of Section 3.3, (x^n) diverges.

Hence (s_n) diverges for $|x| > 1$.

If $x = -1$, the subsequences (s_{2n}) and (s_{2n-1}) converge to different limits, and so (s_n) is divergent by Strategy 1 of Section 3.3.

If $x = 1$, $s_n = na \to \infty$ as $n \to \infty$.

Thus (s_n) converges if and only if $|x| < 1$. This gives the following result.

4.1.1 Result

The geometric series

$$\sum_{r=0}^{\infty} ax^r = a + ax + ax^2 + \ldots \quad (a \neq 0)$$

converges if and only if $|x| < 1$.
Moreover, its sum is then $a/(1 - x)$.

Since the sum of a convergent series is defined to be the limit of the sequence of nth partial sums of the series in question, the results of Section 3.1 can be used to establish theorems concerning series.

The first result provides a useful test for the divergence of series.

4.1.2 The vanishing condition

If $\sum_{r=1}^{\infty} a_r$ is convergent then (a_n) is a null sequence.

Proof

Suppose that (s_n) converges to some limit s. Hence (s_{n-1}) also converges to s. But $a_n = s_n - s_{n-1}$, and so $a_n \to 0$ as $n \to \infty$. $\square$

The contrapositive statement corresponding to 4.1.2 is of most use in applications; namely, if (a_n) is not a null sequence, then $\sum_{r=1}^{\infty} a_r$ is divergent.

■■ EXAMPLE 4

Consider $\sum_{r=1}^{\infty} r/(r+1)$. Now $a_n = n/(n+1)$ and $a_n \to 1 \neq 0$ as $n \to \infty$. By the vanishing condition, $\sum_{r=1}^{\infty} r/(r+1)$ is a divergent series.   ■

It is important to note that the converse of the vanishing condition is false! In other words, there are divergent series whose terms nevertheless tend to zero.

■■ EXAMPLE 5

Consider $\sum_{r=1}^{\infty} (\sqrt{r} - \sqrt{r-1})$. The nth partial sum may be written as

$$s_n = (\sqrt{1} - \sqrt{0}) + (\sqrt{2} - \sqrt{1}) + \ldots$$
$$+ (\sqrt{n} - \sqrt{n-1}) = \sqrt{n}$$

Clearly (s_n) is a divergent sequence, and so

$$\sum_{r=1}^{\infty} (\sqrt{r} - \sqrt{r-1})$$

is a divergent series. However,

$$a_n = \sqrt{n} - \sqrt{n-1}$$
$$= \frac{(\sqrt{n} - \sqrt{n-1})(\sqrt{n} + \sqrt{n-1})}{\sqrt{n} + \sqrt{n-1}}$$
$$= \frac{1}{\sqrt{n} + \sqrt{n-1}} \to 0 \quad \text{as } n \to \infty \qquad ■$$

By considering the nth partial sums of the appropriate series and the sum and scalar product rules for sequences (3.1.2), the following elementary results can easily be proved.

4.1.3   Sum rule

If $\sum_{r=1}^{\infty} a_r$ and $\sum_{r=1}^{\infty} b_r$ are convergent series then $\sum_{r=1}^{\infty} (a_r + b_r)$ is also convergent and

$$\sum_{r=1}^{\infty} (a_r + b_r) = \sum_{r=1}^{\infty} a_r + \sum_{r=1}^{\infty} b_r$$

4.1.4 Scalar product rule

If $\sum_{r=1}^{\infty} a_r$ is convergent then $\sum_{r=1}^{\infty} ka_r$ is convergent for any $k \in \mathbb{R}$ and

$$\sum_{r=1}^{\infty} ka_r = k \sum_{r=1}^{\infty} a_r$$

In the next section, rules will be established that can be used to test whether or not a given series converges. The last of these results, the integral test (see 4.2.5), will be used to prove the following.

4.1.5 Result

The p-series $\sum_{r=1}^{\infty} 1/r^p$ converges if $p > 1$ and diverges if $p \leq 1$.

The p-series, together with the geometric series, gives a fund of known convergent and divergent series. The next two examples examine the p-series with $p = 1$ and $p = 2$.

■■ EXAMPLE 6

The **harmonic series** $\sum_{r=1}^{\infty} 1/r$ diverges.

Solution

If $N = 2^n$ then the Nth partial sum of $\sum_{r=1}^{\infty} 1/r$ can be written as

$$s_N = 1 + \frac{1}{2} + \left(\frac{1}{3} + \frac{1}{4}\right) + \left(\frac{1}{5} + \frac{1}{6} + \frac{1}{7} + \frac{1}{8}\right) + \ldots$$

$$+ \left(\frac{1}{2^{n-1} + 1} + \ldots + \frac{1}{2^n}\right)$$

$$> 1 + \frac{1}{2} + \left(\frac{1}{4} + \frac{1}{4}\right) + \left(\frac{1}{8} + \frac{1}{8} + \frac{1}{8} + \frac{1}{8}\right) + \ldots$$

$$+ \left(\frac{1}{2^n} + \ldots + \frac{1}{2^n}\right)$$

$$= 1 + \frac{1}{2} n$$

Hence (s_N) is a divergent subsequence of (s_n), and so (s_n) is itself divergent by Strategy 2 of Section 3.3. ■

■■   EXAMPLE 7

The series $\sum_{r=1}^{\infty} 1/r^2$ converges.

Solution

If $N = 2^n$ then the Nth partial sum of $\sum_{r=1}^{\infty} 1/r^2$ can be written as

$$
s_N = 1 + \frac{1}{2^2} + \left(\frac{1}{3^2} + \frac{1}{4^2}\right) + \left(\frac{1}{5^2} + \frac{1}{6^2} + \frac{1}{7^2} + \frac{1}{8^2}\right) + \ldots
$$

$$
+ \left(\frac{1}{(2^{n-1} + 1)^2} + \ldots + \frac{1}{(2^n)^2}\right)
$$

$$
< 1 + 1 + \left(\frac{1}{2^2} + \frac{1}{2^2}\right) + \left(\frac{1}{4^2} + \frac{1}{4^2} + \frac{1}{4^2} + \frac{1}{4^2}\right) + \ldots
$$

$$
+ \left(\frac{1}{(2^{n-1})^2} + \ldots + \frac{1}{(2^{n-1})^2}\right)
$$

$$
= 1 + 1 + \frac{1}{2} + \frac{1}{4} + \ldots + \frac{1}{2^{n-1}} < 3
$$

Since $\sum_{r=1}^{\infty} 1/r^2$ is a positive-term series, the sequence (s_n) of partial sums is an increasing sequence. Hence $s_n < s_N < 3$. Therefore (s_n) is an increasing sequence that is bounded above. By 3.4.1, (s_n) converges. In other words, $\sum_{r=1}^{\infty} 1/r^2$ converges. An alternative method of proof is given in Question 2 of Exercises 3.4.   ■

As can be seen from the previous two examples, it is in general not possible to find a simple formula for the nth partial sums of a given series. Also, although having successfully determined the behaviour of $\sum_{r=1}^{\infty} 1/r^2$, the actual sum cannot be determined. For the interested reader, the sum of $\sum_{r=1}^{\infty} 1/r^2$ is $\frac{1}{6}\pi^2$, but the proof of this fact is beyond the scope of this book. The reader will have to be content with tests that establish that certain series converge but do not determine what the sum in question is.

Exercises 4.1

1.   Show that

$$
\frac{1}{r!} - \frac{1}{(r+1)!} = \frac{r}{(r+1)!} \quad \text{for } r \in \mathbb{N}
$$

Hence determine the nth partial sum of $\sum_{r=1}^{\infty} r/(r+1)!$ and show that

$$\sum_{r=1}^{\infty} \frac{r}{(r+1)!} = 1$$

2.   Use the vanishing condition (4.1.2) to show that each of the following series is divergent:

(a) $\sum_{r=1}^{\infty} \dfrac{r}{r+1}$ (b) $\sum_{r=1}^{\infty} [r - \sqrt{r(r-1)}]$

3.   Prove the sum and scalar product rules for series (see 4.1.3 and 4.1.4).

4.   Suppose that $\sum_{r=1}^{\infty} a_r$ is convergent and $\sum_{r=1}^{\infty} b_r$ is divergent. Prove that $\sum_{r=1}^{\infty} (a_r + b_r)$ is divergent.

4.2   Series tests

In this section tests are established that can be used to investigate whether or not a given series is convergent.

4.2.1   First comparison test

If $0 \le a_n \le b_n$ for all $n \in \mathbb{N}$ then

(a) $\sum_{r=1}^{\infty} b_r$ convergent $\Rightarrow \sum_{r=1}^{\infty} a_r$ convergent

(b) $\sum_{r=1}^{\infty} a_r$ divergent $\Rightarrow \sum_{r=1}^{\infty} b_r$ divergent

Proof

(a) Let $s_n = \sum_{r=1}^{n} a_r$ and $t_n = \sum_{r=1}^{n} b_r$. From the given conditions $0 \le s_n \le t_n$ for all $n \in \mathbb{N}$. If $\sum_{r=1}^{\infty} b_r$ converges then $t_n \to t$ as $n \to \infty$, and, since (t_n) is an increasing sequence, $t_n \le t$ for all $n \in \mathbb{N}$. Therefore $s_n \le t$ for all $n \in \mathbb{N}$, and hence (s_n) is a bounded monotone sequence. By 3.4.1, (s_n) converges, and hence $\sum_{r=1}^{\infty} a_r$ converges.

(b) This is just the contrapositive of part (a).   □

■■   EXAMPLE 1

The series

$$\sum_{r=1}^{\infty} \frac{1 + \cos r}{3^r + 2r^3}$$

is convergent.

Solution

Let

$$a_n = \frac{1 + \cos n}{3^n + 2n^3}$$

Then $a_n \geq 0$ since $\cos n \geq -1$. Also, $a_n \leq 2/(3^n + 2n^3)$, since $\cos n \leq 1$. Therefore, since $3^n > 0$,

$$a_n < \frac{2}{2n^3} = \frac{1}{n^3}$$

Let $b_n = 1/n^3$. Then $\sum_{r=1}^{\infty} b_r$ converges, being the *p*-series with $p = 3$. Hence, by 4.2.1, $\sum_{r=1}^{\infty} a_r$ also converges.

Note

At this point, yet another proof can be provided that $\sum_{r=1}^{\infty} 1/r^2$ converges (see Example 7 of Section 4.1). The first step is to establish that

$$0 \leq \frac{1}{n^2} \leq \frac{2}{n(n + 1)} \quad \text{for all } n \in \mathbb{N}$$

Then, by Example 3 of Section 4.1 and 4.1.4, $\sum_{r=1}^{\infty} 2/r(r + 1)$ converges. Hence, by 4.2.1, $\sum_{r=1}^{\infty} 1/r^2$ also converges. Incidentally, 4.2.1 now allows the deduction that $\sum_{r=1}^{\infty} 1/r^p$ converges for $p \geq 2$.   ■

4.2.2   Second comparison test

Let $\sum_{r=1}^{\infty} a_r$ and $\sum_{r=1}^{\infty} b_r$ be positive-term series such that

$$\lim_{n \to \infty} (a_n/b_n) = L \neq 0$$

Then $\sum_{r=1}^{\infty} a_r$ converges if and only if $\sum_{r=1}^{\infty} b_r$ converges.

Proof

Suppose that $\sum_{r=1}^{\infty} b_r$ is convergent and let s_n and t_n be as in the proof of 4.2.1. Since $\lim_{n\to\infty}(a_n/b_n) = L$, for $\varepsilon = 1$ there is an N such that

$$\left| \frac{a_n}{b_n} - L \right| < 1 \quad \text{for all } n > N$$

Hence

$$\frac{a_n}{b_n} = \left| \frac{a_n}{b_n} \right|$$

$$= \left| \frac{a_n}{b_n} - L + L \right| \leq \left| \frac{a_n}{b_n} - L \right| + |L| < 1 + |L|$$

$$= k \quad \text{for } n > N$$

Now consider the positive-term series $\sum_{r=1}^{\infty} \alpha_r$ and $\sum_{r=1}^{\infty} \beta_r$, where $\alpha_r = a_{r+N}$ and $\beta_r = kb_{r+N}$. Hence $0 \leq \alpha_n \leq \beta_n$ for all $n \in \mathbb{N}$. Since $\sum_{r=1}^{\infty} b_r$ converges, so too does $\sum_{r=N+1}^{\infty} b_r$ (why?), and hence $\sum_{r=1}^{\infty} \beta_r$ converges by the scalar product rule for series (4.1.4). By 4.2.1, $\sum_{r=1}^{\infty} \alpha_r$ converges, and, since the addition of a finite number of terms to a convergent series produces another convergent series, $\sum_{r=1}^{\infty} a_r$ converges. This proves that $\sum_{r=1}^{\infty} b_r$ is convergent implies that $\sum_{r=1}^{\infty} a_r$ is convergent. The converse of this statement can be proved by reversing the roles of a_n and b_n in the above argument and observing that $b_n/a_n \to 1/L$ as $n \to \infty$. $\qquad \square$

■■ EXAMPLE 2

The series

$$\sum_{r=1}^{\infty} \frac{2r}{r^2 - 5r + 8}$$

is divergent.

Solution

Let

$$a_n = \frac{2n}{n^2 - 5n + 8} \quad \text{and} \quad b_n = \frac{1}{n}$$

Then

$$\frac{a_n}{b_n} = \frac{2n^2}{n^2 - 5n + 8}$$

Now $a_n/b_n \to 2 \neq 0$ as $n \to \infty$, and so $\sum_{r=1}^{\infty} a_r$ diverges by comparison with the divergent harmonic series. ∎

4.2.3 D'Alembert's ratio test

Let $\sum_{r=1}^{\infty} a_r$ be a series of positive terms and for each $n \in \mathbb{N}$ let $\alpha_n = a_{n+1}/a_n$. Suppose that (α_n) converges to some limit L. If $L > 1$ then $\sum_{r=1}^{\infty} a_r$ diverges; if $L < 1$ then $\sum_{r=1}^{\infty} a_r$ converges; and if $L = 1$ the test gives no information.

Proof

Suppose that $L < 1$ and let $\varepsilon = \frac{1}{2}(1 - L)$. Now $\varepsilon > 0$ and $L + \varepsilon = k < 1$. Since $\lim_{n \to \infty} \alpha_n = L$, there is a value of N such that $\alpha_n = |\alpha_n - L + L| \leq \varepsilon + L = k < 1$ for all $n > N$. Therefore $a_{n+1} \leq k a_n$ for all $n > N$. Let $\beta_r = a_{r+N}$. Then

$$\beta_{n+1} \leq k \beta_n \quad \text{for all } n \in \mathbb{N}$$

and so (by induction on n)

$$\beta_{n+1} \leq k^n \beta_1 \quad \text{for all } n \in \mathbb{N}$$

Now $\sum_{r=0}^{\infty} k^r \beta_1$ is a convergent geometric series since $k < 1$. By 4.2.1, $\sum_{r=1}^{\infty} \beta_r$ converges, and hence $\sum_{r=1}^{\infty} a_r$ also converges. Suppose now that $L > 1$ and let $\varepsilon = L - 1$. Now $\varepsilon > 0$ and, since $\lim_{n \to \infty} \alpha_n = L$, there is a value of N such that

$$\alpha_n > L - \varepsilon = 1 \quad \text{for all } n > N$$

Hence

$$a_{n+1} > a_n \quad \text{for all } n > N$$

and so

$$a_n > a_N \quad \text{for all } n > N$$

Since $a_N \neq 0$, (a_n) is not null. By the vanishing condition (4.1.2), $\sum_{r=1}^{\infty} a_r$ diverges. □

Jean-le-Rond d'Alembert (1717–1783) was one of the leading French mathematicians of the mid eighteenth century. The son of the aristocratic sister of a cardinal, d'Alembert was abandoned near the church of Saint Jean-le-Rond when a newly-born infant. He is renowned for his contributions to kinetics and his work on the solution of partial differential equations. He showed interest in the foundations of analysis, and was responsible, in 1754, for suggesting that analysis needed to be placed on firm foundations by the development of a sound theory of limits. Little heed was paid to this suggestion during his lifetime. However, in 1797 Lagrange (who ranked with Euler as an outstanding eighteenth-century mathematician) attempted to provide a rigorous foundation for analysis, and in 1821 Augustin-Louis Cauchy successfully executed d'Alembert's suggestion.

■■ **EXAMPLE 3**

Determine those values of x for which $\sum_{r=1}^{\infty} r(4x^2)^r$ is convergent.

Solution

Here $a_n = n(4x^2)^n$, and so for $x \neq 0$

$$\alpha_n = \frac{(n+1)(4x^2)^{n+1}}{n(4x^2)^n} = 4x^2\left(1 + \frac{1}{n}\right)$$

Now $\alpha_n \to 4x^2 = L$ as $n \to \infty$. By 4.2.3, $\sum_{r=1}^{\infty} a_r$ diverges if $4x^2 > 1$ (in other words, $|x| > \frac{1}{2}$), $\sum_{r=1}^{\infty} a_r$ converges if $|x| < \frac{1}{2}$, and no information is gained if $|x| = \frac{1}{2}$. If $|x| = \frac{1}{2}$ then $\sum_{r=1}^{\infty} a_r = \sum_{r=1}^{\infty} r$, a divergent series. The series clearly converges for $x = 0$, and therefore $\sum_{r=1}^{\infty} r(4x^2)^r$ converges if and only if $|x| < \frac{1}{2}$. ■

4.2.4 Altenating series test

The alternating series

$$\sum_{r=1}^{\infty} (-1)^{r-1} b_r$$

where $b_n > 0$, converges if (b_n) is a monotone decreasing sequence with limit zero.

Proof

Let

$$s_n = \sum_{r=1}^{n} (-1)^{r-1} b_r$$

Then

$$s_{2m} = b_1 - (b_2 - b_3) - \ldots - (b_{2m-2} - b_{2m-1}) - b_{2m} \leqslant b_1$$

Hence (s_{2m}) is bounded above. Since $s_{2m} = (b_1 - b_2) + (b_3 - b_4) + \ldots + (b_{2m-1} - b_{2m})$ and (b_n) is decreasing, (s_{2m}) is increasing. By 3.4.1, $s_{2m} \to s$ as $n \to \infty$. Similarly, (s_{2m+1}) is a decreasing sequence that is bounded below (by $b_1 - b_2$), and so $s_{2m+1} \to t$ as $n \to \infty$. Now

$$t - s = \lim_{m \to \infty} (s_{2m+1} - s_{2m})$$

$$= \lim_{m \to \infty} b_{2m+1} = 0$$

so $t = s$. Then, arguing as in the solution to Question 4 of Exercises 3.3, (s_n) is convergent with limit s. Hence $\sum_{r=1}^{\infty} (-1)^{r-1} b_r$ is convergent. $\quad\square$

■■ EXAMPLE 4

$\sum_{r=1}^{\infty} (-1)^{r-1}/r$ is convergent since $(1/n)$ is a decreasing sequence with limit zero. $\quad\blacksquare$

4.2.5   Integral test

Let $f: \mathbb{R}^+ \to \mathbb{R}^+$ be a decreasing function and let $a_n = f(n)$ for each $n \in \mathbb{N}$. Let $j_n = \int_1^n f(x)\,dx$. Then the series $\sum_{r=1}^{\infty} a_r$ converges if and only if (j_n) converges.

The proof of 4.2.5 is given in Section 7.1.

■■ EXAMPLE 5

Establish the previously stated fact that the p-series $\sum_{r=1}^{\infty} 1/r^p$ converges if and only if $p > 1$.

Solution

Consider the function $f_p: \mathbb{R}^+ \to \mathbb{R}^+$ given by $f_p(x) = 1/x^p$. When $p > 0$, this is a decreasing function of x, and $a_n = f_p(n) = 1/n^p$ is the nth term of the p-series $\sum_{r=1}^{\infty} 1/r^p$. For $p \neq 1$,

$$j_n = \int_1^n \frac{1}{x^p}\,dx = \left[\frac{x^{1-p}}{1-p}\right]_1^n = \frac{1}{1-p}(n^{1-p} - 1)$$

So, for $p > 1$, $j_n \to 1/(p-1)$ as $n \to \infty$, and, for $p < 1$, (j_n) is divergent. For $p = 1$

$$j_n = \int_1^n \frac{1}{x} \, dx = [\log_e x]_1^n = \log_e n$$

and so (j_n) again diverges.

When $p \leqslant 0$, $\sum_{r=1}^{\infty} 1/r^p$ diverges by the vanishing condition. ∎

The series $1 - \frac{1}{2} + \frac{1}{3} - \frac{1}{4} + \ldots$ is convergent by virtue of Example 4. Denote the sum of this series by s. Consider the following plausible manipulations:

$$s = 1 - \tfrac{1}{2} + \tfrac{1}{3} - \tfrac{1}{4} + \tfrac{1}{5} - \tfrac{1}{6} + \ldots$$
$$= 1 - \tfrac{1}{2} - \tfrac{1}{4} + \tfrac{1}{3} - \tfrac{1}{6} - \tfrac{1}{8} + \tfrac{1}{5} - \tfrac{1}{10} - \tfrac{1}{12} + \ldots$$

where precisely the same terms are used, but each positive term is followed by two negative ones. Hence

$$s = (1 - \tfrac{1}{2}) - \tfrac{1}{4} + (\tfrac{1}{3} - \tfrac{1}{6}) - \tfrac{1}{8} + (\tfrac{1}{5} - \tfrac{1}{10}) - \tfrac{1}{12} + \ldots$$

where the terms have been bracketed. Thus

$$s = \tfrac{1}{2} - \tfrac{1}{4} + \tfrac{1}{6} - \tfrac{1}{8} + \tfrac{1}{10} - \tfrac{1}{12} + \ldots$$
$$= \tfrac{1}{2}(1 - \tfrac{1}{2} + \tfrac{1}{3} - \tfrac{1}{4} + \tfrac{1}{5} - \tfrac{1}{6} + \ldots)$$
$$= \tfrac{1}{2}s$$

Hence $s = 0$. However, rebracketing the terms gives

$$s = (1 - \tfrac{1}{2}) + (\tfrac{1}{3} - \tfrac{1}{4}) + (\tfrac{1}{5} - \tfrac{1}{6}) + \ldots$$
$$= \tfrac{1}{2} + \tfrac{1}{12} + \tfrac{1}{30} + \ldots$$

and so the $(2n)$th partial sums of $\sum_{r=1}^{\infty} (-1)^{r-1}/r$ form an increasing sequence and so $s \geqslant \frac{1}{2}$ (the second partial sum). This paradoxical result arises because it has been taken for granted that operations that are valid for finite sums also work for infinite sums. In particular, the first step above rearranges the terms of the original series. A **rearrangement** of a given series is another series containing precisely the same terms as the original series, but in a different order. There is no reason to expect that the rearranged series has the same sum as the original series, nor even that the rearranged series is convergent! Luckily, as will be seen, a large class of series can be rearranged without affecting their sum.

A series $\sum_{r=1}^{\infty} a_r$ is called **absolutely convergent** if the associated series $\sum_{r=1}^{\infty} |a_r|$ is convergent.

4.2.6 Theorem

$$\sum_{r=1}^{\infty} a_r \text{ absolutely convergent} \Rightarrow \sum_{r=1}^{\infty} a_r \text{ convergent}$$

Proof

Let

$$a_n^+ = \begin{cases} a_n & \text{if } a_n \geqslant 0 \\ 0 & \text{if } a_n \leqslant 0 \end{cases} \quad \text{and} \quad a_n^- = \begin{cases} -a_n & \text{if } a_n \leqslant 0 \\ 0 & \text{if } a_n \geqslant 0 \end{cases}$$

By definition, $a_n^+ \geqslant 0$ and $a_n^- \geqslant 0$ for all $n \in \mathbb{N}$.

Also $|a_n| = a_n^+ + a_n^-$. If $\sum_{r=1}^{\infty} |a_r|$ converges then, by 4.2.1, both $\sum_{r=1}^{\infty} a_r^+$ and $\sum_{r=1}^{\infty} a_r^-$ converge. But $a_n = a_n^+ - a_n^-$, and so $\sum_{r=1}^{\infty} a_r$ converges by 4.1.3 and 4.1.4.   □

A series $\sum_{r=1}^{\infty} a_r$ for which $\sum_{r=1}^{\infty} a_r$ converges but $\sum_{r=1}^{\infty} |a_r|$ diverges is called a **conditionally convergent** series. By virtue of the proof of 4.2.6, at least one, and in fact both, of $\sum_{r=1}^{\infty} a_r^+$ and $\sum_{r=1}^{\infty} a_r^-$ must diverge. The paradox concerning the conditionally convergent series $\sum_{r=1}^{\infty} (-1)^{r-1}/r$ that was seen earlier can now be explained by the next result.

4.2.7 Theorem

If $\sum_{r=1}^{\infty} a_r$ is conditionally convergent (with sum s) and $s' \in \mathbb{R}$ then the series can be rearranged to converge to s'.

Proof

Since $\sum_{r=1}^{\infty} a_r$ is conditionally convergent, both of the series $\sum_{r=1}^{\infty} a_r^+$ and $\sum_{r=1}^{\infty} a_r^-$ are divergent, and both consist of positive terms only. Hence the partial sums of $\sum_{r=1}^{\infty} a_r^+$ increase without bound, and so for any real number s' there exists a positive integer n_1 such that

$$T_1 = a_1^+ + a_2^+ + \ldots + a_{n_1}^+ > s'$$

Moreover, if n_1 is chosen to be the *least* positive integer with this property then $a_1^+ + a_2^+ + \ldots + a_{n_1-1}^+ \leqslant s'$, and so $T_1 - a_{n_1}^+ \leqslant s'$. Therefore $0 < T_1 - s' \leqslant a_{n_1}^+$, and so $|T_1 - s'| \leqslant a_{n_1}^+$. Since the partial sums of $\sum_{r=1}^{\infty} a_r^-$ also increase without bound, there exists a positive integer n_2 such that

$$T_2 = T_1 - (a_1^- + a_2^- + \ldots + a_{n_2}^-)$$
$$= (a_1^+ + a_2^+ + \ldots + a_{n_1}^+) - (a_1^- + a_2^- + \ldots + a_{n_2}^-)$$
$$< s'$$

If n_2 is chosen to be the *least* positive integer with this property then $T_2 + a_{n_2}^- \geqslant s'$. Therefore $0 < s - T_2 \leqslant a_{n_2}^-$, and so $|T_2 - s'| \leqslant a_{n_2}^-$. Now choose n_3 to be the *least* positive integer such that

$$
\begin{aligned}
T_3 &= T_2 + (a_{n_1+1}^+ + a_{n_1+2}^+ + \ldots + a_{n_3}^+) \\
&= (a_1^+ + a_2^+ + \ldots + a_{n_1}^+) - (a_1^- + a_2^- + \ldots + a_{n_2}^-) \\
&\quad + (a_{n_1+1}^+ + a_{n_1+2}^+ + \ldots + a_{n_3}^+) \\
&> s'
\end{aligned}
$$

This choice of n_3 guarantees that $|T_3 - s'| \leqslant a_{n_3}^+$.

This process of adding on sufficiently many positive terms from the original series to once again produce a total exceeding s', followed by the addition of sufficiently many negative terms to take the accumulated total below s' again, can be continued. The end result is a rearrangement of $\sum_{r=1}^{\infty} a_r$ in which the particular partial sums T_1, T_2, T_3, ... are alternately greater than and then less than s', and in which $|T_n - s'|$ is less than or equal to some term of $\sum_{r=1}^{\infty} |a_r|$. By construction, any partial sum S_n of this rearranged series lies between T_m and T_{m+1} for a suitable integer m (or equals one of T_m or T_{m+1}). If m is even then $T_m \leqslant S_n \leqslant T_{m+1}$, and if m is odd then the inequalities are reversed. In either case, arguing as in the proof of the sandwich rule for sequences (3.1.3), $|S_n - s'|$ is less than or equal to the maximum of $|T_m - s'|$ and $|T_{m+1} - s'|$. Hence the partial sums S_1, S_2, S_3, ... of the rearranged series are such that $|S_n - s'|$ is less than or equal to some term of $\sum_{r=1}^{\infty} |a_r|$. But $\sum_{r=1}^{\infty} |a_r|$ is a convergent series, and so, by the vanishing condition (4.1.2), $|a_r| \to 0$ as $n \to \infty$. Therefore $|S_n - s'| \to 0$ as $n \to \infty$, and so the partial sums of the rearranged series converge to s'.

In other words, $\sum_{r=1}^{\infty} a_r$ can be rearranged to produce a series converging to any real number s'. $\qquad\square$

In complete contrast, rearrangement of an absolutely convergent series produces an absolutely convergent series whose sum is the same as the sum of the original series.

4.2.8 The rearrangement rule

If $\sum_{r=1}^{\infty} a_r$ is absolutely convergent and $\sum_{r=1}^{\infty} b_r$ is any rearrangement of $\sum_{r=1}^{\infty} a_r$ then $\sum_{r=1}^{\infty} b_r$ is absolutely convergent and

$$
\sum_{r=1}^{\infty} b_r = \sum_{r=1}^{\infty} a_r
$$

Proof

Assume that $a_n \geq 0$ for all $n \in \mathbb{N}$. Choose M such that $b_1, b_2, \ldots,$ b_n occur among $a_1, a_2, \ldots, a_M$. Hence

$$\sum_{r=1}^{n} b_r \leq \sum_{r=1}^{M} a_r \leq \sum_{r=1}^{\infty} a_r$$

Therefore the nth partial sums of $\sum_{r=1}^{\infty} b_r$ are bounded above, and so $\sum_{r=1}^{\infty} b_r$ converges. In addition,

$$\sum_{r=1}^{\infty} b_r \leq \sum_{r=1}^{\infty} a_r$$

and, by reversing the roles of a_n and b_n in the above argument,

$$\sum_{r=1}^{\infty} a_r \leq \sum_{r=1}^{\infty} b_r$$

Hence the two series have the same sum. Now the condition that $a_n \geq 0$ can be relaxed and a_n^+ and a_n^- can be defined as in the proof of 4.2.6. Similarly define b_n^+ and b_n^-. Now $\sum_{r=1}^{\infty} b_r^+$ is a rearrangement of $\sum_{r=1}^{\infty} a_r^+$, and both series have non-negative terms. So, by the first part of the proof, $\sum_{r=1}^{\infty} b_r^+$ converges. Similarly, $\sum_{r=1}^{\infty} b_r^-$ converges, and hence

$$\sum_{r=1}^{\infty} |b_r| = \sum_{r=1}^{\infty} b_r^+ + \sum_{r=1}^{\infty} b_r^-$$

converges. In other words, $\sum_{r=1}^{\infty} b_r$ is absolutely convergent. Finally,

$$\sum_{r=1}^{\infty} b_r = \sum_{r=1}^{\infty} b_r^+ - \sum_{r=1}^{\infty} b_r^-$$

$$= \sum_{r=1}^{\infty} a_r^+ - \sum_{r=1}^{\infty} a_r^-$$

$$= \sum_{r=1}^{\infty} a_r \qquad \square$$

The final result in this section is further confirmation that absolutely convergent series behave like finite sums.

4.2.9 Cauchy product of series

If $\sum_{r=1}^{\infty} a_r$ and $\sum_{r=1}^{\infty} b_r$ are absolutely convergent series and

$$c_n = a_1 b_n + a_2 b_{n-1} + \ldots + a_n b_1$$

then $\sum_{r=1}^{\infty} c_r$ is absolutely convergent and

$$\sum_{r=1}^{\infty} c_r = \left(\sum_{r=1}^{\infty} a_r \right) \left(\sum_{r=1}^{\infty} b_r \right)$$

Proof

Suppose first that $\sum_{r=1}^{\infty} a_r$ and $\sum_{r=1}^{\infty} b_r$ are positive-term series and consider the array

$$
\begin{array}{cccc}
a_1 b_1 & a_1 b_2 & a_1 b_3 & \ldots \\
a_2 b_1 & a_2 b_2 & a_2 b_3 & \ldots \\
a_3 b_1 & a_3 b_2 & a_3 b_3 & \ldots \\
\vdots & \vdots & \vdots &
\end{array}
$$

If w_n is the sum of the terms in the array that lie in the $n \times n$ square with $a_1 b_1$ at one corner then $w_n = s_n t_n$, where s_n and t_n are the nth partial sums of $\sum_{r=1}^{\infty} a_r$ and $\sum_{r=1}^{\infty} b_r$ respectively. Hence

$$w_n \to \left(\sum_{r=1}^{\infty} a_r \right) \left(\sum_{r=1}^{\infty} b_r \right) \quad \text{as } n \to \infty$$

Now $\sum_{r=1}^{\infty} c_r$ is the sum of the terms in the array summed 'by diagonals' and so if u_n is the nth partial sum of $\sum_{r=1}^{\infty} c_r$ then

$$w_{[n/2]} \leq u_n \leq w_n$$

where $[x]$ denotes the largest integer not exceeding x (see Example 11 of Section 5.1). By the sandwich rule for sequences (3.1.3),

$$u_n \to \left(\sum_{r=1}^{\infty} a_r \right) \left(\sum_{r=1}^{\infty} b_r \right) \quad \text{as } n \to \infty$$

as required.

For the general case the above argument can be applied to $\sum_{r=1}^{\infty} |a_r|$ and $\sum_{r=1}^{\infty} |b_r|$ to deduce that $\sum_{r=1}^{\infty} c_r$ is absolutely convergent. Since $\sum_{r=1}^{\infty} a_r$ and $\sum_{r=1}^{\infty} b_r$ are linear combinations of $\sum_{r=1}^{\infty} a_r^+$, $\sum_{r=1}^{\infty} a_r^-$, and $\sum_{r=1}^{\infty} b_r^+$, $\sum_{r=1}^{\infty} b_r^-$ respectively, it is readily established that

$$\sum_{r=1}^{\infty} c_r = \left(\sum_{r=1}^{\infty} a_r\right)\left(\sum_{r=1}^{\infty} b_r\right)$$

by 4.1.3 and 4.1.4.   □

The exercises for this section provide practice in applying a specified test to a given series. However, when faced with a series, the choice of which test to try may not be immediately obvious. In such cases the procedure illustrated in Figure 4.2 is a useful guide as to what strategy to adopt.

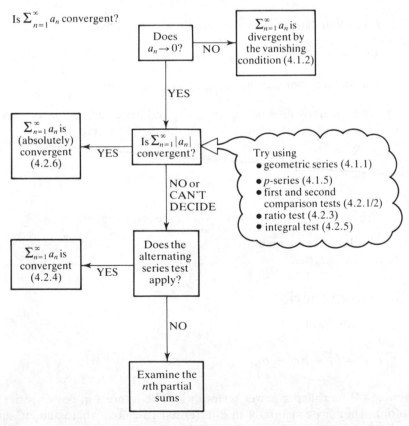

Figure 4.2

Exercises 4.2

1. Use the first comparison test (4.2.1) to show that

    (a) $\sum_{r=1}^{\infty} \dfrac{1}{2^r r!}$ is convergent   (b) $\sum_{r=2}^{\infty} \dfrac{1}{\log_e r}$ is divergent

2. Use the second comparison test (4.2.2) to show that

    $$\sum_{r=1}^{\infty} \frac{r+3}{r^2+r}$$

    diverges.

3. Decide which of the following are convergent using the ratio test (4.2.3):

    (a) $\sum_{r=1}^{\infty} \dfrac{r}{3^r}$ (b) $\sum_{r=1}^{\infty} \dfrac{r^3}{r!}$ (c) $\sum_{r=1}^{\infty} \dfrac{3^r}{2^r+1}$

4. Show that each of the following series converges:

    (a) $1 - \frac{1}{3} + \frac{1}{5} - \frac{1}{7} + \ldots$ (b) $\sum_{r=1}^{\infty} \dfrac{\cos r\pi}{r \sqrt{r}}$

    and say whether they are absolutely convergent.

5. Find the derivative of $\log_e(\log_e x)$, and hence use the integral test (4.2.5) to show that the following series is divergent.

    $$\sum_{r=2}^{\infty} \frac{1}{r \log_e r}$$

6. Prove that if $\sum_{r=1}^{\infty} a_r^+$ and $\sum_{r=1}^{\infty} a_r^-$ are convergent, then $\sum_{r=1}^{\infty} a_r$ is absolutely convergent

4.3   Power series

A series of the form

$$\sum_{r=0}^{\infty} b_r x^r = b_0 + b_1 x + b_2 x^2 + \ldots$$

where $x \in \mathbb{R}$, is called a **power series** in x. The interest in power series is twofold. They arise naturally in differential calculus, where, under suitable conditions, a function f can be written as

$$f(x) = f(0) + f'(0)x + \frac{f''(0)}{2!}x^2 + \ldots$$

This particular power series is called the Maclaurin series for f and is discussed fully in Section 6.3. Power series are also used to give rigorous definitions of the elementary functions. This is discussed at the end of this section.

Since a power series gives a particular series for each real value of x, its convergence will depend on x. Clearly any power series converges when $x = 0$.

4.3.1   Theorem

> The set of values of x for which the power series $\sum_{r=0}^{\infty} b_r x^r$ converges is an interval with midpoint zero.

Proof

There are three mutually exclusive possibilities:

(i) $\sum_{r=0}^{\infty} b_r x^r$ converges only for $x = 0$;

(ii) $\sum_{r=0}^{\infty} b_r x^r$ converges for all real x;

(iii) $\sum_{r=0}^{\infty} b_r x^r$ converges for some $x \neq 0$ and diverges for some $x \neq 0$.

The result follows immediately in cases (i) and (ii). Now consider case (iii) and suppose that $\sum_{r=0}^{\infty} b_r x^r$ converges for $x = x_1$. By the vanishing condition (4.1.2), $b_n x_1^n \to 0$ as $n \to \infty$. Hence $(b_n x_1^n)$ is bounded, and so there is some number M for which

$$|b_n x_1^n| \leqslant M \quad \text{for all } n \in \mathbb{N}$$

If x_2 satisfies $|x_2| < |x_1|$ then

$$t = \frac{|x_2|}{|x_1|} < 1$$

Now

$$|b_n x_2^n| = |b_n x_1^n| t^n \leqslant M t^n$$

But $\sum_{r=0}^{\infty} M t^r$ is a convergent geometric series (4.1.1), and so, by the first comparison test (4.2.1), $\sum_{r=0}^{\infty} |b_r x_2^r|$ is convergent. Hence $\sum_{r=0}^{\infty} b_r x_2^r$ is absolutely convergent and hence convergent. Therefore $\sum_{r=0}^{\infty} b_r x^r$ converges for $|x| < |x_1|$. Let $S = \{c : c > 0$ and $\sum_{r=0}^{\infty} b_r x^r$ converges for $|x| < c\}$. Since $|x_1| \in S$, S is non-empty. Also, since there exists a $y \neq 0$ such that $\sum_{r=0}^{\infty} b_r y^r$ is divergent, $d \notin S$ for all $d > |y|$ by the first comparison test (4.2.1). Therefore S is bounded above. Let $k = \sup S$. Then $k \geq |x_1| > 0$. Now suppose $|x| < k$, so that, by the definition of k, there is an $l \in S$ with $|x| < l < k$ and such that $\sum_{r=0}^{\infty} b_r x^r$ is convergent. Hence $k \in S$. In other words, $\sum_{r=0}^{\infty} b_r x^r$ converges for $|x| < k$. If it were possible for $\sum_{r=0}^{\infty} b_r y^r$ to converge for $|y| > k$ then, as previously demonstrated, $\sum_{r=0}^{\infty} b_r x^r$ would converge for $|x| < |y|$, and so $|y| \in S$, which would contradict the fact that $k = \sup S$. Therefore $\sum_{r=0}^{\infty} b_r x^r$ diverges for $|x| > k$. This establishes the theorem. $\square$

The set of real numbers x for which $\sum_{r=0}^{\infty} b_r x^r$ converges is called the **interval of convergence**, which (excluding the cases where the given series converges only for $x = 0$, or converges for all x) is one of the intervals $[-R, R]$, $(-R, R]$, $[-R, R)$ or $(-R, R)$, where R is called the **radius of convergence**. If the power series converges only for $x = 0$, the interval of convergence is $\{0\}$ and the radius of convergence is taken to be $R = 0$; if the power series converges for all real values of x, the interval of convergence is $\mathbb{R}$ and the radius of convergence is, by convention, $R = \infty$. Now the proof of 4.3.1 shows that a power series is absolutely convergent for $|x| < R$. It could be absolutely convergent at $x = \pm R$, but may be conditionally convergent, or even divergent, at one or both of the endpoints of the interval of convergence.

■■   EXAMPLE 1

The power series $\sum_{r=1}^{\infty} x^r/r$ has interval of convergence $[-1, 1)$.

Solution

Apply the ratio test to

$$\sum_{r=1}^{\infty} \left| \frac{x^r}{r} \right| = \sum_{r=1}^{\infty} \frac{|x|^r}{r} \quad \text{for } x \neq 0$$

Since $a_n = |x|^n/n$,

$$\alpha_n = \frac{a_{n+1}}{a_n} = \frac{|x|^{n+1}}{n+1} \frac{n}{|x|^n}$$

$$= \left(1 + \frac{1}{n} \right)^{-1} |x|$$

Now $\alpha_n \to |x|$ as $n \to \infty$. By 4.2.3, the power series converges (absolutely) for $|x| < 1$ and diverges for $|x| > 1$. Hence its radius of convergence is $R = 1$. To determine the interval of convergence, examine the behaviour of the series at the points $x = \pm 1$ (where the ratio test gives no information). At $x = 1$ the series is $\sum_{r=1}^{\infty} 1/r$, the divergent harmonic series. At $x = -1$ the series is $\sum_{r=1}^{\infty} (-1)^r/r$, a convergent alternating series. Hence the required interval of convergence is $[-1, 1)$. ∎

4.3.2 Arithmetic of power series

Let $\sum_{r=0}^{\infty} a_r x^r$ and $\sum_{r=0}^{\infty} b_r x^r$ be power series with radii of convergence R_1 and R_2 respectively, where $0 < R_1 \leqslant R_2$. Then the

sum $$\sum_{r=0}^{\infty} (a_r + b_r)x^r$$

scalar product $$\sum_{r=0}^{\infty} (ka_r)x^r$$

and

Cauchy product $$\sum_{r=0}^{\infty} c_r x^r \quad \left(\text{with } c_n = \sum_{r=0}^{n} a_r b_{n-r} \right)$$

all have radius of convergence at least R_1.

Moreover, if $\sum_{r=0}^{\infty} a_r x^r$ has sum $f(x)$ and $\sum_{r=0}^{\infty} b_r x^r$ has sum $g(x)$ then

$$\sum_{r=0}^{\infty} (a_r + b_r)x^r = f(x) + g(x)$$

$$= \sum_{r=0}^{\infty} a_r x^r + \sum_{r=0}^{\infty} b_r x^r$$

$$\sum_{r=0}^{\infty} (ka_r)x^r = kf(x)$$

$$= k \sum_{r=0}^{\infty} a_r x^r$$

and

$$\sum_{r=0}^{\infty} c_r x^r = f(x)g(x)$$

$$= \left(\sum_{r=0}^{\infty} a_r x^r \right)\left(\sum_{r=0}^{\infty} b_r x^r \right)$$

Proof

These claims concerning the sum and scalar product follow from the sum and scalar product rules for series (4.1.3 and 4.1.4). To establish the Cauchy product result, note that $\sum_{r=0}^{\infty} a_r x^r$ and $\sum_{r=0}^{\infty} b_r x^r$ are absolutely convergent for $|x| < R_1$. Since

$$c_n x^n = \sum_{r=0}^{n} (a_r x^r)(b_{n-r} x^{n-r})$$

$\sum_{r=0}^{\infty} c_r x^r$ is absolutely convergent for $|x| < R_1$ by 4.2.9 and has the sum stated. □

Much of the preceding discussion can be modified to apply to series of the form $\sum_{r=0}^{\infty} a_r (x - a)^r$. For example, the ratio test (4.2.3) may often be employed to determine the interval of convergence, an interval with a as its midpoint. Such series arise in Section 6.3.

Finally, this chapter briefly describes one approach to defining the elementary functions sine, cosine and exponential. It is assumed that the reader is familiar with these functions and their properties only in order to motivate and illustrate the theoretical results arising in later chapters. In fact, they have already been used in this chapter. However, they have not been used in the logical development of the theory. This ensures that the theoretical results have been derived in a rigorous manner. Naturally, once the elementary functions have been formally defined, a start can be made to deduce their many and varied properties.

Suppose then that $\sum_{r=0}^{\infty} a_n x^n$ is a power series with radius of convergence $R > 0$, and that $f(x)$ is its sum for $|x| < R$. This defines a function $f: (-R, R) \to \mathbb{R}$ given by $f(x) = \sum_{n=0}^{\infty} a_n x^n$. Note that, for each x in the domain of f, $f(x)$ is the limit of the sequence $\{f_N(x)\}$ of Nth partial sums, where $f_N(x) = \sum_{n=0}^{N} a_n x^n$ is a polynomial in x. The Appendix shows that the sum function $f(x)$ inherits important analytical properties of the polynomials $f_N(x)$. The exponential, sine and cosine functions can now be defined as follows.

4.3.3 Definition

The **exponential** function $\exp: \mathbb{R} \to \mathbb{R}$ is given by

$$\exp x = \sum_{n=0}^{\infty} \frac{x^n}{n!}$$

4.3.4   Definition

The **sine** function sin: $\mathbb{R} \to \mathbb{R}$ is given by

$$\sin x = \sum_{n=0}^{\infty} \frac{(-1)^n x^{2n+1}}{(2n+1)!}$$

4.3.5   Definition

The **cosine** function cos: $\mathbb{R} \to \mathbb{R}$ is given by

$$\cos x = \sum_{n=0}^{\infty} \frac{(-1)^n x^{2n}}{(2n)!}$$

The ratio test establishes immediately that these power series converge (absolutely) for all $x \in \mathbb{R}$, After reading Chapters 5–7, the reader can consult the Appendix, where the above power series definitions of the elementary functions are used to establish the analytic properties of those functions that are assumed in Chapters 5–7. For the time being, the reader can be content with the derivation of some arithmetic properties of the elementary functions.

■■   EXAMPLE 2

$$\exp(x + y) = \exp x \exp y \quad \text{for all } x, y \in \mathbb{R}$$

Solution

By definition,

$$\exp x = \sum_{n=0}^{\infty} \frac{x^n}{n!} \quad \text{and} \quad \exp y = \sum_{n=0}^{\infty} \frac{y^n}{n!}$$

By 4.2.9, $\exp x \exp y = \sum_{n=0}^{\infty} \alpha_n$, where

$$\alpha_n = \sum_{r=0}^{n} \frac{x^r}{r!} \frac{y^{n-r}}{(n-r)!}$$

Now, by the binomial expansion (see Section 3.2, after 3.2.1),

$$\alpha_n = \sum_{r=0}^{n} \binom{n}{r} \frac{x^r y^{n-r}}{n!} = \frac{(x+y)^n}{n!}$$

Hence $\exp(x + y) = \exp x \exp y$.   ■

■■ **EXAMPLE 3**

$$\cos^2 x + \sin^2 x = 1 \quad \text{for all } x \in \mathbb{R}$$

Solution

Now

$$\cos^2 x + \sin^2 x = \left[\sum_{n=0}^{\infty} \frac{(-1)^n x^{2n}}{(2n)!}\right]\left[\sum_{n=0}^{\infty} \frac{(-1)^n x^{2n}}{(2n)!}\right]$$

$$+ x^2\left[\sum_{n=0}^{\infty} \frac{(-1)^n x^{2n}}{(2n+1)!}\right]\left[\sum_{n=0}^{\infty} \frac{(-1)^n x^{2n}}{(2n+1)!}\right]$$

$$= \sum_{n=0}^{\infty} \alpha_n x^{2n} + x^2 \sum_{n=0}^{\infty} \beta_n x^{2n}, \quad \text{by 4.3.2}$$

where

$$\alpha_n = \sum_{r=0}^{n} \frac{(-1)^r}{(2r)!} \frac{(-1)^{n-r}}{(2n-2r)!}$$

and

$$\beta_n = \sum_{r=0}^{n} \frac{(-1)^r}{(2r+1)!} \frac{(-1)^{n-r}}{(2n-2r+1)!}$$

Hence

$$\cos^2 x + \sin^2 x = 1 + \sum_{n=1}^{\infty} \omega_n x^{2n}, \quad \text{where } \omega_n = \alpha_n + \beta_{n-1}$$

Now

$$\alpha_n + \beta_{n-1} = \frac{(-1)^n}{(2n)!} \sum_{r=0}^{n} \binom{2n}{2r} + \frac{(-1)^{n-1}}{(2n)!} \sum_{r=0}^{n-1} \binom{2n}{2r+1}$$

$$= \frac{(-1)^n}{(2n)!} \left\{ \sum_{r=0}^{n-1}\left[\binom{2n}{2r} - \binom{2n}{2r+1}\right] + \binom{2n}{2n} \right\}$$

$$= \frac{(-1)^n}{(2n)!} \left[\binom{2n}{0} - \binom{2n}{1} + \binom{2n}{2} - \binom{2n}{3} + \cdots \right.$$

$$\left. + \binom{2n}{2n-2} - \binom{2n}{2n-1} + \binom{2n}{2n} \right]$$

$$= \frac{(-1)^n}{(2n)!} \sum_{r=0}^{2n} \binom{2n}{r} x^r, \quad \text{where } x = -1$$

Since, by the binomial expansion,

$$\sum_{r=0}^{2n} \binom{2n}{r} x^r = (1 + x)^{2n}$$

$\omega_n = \alpha_n + \beta_{n-1} = 0$, and hence $\cos^2 x + \sin^2 x = 1$. ∎

Exercises 4.3

1. Determine the radius of convergence of each of the following power series:

    (a) $\displaystyle\sum_{n=0}^{\infty} \frac{(n!)^2 x^n}{(2n)!}$ (b) $\displaystyle\sum_{n=0}^{\infty} \frac{n^3 x^n}{n!}$

2. Show that the Cauchy product of $\sum_{n=0}^{\infty} x^n$ with itself is $\sum_{n=0}^{\infty} (n + 1)x^n$, and deduce that

    $$\sum_{n=0}^{\infty} (n + 1)x^n = (1 - x)^{-2} \quad \text{for } |x| < 1$$

3. Verify that the power series defining the sine, cosine and exponential functions (see 4.3.3, 4.3.4 and 4.3.5) are (absolutely) convergent for all x.

4. Show that

    (a) $\cos(-x) = \cos x$ and $\sin(-x) = -\sin x$ for all $x \in \mathbb{R}$,
    (b) $|\cos x| \leq 1$ and $|\sin x| \leq 1$ for all $x \in \mathbb{R}$.

5. If $\sum_{n=0}^{\infty} a_n x^n$ has radius of convergence $R > 0$ and $\lim_{n \to \infty} |a_{n+1}/a_n|$ exists, show that

    $$\sum_{n=1}^{\infty} na_n x^{n-1} \quad \text{converges for } |x| < R$$

    Verify that if $f(x) = \sum_{n=0}^{\infty} x^n$ then the derivative $f'(x) = \sum_{n=1}^{\infty} nx^{n-1}$ $(|x| < 1)$

Problems 4

1. Show that

    $$\frac{1}{4r^2 - 1} = \frac{1}{2}\left(\frac{1}{2r - 1} - \frac{1}{2r + 1}\right)$$

Deduce that

$$\sum_{r=1}^{\infty} \frac{1}{4r^2 - 1} = \frac{1}{2}$$

2. Let $\sum_{r=1}^{\infty} a_r$ and $\sum_{r=1}^{\infty} b_r$ be positive-term series such that $a_n \geqslant k b_n$ for all $n \in \mathbb{N}$, where $k > 0$ is a constant. Prove that

$$\sum_{r=1}^{\infty} b_r \text{ divergent} \Rightarrow \sum_{r=1}^{\infty} a_r \text{ divergent}$$

3. Determine which of the following series are convergent.

(a) $\displaystyle\sum_{r=1}^{\infty} \left[\left(\frac{3}{4}\right)^r - \frac{2}{r(r+1)} \right]$

(b) $\displaystyle\sum_{r=1}^{\infty} \left[\frac{1}{\sqrt{r}} - \frac{1}{\sqrt{r+1}} \right]$

(c) $\displaystyle\sum_{r=1}^{\infty} \frac{10^r}{r!}$

(d) $\displaystyle\sum_{r=1}^{\infty} \frac{(-1)^r r^2}{r^2 + 1}$

(e) $\displaystyle\sum_{r=1}^{\infty} \frac{(-1)^r r}{r^2 + 2}$

(f) $\displaystyle\sum_{r=1}^{\infty} \frac{5r + 2^r}{3^r}$

(g) $\displaystyle\sum_{r=1}^{\infty} \frac{2^r}{r^6}$

(h) $\displaystyle\sum_{r=1}^{\infty} \frac{(2r)!}{r^r}$

4. For which values of x is $\sum_{r=1}^{\infty} x^r r! / r^r$ convergent?

5. (a) Show that

$$\frac{1}{3n-2} + \frac{1}{3n-1} - \frac{1}{3n} \geqslant \frac{1}{3n}$$

for $n \geqslant 1$, and hence deduce that

$$1 + \tfrac{1}{2} - \tfrac{1}{3} + \tfrac{1}{4} + \tfrac{1}{5} - \tfrac{1}{6} + \dots$$

is divergent.

(b) Show that

$$\frac{1}{\sqrt{2n-1}} - \frac{1}{2n} \geqslant \frac{1}{2n}$$

for $n \geqslant 1$, and hence deduce that

$$1 - \tfrac{1}{2} + \sqrt{\tfrac{1}{3}} - \tfrac{1}{4} + \sqrt{\tfrac{1}{5}} - \tfrac{1}{6} + \dots$$

is divergent.

6. Test each of the following series for absolute or conditional convergence:

(a) $\sum_{r=1}^{\infty} \frac{(-1)^r}{\sqrt{r}}$ (b) $\sum_{r=1}^{\infty} (-r)^{-r}$

(c) $\sum_{r=1}^{\infty} r(-\tfrac{1}{2})^r$ (d) $\sum_{r=1}^{\infty} \frac{\tfrac{1}{2} + (-1)^r}{r}$

7. Let $\sum_{r=1}^{\infty} a_r$ be a positive-term series such that $(a_n)^{1/n} \to L$ as $n \to \infty$. By comparing $\sum_{r=1}^{\infty} a_r$ with a suitable geometric series, show that $\sum_{r=1}^{\infty} a_r$ converges if $L < 1$ and diverges for $L > 1$. This result is known as the **nth-root test**.

8. Use the nth-root test to determine which of the following converge:

(a) $\sum_{r=1}^{\infty} \left(2 + \frac{4}{r}\right)^r$ (b) $\sum_{r=1}^{\infty} r^{-\sqrt{r}}$ (c) $\sum_{r=2}^{\infty} \frac{1}{(\log_e r)^r}$

9. Determine the radius of convergence of each of the following power series:

(a) $\sum_{n=0}^{\infty} (2^n + 3^n) x^n$ (b) $\sum_{n=0}^{\infty} n^2 x^n$

(c) $\sum_{n=0}^{\infty} \frac{n!(2n)!}{(3n)!} x^n$

10. Show that the Cauchy product of $\sum_{n=1}^{\infty} x^n/n$ with itself is

$$\sum_{n-1}^{\infty} \frac{2}{n+1} \left(1 + \frac{1}{2} + \frac{1}{3} + \ldots + \frac{1}{n}\right) x^{n+1}$$

CHAPTER FIVE

5

Continuous Functions

5.1   Limits
5.2   Continuity
5.3   Theorems

In this chapter the important concept of a continuous function will be introduced. The reader is no doubt familiar with sketch-graphs of basic functions such as $f(x) = x^2$, $g(x) = 1 - 1/x$ and $h(x) = \sin x$ (see Figure 5.1). Underlying the sketching of these graphs is the plausible assumption that, except for points where the function is undefined, the graphs consist of smooth unbroken curves. This is frequently taken for granted, but can it be justified?

In Section 5.2 a precise definition of a continuous function will be given. The formulation given is expressed in terms of the concept of the limit of a function; this is covered in Section 5.1. The advantage of having a formal definition of a continuous function is that it is then possible to *prove* that graphs such as those in Figure 5.1 do consist of 'continuous curves'. In addition, the rules established in Section 5.2 can be used to construct new continuous functions from more basic ones. This results in quite complicated functions that, by virtue of the fact that they are continuous in the sense of the formal definition, are guaranteed to have graphs consisting of 'continuous curves' in the intuitive sense.

Firm evidence that the mathematically precise definition of continuity adopted does accord well with intuitive geometrical ideas is given in Section 5.3. In this section several fundamental properties possessed by all continuous functions are proved. The theorems themselves are of independent interest, since they form part of the theoretical basis for

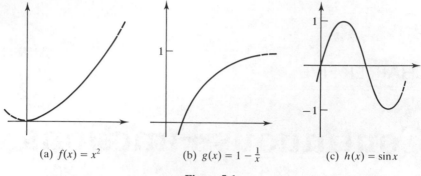

(a) $f(x) = x^2$ (b) $g(x) = 1 - \frac{1}{x}$ (c) $h(x) = \sin x$

Figure 5.1

many of the practical applications of **numerical analysis**; a subject concerned with the development of efficient algorithms for solving a wide range of mathematical problems. In particular, the intermediate value property (5.3.2) leads directly to the bisection method for locating roots of a real polynomial (see Question 1 of Exercises 5.3). The intermediate value property was first formulated and proved by Bolzano (1781–1848), one of the founders of modern analysis.

5.1 Limits

The first task in this section is to extend the idea of the limit of a convergent sequence in order to analyse the behaviour of a function $f: A \to \mathbb{R}$ whose domain A contains arbitrarily large positive real numbers x. First note that a sequence (a_n) may be regarded as a function $f: \mathbb{N} \to \mathbb{R}$, with domain $\mathbb{N}$, given by the formula $f(n) = a_n$. The graph of such a function consists of an infinite number of isolated points, as indicated in Figures 3.1 and 3.3. Hence in Chapter 3 the behaviour of $f(n)$ was investigated for increasingly large positive integral values of n. In contrast, the functions graphed in Figure 5.1 have domains that include an interval of real numbers of the form $[a, \infty)$ and whose graphs are (or appear to be) unbroken curves. As x increases through real values, the behaviour of these three functions is very reminiscent of the behaviour of sequences as n tends to infinity through integral values. This similarity is examined in the following example.

■■   EXAMPLE 1

Figure 5.1(a) gives the graph of the function $f: \mathbb{R} \to \mathbb{R}$ where $f(x) = x^2$; as x increases, the graph rises more and more steeply and the values $f(x)$ become larger and larger.

Figure 5.1(b) gives the graph of the function $g: \mathbb{R}^+ \to \mathbb{R}$ where $g(x) = 1 - 1/x$; as x increases, the graph rises but the values $g(x)$ always remain less than one.

Figure 5.1(c) gives the graph of the function $h: \mathbb{R} \to \mathbb{R}$ where $h(x) = \sin x$; as x increases, the graph oscillates between the values ± 1.

If the domains of f, g and h are restricted to the set $\mathbb{N}$ of natural numbers then the resulting sequences are respectively divergent (to infinity), convergent to 1 and divergent (though bounded). ∎

It is now tempting, for example, to write $1 - 1/x \to 1$ as $x \to \infty$ to describe the behaviour of the function g as x increases. Since x is now a real variable, this will be permissible once the following formal definition has been made.

5.1.1   Definition

Let $f: A \to \mathbb{R}$ be a function whose domain A contains the interval $[a, \infty)$ for some fixed real value of a. Then $f(x)$ **tends to** L as x **tends to infinity**, if for every $\varepsilon > 0$ there exists a real number X such that $x > X \Rightarrow |f(x) - L| < \varepsilon$.

This is denoted by $\lim_{x \to \infty} f(x) = L$ or $f(x) \to L$ as $x \to \infty$.

■■   EXAMPLE 2

Prove that $\lim_{x \to \infty} (1 - 1/x) = 1$.

Solution

For positive x

$$\left| \left(1 - \frac{1}{x} \right) - 1 \right| = \frac{1}{x}$$

Let $X = 1/\varepsilon$. Then $x > X$ implies that

$$\left| \left(1 - \frac{1}{x} \right) - 1 \right| < \varepsilon$$

In other words, $1 - 1/x \to 1$ as $x \to \infty$. ∎

If the domain of f contains an interval of the form $(-\infty, b]$ for some $b \in \mathbb{R}$ then its 'mirror image' g given by $g(x) = f(-x)$ contains the interval $[b, \infty)$. See Figure 5.2. Then $f(x)$ **tends to** L **as** x **tends to minus**

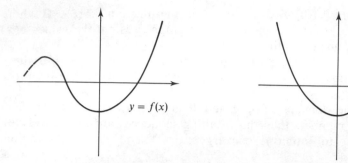

Figure 5.2

infinity if and only if $g(x) = f(-x) \to L$ as $x \to \infty$. This is denoted by $\lim_{x \to -\infty} f(x) = L$ or $f(x) \to L$ as $x \to -\infty$.

Having extended the language of limits to analyse the behaviour of a function $f: A \to \mathbb{R}$ of a real variable x as x takes increasingly large positive or large negative values, it is now natural to focus attention on the function values $f(x)$ for x close to a fixed real number a. To this end, assume that $f(x)$ is defined for all real values of x close to a, but not necessarily at a. In other words, assume that the domain of f contains the set $(a - \delta, a) \cup (a, a + \delta)$ for some $\delta > 0$. The reason for excluding $x = a$ is that it is the behaviour of $f(x)$ as x tends to a that is required, and not the particular value of $f(x)$ at $x = a$. Indeed, as the following example demonstrates, $f(x)$ may tend to a limiting value as x tends to a, even though $f(a)$ is undefined.

■■ EXAMPLE 3

Let

$$f(x) = \frac{\sin x}{x}$$

a function defined only for $x \neq 0$. Calculation gives

$$f(0.1) = 0.998\,334\,17$$

$$f(0.01) = 0.999\,983\,33$$

$$f(0.001) = 0.999\,999\,83$$

It is the case that, as $x \to 0$, $(\sin x)/x \to 1$ (see Example 3, Section 6.2), even though $f(0)$ is undefined. ■

Hence in the definition of the limiting value of $f(x)$ as x tends to a care must be taken to exclude $x = a$ when x tends to a. Now the condi-

$a-\delta$ a $a+\delta$ $|x-a|<\delta$

$a-\delta$ a $a+\delta$ $0<|x-a|<\delta$

Figure 5.3

tion $|x - a| < \delta$ describes all x in the interval $(a - \delta, a + \delta)$, called a **neighbourhood** of a. To exclude $x = a$, it must be written that $0 < |x - a| < \delta$, and this describes the **deleted neighbourhood** $(a - \delta, a) \cup (a, a + \delta)$ of a. See Figure 5.3.

Let $f: A \to \mathbb{R}$ be a function whose domain contains a deleted neighbourhood of a.

5.1.2 Definition

> $f(x)$ tends to the **limit** L as x tends to a if and only if for every $\varepsilon > 0$ there exists a $\delta > 0$ such that
>
> $$0 < |x - a| < \delta \Rightarrow |f(x) - L| < \varepsilon$$

This is denoted by $\lim_{x \to a} f(x) = L$ or $f(x) \to L$ as $x \to a$.

If for some given function f it is required to prove that $f(x) \to L$ as $x \to a$ then for each positive ε, however small, a $\delta > 0$ (depending on ε) must be produced such that $f(x)$ is within a distance ε of L for all x within a distance δ of a (excluding $x = a$, of course). In Figure 5.4 the points of the graph lying in the horizontal strip from $y = L - \varepsilon$ to $y = L + \varepsilon$ all satisfy $|f(x) - L| < \varepsilon$. Since the graph is a continuous curve,

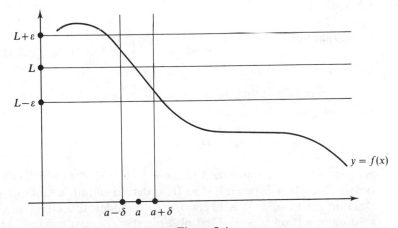

Figure 5.4

these points also lie in a vertical strip of width 2δ centred on $x = a$. In other words, there is a $\delta > 0$ such that $|x - a| < \delta$ forces $|f(x) - L| < \varepsilon$.

■■ **EXAMPLE 4**

Prove that $\lim_{x \to 2} (3x - 1) = 5$.

Solution

What is required is that for every $\varepsilon > 0$ there can be found a $\delta > 0$ such that

$$0 < |x - 2| < \delta \Rightarrow |(3x - 1) - 5| < \varepsilon$$

But $|(3x - 1) - 5| < \varepsilon$ is equivalent to $|3x - 6| < \varepsilon$, which can be rewritten as $|x - 2| < \frac{1}{3}\varepsilon$. Hence, whatever value is assigned to $\varepsilon > 0$, it suffices to choose $\delta = \frac{1}{3}\varepsilon$. Then

$$0 < |x - 2| < \delta \Rightarrow 0 < |x - 2| < \tfrac{1}{3}\varepsilon$$
$$\Rightarrow |3x - 6| < \varepsilon$$
$$\Rightarrow |(3x - 1) - 5| < \varepsilon$$

In other words, $\lim_{x \to 2} (3x - 1) = 5$. ■

In the preceding example it was relatively easy to obtain the restriction on $|x - 2|$ guaranteeing that $|(3x - 1) - 5|$ was less than ε. It is not always so straightforward, as the following two examples illustrate.

■■ **EXAMPLE 5**

Prove that $\lim_{x \to 3} (x^2) = 9$

Solution

What is required is that for every $\varepsilon > 0$ there can be found a $\delta > 0$ such that

$$0 < |x - 3| < \delta \Rightarrow |x^2 - 9| < \varepsilon$$

Now $|x^2 - 9| < \varepsilon$ is equivalent to $|(x + 3)(x - 3)| < \varepsilon$, which can be rewritten as $|x + 3||x - 3| < \varepsilon$. In order to obtain a suitable restriction on $|x - 3|$, it is necessary to consider the effect such a restriction has on $|x + 3|$. First observe that, once a suitable δ has been obtained, any positive number less than δ could also be substituted for δ. It is thus permissible to assume that $\delta \leqslant 1$. If $0 < |x - 3| < 1$ then $2 < x < 4$, and so $5 < x + 3 < 7$. Hence $|x + 3| < 7$. In other words,

$$0 < |x - 3| < 1 \Rightarrow |x + 3| |x - 3| < 7|x - 3|$$
$$\Rightarrow |x^2 - 9| < 7|x - 3|$$

Therefore the inequality $|x^2 - 9| < \varepsilon$ will hold provided that $7|x - 3| < \varepsilon$. This latter inequality is clearly equivalent to $|x - 3| < \frac{1}{7}\varepsilon$. So, given any $\varepsilon > 0$, choose δ to be the minimum of 1 and $\frac{1}{7}\varepsilon$. Then

$$0 < |x - 3| < \delta \Rightarrow |x + 3| |x - 3| < 7|x - 3|$$
$$\Rightarrow |x^2 - 9| < 7|x - 3|$$
$$\Rightarrow |x^2 - 9| < 7(\tfrac{1}{7}\varepsilon)$$
$$\Rightarrow |x^2 - 9| < \varepsilon$$

In other words, $\lim_{x \to 3} (x^2) = 9$ ■

■■ EXAMPLE 6

Prove that $\lim_{x \to 1} (x^2 - 5x + 7) = 3$

Solution

What is required is that for every $\varepsilon > 0$ there can be found a $\delta > 0$ such that

$$0 < |x - 1| < \delta \Rightarrow |(x^2 - 5x + 7) - 3| < \varepsilon$$

Now

$$|(x^2 - 5x + 7) - 3| = |x^2 - 5x + 4|$$
$$= |(x - 1)^2 - 3(x - 1)|$$
$$\leqslant |x - 1|^2 + 3|x - 1|,$$

using the triangle inequality

Hence, if $0 < |x - 1| < 1$,

$$|x - 1|^2 + 3|x - 1| < 4|x - 1|$$

So, given $\varepsilon > 0$, choose δ to be the minimum of 1 and $\frac{1}{4}\varepsilon$. Then

$$0 < |x - 1| < \delta \Rightarrow |(x^2 - 5x + 7) - 3| < \varepsilon$$

In other words, $\lim_{x \to 1} (x^2 - 5x + 7) = 3$.   ■

Resorting to εs and δs in order to establish limits such as those in Examples 5 and 6 is clearly technically demanding. A far better approach is to derive rules that can be used to evaluate such limits more easily. Application of these rules will require knowledge of the limits of certain simple functions – limits that must be verified by appealing to Definition 5.1.1. The definition will of course be instrumental in proving the rules.

5.1.3 Rules

Suppose that $\lim_{x \to a} f(x) = L$ and $\lim_{x \to a} g(x) = M$. Then the following rules apply.

Sum rule

$$\lim_{x \to a} [f(x) + g(x)] = L + M$$

Product rule

$$\lim_{x \to a} f(x)g(x) = LM$$

Quotient rule

$$\lim_{x \to a} \frac{f(x)}{g(x)} = \frac{L}{M}, \quad \text{provided that } M \neq 0$$

The sum and product rules are proved below, the proof of the quotient rule being left as an exercise.

Proof of the sum rule

Given $\varepsilon > 0$, there exist $\delta_1, \delta_2 > 0$ such that

$$0 < |x - a| < \delta_1 \Rightarrow |f(x) - L| < \tfrac{1}{2}\varepsilon$$

and

$$0 < |x - a| < \delta_2 \Rightarrow |g(x) - L| < \tfrac{1}{2}\varepsilon$$

Let δ be the minimum of δ_1 and δ_2. Then for x satisfying $0 < |x - a| < \delta$ it follows that

$$|f(x) + g(x) - (L + M)| \leqslant |f(x) - L| + |g(x) - M|$$
$$< \tfrac{1}{2}\varepsilon + \tfrac{1}{2}\varepsilon = \varepsilon \qquad \square$$

Proof of the product rule

If $\varepsilon = 1$, there exists a $\delta > 0$ such that $|g(x) - M| < 1$ for all x satisfying $0 < |x - a| < \delta$. Hence for such x

$$|g(x)| = |g(x) - M + M| \leq |g(x) - M| + |M| < 1 + |M|$$

Given any $\varepsilon > 0$, let

$$\varepsilon_1 = \frac{\varepsilon}{2(1 + |M|)} \quad \text{and} \quad \varepsilon_2 = \frac{\varepsilon}{2(|L| + 1)}$$

Now there exist $\delta_1, \delta_2 > 0$ such that

$$0 < |x - a| < \delta_1 \Rightarrow |f(x) - L| < \varepsilon_1 = \frac{\varepsilon}{2(1 + |M|)}$$

and

$$0 < |x - a| < \delta_2 \Rightarrow |g(x) - M| < \varepsilon_2 = \frac{\varepsilon}{2(|L| + 1)}$$

Now choose δ_3 to be the minimum of δ_1, δ_2 and δ. For all x satisfying $0 < |x - a| < \delta_3$ it follows that:

$$
\begin{aligned}
|f(x)g(x) - LM| &= |f(x)g(x) - Lg(x) + Lg(x) - LM| \\
&\leq |g(x)| |f(x) - L| + |L| |g(x) - M| \\
&< (1 + |M|)\varepsilon_1 + |L|\varepsilon_2 \\
&= (1 + |M|) \frac{\varepsilon}{2(1 + |M|)} \\
&\quad + |L| \frac{\varepsilon}{2(|L| + 1)} < \varepsilon
\end{aligned}
$$

Therefore $\lim_{x \to a} f(x)g(x) = LM$.   □

■■   **EXAMPLE 7**

Evaluate

$$\lim_{x \to 1} \frac{2x^2 - 3x + 4}{x^3 + 5x + 1}$$

Solution

Since $\lim_{x \to 1} x = 1$, the product rule gives $\lim_{x \to 1} x^2 = 1$ and $\lim_{x \to 1} x^3 = 1$. Since the constant function $f(x) = k$, for all x, is such that $\lim_{x \to 1} f(x) = k$, the sum and product rules give

$$\lim_{x \to 1} (2x^2 - 3x + 4) = 3$$

and

$$\lim_{x \to 1} (x^3 + 5x + 1) = 7$$

Finally, the quotient rule gives

$$\lim_{x \to 1} \frac{2x^2 - 3x + 4}{x^3 + 5x + 1} = \frac{3}{7}$$ ∎

5.1.4   Sandwich rule

Let f, g and h be three functions satisfying $h(x) \leqslant f(x) \leqslant g(x)$ in some deleted neighbourhood of $x = a$. If

$$\lim_{x \to a} h(x) = \lim_{x \to a} g(x) = L$$

then

$$\lim_{x \to a} f(x) = L$$

Proof of the sandwich rule

Given $\varepsilon > 0$, there exist δ_1, $\delta_2 > 0$ such that

$$0 < |x - a| < \delta_1 \Rightarrow |h(x) - L| < \varepsilon$$

and

$$0 < |x - a| < \delta_2 \Rightarrow |g(x) - L| < \varepsilon$$

Now $|f(x) - L|$ does not exceed the maximum of $|h(x) - L|$ and $|g(x) - L|$, and so, with δ equal to the minimum of δ_1 and δ_2,

$$0 < |x - a| < \delta \Rightarrow |f(x) - L| < \varepsilon$$

as required. ☐

■■   EXAMPLE 8

Consider $f(x) = x^2 (\sin 1/x)$, $x \neq 0$. Since $|\sin t| \leqslant 1$ for all real t,

$$\left| x^2 \sin\left(\frac{1}{x}\right) \right| \leqslant |x^2| \quad \text{for all } x$$

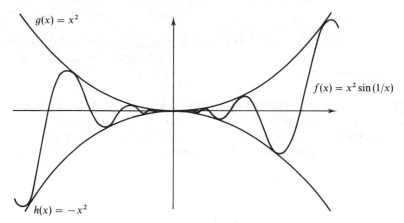

Figure 5.5

in any deleted neighbourhood of 0. If $h(x) = -x^2$ and $g(x) = x^2$ then $h(x) \leqslant f(x) \leqslant g(x)$ for all x in any deleted neighbourhood of 0. Since

$$\lim_{x \to 0} h(x) = \lim_{x \to 0} g(x) = 0$$

then, by the sandwich rule,

$$\lim_{x \to 0} \left[x^2 \sin\left(\frac{1}{x}\right) \right] = 0$$

The graph of f appears in Figure 5.5, and it is clear that $f(x)$ is indeed 'sandwiched' between $\pm x^2$ at the origin.   ■

The final rule involves the concept of a continuous function, and so its proof is deferred until Section 5.2, where continuous functions are discussed.

5.1.5 Composite rule

If $\lim_{x \to a} f(x) = L$ and g is a function that is continuous in some neighbourhood of $x = L$ then

$$\lim_{x \to a} (g \circ f)(x) = g(L)$$

■■   EXAMPLE 9

Show that $\lim_{x \to 1} \sqrt{x^2 + 1} = \sqrt{2}$.

Solution

From the sum and product rules, $\lim_{x \to 1} (x^2 + 1) = 2$. Assuming that the function $g(x) = \sqrt{x}$ is continuous for positive x, the composite rule gives $\lim_{x \to 1} \sqrt{x^2 + 1} = \sqrt{2}$. ∎

This section concludes with a discussion of the notion of **one-sided limits**. If x is allowed to tend to a while $x > a$ throughout then the **right-hand limit**, $\lim_{x \to a+} f(x)$ is obtained. In ε-δ terms, $\lim_{x \to a+} f(x) = L$ means that for every $\varepsilon > 0$ there exists a $\delta > 0$ such that

$$a < x < a + \delta \Rightarrow |f(x) - L| < \varepsilon$$

If, on the other hand, x is allowed to tend to a while $x < a$, the **left-hand limit**, $\lim_{x \to a-} f(x)$ is obtained. This has the obvious ε-δ definition. These definitions reveal the following result.

5.1.6 Theorem

$$\lim_{x \to a} f(x) = L \Leftrightarrow \lim_{x \to a-} f(x) = \lim_{x \to a+} f(x) = L$$

■■ EXAMPLE 10

Consider $f(x) = |x|$. Since

$$\lim_{x \to 0+} |x| = \lim_{x \to 0+} x = 0$$

and

$$\lim_{x \to 0-} |x| = \lim_{x \to 0-} (-x) = 0$$

it can be deduced that $\lim_{x \to 0} |x| = 0$. ∎

■■ EXAMPLE 11

If x is a real number then there exists an integer n such that $x - 1 < n \leqslant x$; n is called the **integer part** of x and is denoted by $[x]$. Consider $f(x) = x - [x]$. The graph of f appears in Figure 5.6. Clearly

$$\lim_{x \to n+} f(x) = \lim_{x \to n+} (x - n) = 0$$

and

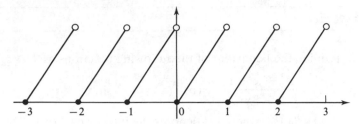

Figure 5.6

$$\lim_{x \to n-} f(x) = \lim_{x \to n-} (x - n + 1) = 1$$

Hence $\lim_{x \to n} f(x)$ does not exist.   ∎

One-sided limits arise naturally when investigating the behaviour of functions as x tends to $\pm\infty$, as the following example illustrates.

■■ **EXAMPLE 12**

Consider the function given by

$$f(x) = \frac{3x^2 + 1}{x^2 - 1} \quad (x \neq \pm 1)$$

Let $t = 1/x$ and $g(t) = f(x) = f(1/t)$, where x, $t \neq 0$. Then

$$g(t) = \frac{3/t^2 + 1}{1/t^2 - 1} = \frac{t^2 + 3}{1 - t^2} \quad \text{for } t \neq 0, \pm 1$$

and so $\lim_{t \to 0} g(t) = 3$ via 5.1.3. By 5.1.6, $\lim_{t \to 0+} g(t) = 3$. In other words, for any $\varepsilon > 0$ there exists a $\delta > 0$ such that

$$0 < t < \delta \Rightarrow |g(t) - 3| < \varepsilon$$

Let $X = 1/\delta$. Then

$$x > X \Rightarrow \frac{1}{t} > \frac{1}{\delta} > 0$$

$$\Rightarrow 0 < t < \delta$$

$$\Rightarrow |g(t) - 3| < \varepsilon$$

$$\Rightarrow |f(x) - 3| < \varepsilon$$

Hence $\lim_{x \to \infty} f(x) = 3$. A similar analysis using the left-hand limit $\lim_{t \to 0-} g(t)$ soon gives $\lim_{x \to -\infty} f(x) = 3$.   ∎

Exercises 5.1

1. Determine the behaviour of the following functions as $x \to \pm\infty$:

   (a) $f(x) = \dfrac{x^3 + x + 1}{2 - 3x^3}$ (b) $f(x) = e^{-x} \sin x$

   (c) $f(x) = [x]/x$, where $[x]$ denotes the integer part of x

2. Given an $\varepsilon{-}\delta$ proof of the fact that

$$\lim_{x \to 0} 2x \sin\left(\frac{1}{x}\right) = 0$$

3. Prove the quotient rule for limits of functions (see 5.1.3).

4. Use the rules for limits to evaluate the following.

   (a) $\displaystyle\lim_{x \to 0} \dfrac{x^3 - 1}{x^2 + 1}$ (b) $\displaystyle\lim_{x \to 1} \dfrac{x - 1}{x^3 - 1}$

   (c) $\displaystyle\lim_{x \to 1} \cos\left(\dfrac{\pi x}{x + 1}\right)$ (d) $\displaystyle\lim_{x \to 0} x^2 \cos\left(\dfrac{1}{x}\right)$

5. Evaluate the following one-sided limits.

   (a) $\displaystyle\lim_{x \to 1+} \dfrac{1}{[x] + 1}$ (b) $\displaystyle\lim_{x \to 4+} [\sqrt{x}\,]$ (c) $\displaystyle\lim_{x \to 4-} [\sqrt{x}\,]$

6. Let $f(x)$ be defined on an interval $(0, a)$ and let $t = 1/x$. Prove that if either of the limits $\lim_{x \to 0+} f(x)$ or $\lim_{t \to \infty} f(1/t)$ exists then both exist and have the same value.

5.2 Continuity

Naively, a function $f: A \to \mathbb{R}$ is continuous if its graph is a continuous curve. In particular, if the domain of f contains a neighbourhood of a fixed real number c then the graph of f can be drawn through the point $(c, f(c))$ without removing the pen from the paper. The desired behaviour at $(c, f(c))$ can be arranged by insisting that, for all values of x sufficiently close to c, $f(x)$ is close to $f(c)$. Before giving the official definition of continuity at $x = c$, it is helpful to look at the graphs of a few functions.

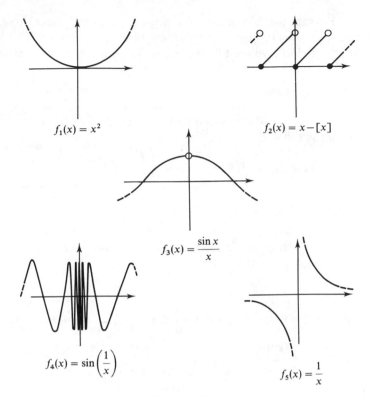

$$f_1(x) = x^2$$

$$f_2(x) = x - [x]$$

$$f_3(x) = \frac{\sin x}{x}$$

$$f_4(x) = \sin\left(\frac{1}{x}\right)$$

$$f_5(x) = \frac{1}{x}$$

Figure 5.7

■■   EXAMPLE 1

Refer to Figure 5.7.

(1) $f_1(x) = x^2$ appears to be a continuous curve for all x.

(2) $f_2(x) = x - [x]$ has finite jumps at integer values of x.

(3) $f_3(x) = (\sin x)/x$, $f_4(x) = \sin(1/x)$ and $f_5(x) = 1/x$ $(x \neq 0)$ are all undefined at $x = 0$; however, they exhibit completely different sorts of behaviour near $x = 0$.   ■

If $f: A \to \mathbb{R}$ is a function whose domain contains a neighbourhood of c then the following definition is made.

5.2.1   Definition

> f is **continuous** at c if and only if $\lim_{x \to c} f(x) = f(c)$.

Note that this definition demands three things; first that $\lim_{x \to c} f(x)$ exists, secondly that $f(c)$ is defined, and finally that the previous two values are equal. Look at the functions in Example 1 and think about these three requirements as $x \to 0$. Our ε-δ definition of $\lim_{x \to a} f(x) = L$ (5.1.2) can easily be adapted to give the following ε-δ definition of continuity at c.

5.2.2 Definition

> f is **continuous** at c if and only if for every $\varepsilon > 0$ there exists a $\delta > 0$ such that
>
> $$|x - c| < \delta \Rightarrow |f(x) - f(c)| < \varepsilon$$

■■ **EXAMPLE 2**

Use the ε-δ definition of continuity to prove that $f(x) = x^2$ is continuous at $c = 0$.

Solution

For any $\varepsilon > 0$ determine those x for which $|f(x) - f(0)| < \varepsilon$. Now

$$|f(x) - f(0)| = |x^2 - 0| = |x^2| < \varepsilon$$

provided that $|x| < \sqrt{\varepsilon}$. So let $\delta = \sqrt{\varepsilon}$. If $|x - 0| < \delta$ then $|f(x) - f(0)| < \varepsilon$. In other words, $\lim_{x \to 0} f(x) = 0$, and, since $f(0) = 0$, $\lim_{x \to 0} f(x) = f(0)$. Hence f is continuous at 0. ■

If a function f is continuous for all x in the range $a < x < b$ then it can be said that f is **continuous on the interval** (a, b). If f is continuous for all x in its domain, it can be simply said that f is **continuous**. If $\lim_{x \to c+} f(x)$ exists and equals $f(c)$ then f is called **right-continuous** at c, and if $\lim_{x \to c-} f(x)$ exists and equals $f(c)$ then f is called **left-continuous** at c. The ε-δ formulations of the last two definitions are not difficult to write down. Moreover, the following result holds.

5.2.3 Theorem

> A function f is continuous at c if and only if f is both left-continuous and right-continuous at c.

Note

If a function f is only defined on the closed interval $[a, b]$ and it is claimed that f is 'continuous on $[a, b]$', what is meant is that f is continuous on (a, b), right-continuous at a and left-continuous at b.

Soon, rules that enable the building up of complicated continuous functions from a given list of standard continuous functions such as sine, cosine, exponential and so on will be established. That these standard functions are indeed continuous follows from their power series definitions and the fact that the sum function of a power series is continuous for x inside the interval of convergence of the defining power series. See the Appendix.

Returning to Example 1, the following comment applies to the discontinuities observed there. The following example distinguishes between a **discontinuity**, where a function is defined but is not continuous, and a **singularity**, where a function is undefined.

■■   EXAMPLE 1 (Revisited)

(1) $f_3(x) = (\sin x)/x$ for $x \neq 0$ has what is called a **removable singularity** at $x = 0$, since, defining further $f_3(0) = 1$, the new f_3 is continuous not only for $x \neq 0$ but also at $x = 0$.

(2) The discontinuities of $f_2(x) = x - [x]$ are called **jump discontinuities**. This is because the left- and right-hand limits of f_2 at integer values of x exist but are unequal. In fact, f_2 is right-continuous but not left-continuous at integer values of x.

(3) $f_4(x) = \sin(1/x)$, $x \neq 0$ has an **oscillating singularity** at $x = 0$. The oscillations are bounded.

(4) $f_5(x) = 1/x$ for $x \neq 0$ exhibits what is termed an **infinite singularity** at $x = 0$.

Neither f_4 nor f_5 can be extended to give functions continuous on the whole of $\mathbb{R}$.   ■

■■   EXAMPLE 3

Dirichlet's function is defined by

$$f(x) = \begin{cases} \dfrac{1}{q} & \text{if } x = \dfrac{p}{q} \text{ is a rational in its lowest form} \\ 0 & \text{if } x \text{ is irrational} \end{cases}$$

Restricting our attention to the interval $(0, 1)$, and starting to construct the rather bizarre graph of Figure 5.8 gives the following:

$f(x) = \frac{1}{2}$ when $x = \frac{1}{2}$

$f(x) = \frac{1}{3}$ when $x = \frac{1}{3}$ or $\frac{2}{3}$

$f(x) = \frac{1}{4}$ when $x = \frac{1}{4}$ or $\frac{3}{4}$

$f(x) = \frac{1}{5}$ when $x = \frac{1}{5}$ or $\frac{2}{5}$ or $\frac{3}{5}$ or $\frac{4}{5}$

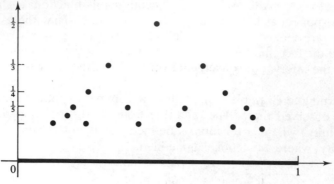

Figure 5.8

and so on. Now $f(x) \geqslant 1/q$ for fractions of the form $x = m/q$, $0 < m < q$. There are at most $\frac{1}{2}q(q - 1)$ such values of x. For any $\varepsilon > 0$ let q be a positive integer satisfying $1/q < \varepsilon$. For any c, $0 < c < 1$, there exists a $\delta > 0$ such that the interval $(c - \delta, c + \delta)$ contains none of the finitely many x satisfying $f(x) \geqslant 1/q$ (apart possibly from $x = c$ itself). Hence

$$|f(x)| = f(x) < \frac{1}{q} < \varepsilon$$

for all x satisfying $0 < |x - c| < \delta$. Thus $\lim_{x \to c} f(x) = 0$ for all $c \in (0, 1)$. So f is *continuous at every irrational and discontinuous at every rational in* $(0, 1)$! ∎

5.2.4 Rules

> **Sum rule**
>
> If f and g are continuous at c then $f + g$ is continuous at c.
>
> **Product rule**
>
> If f and g are continuous at c then $f \cdot g$ is continuous at c.
>
> **Reciprocal rule**
>
> If f is continuous at c and $f(c) \neq 0$ then $1/f$ is continuous at c.

Peter Gustav Lejeune Dirichlet (1805–1859) was principally interested in number theory, and his name arises in connection with many areas of both pure and applied mathematics. His main interest in analysis was in its applications to physical problems and number theory. As early as 1837, Dirichlet suggested a very broad definition of a function, not dissimilar to that presented in Chapter 1.3. To illustrate the generality of this definition and to exhibit very 'badly behaved' functions, he invented functions such as that in Example 3.

The proofs of the above rules and of the following sandwich rule are left as exercises. They all follow immediately from the corresponding rules for limits of functions and the definition of continuity (see 5.1.3, 5.1.4 and 5.2.1).

5.2.5   Sandwich rule

Let f, g and h be functions such that $h(x) \leqslant f(x) \leqslant g(x)$ for all x in some neighbourhood of c and such that $h(c) = f(c) = g(c)$. If h and g are continuous at c then so is f.

5.2.6   Composite rule

Let f and g be continuous at c and $f(c)$ respectively, such that the composite $g \circ f$ is defined. Then $g \circ f$ is continuous at c.

Proof of the composite rule

Let $f(c) = d$. Since g is continuous at d, for every $\varepsilon > 0$ there exists a $\delta_1 > 0$ such that

$$|t - d| < \delta_1 \Rightarrow |g(t) - g(d)| < \varepsilon \tag{1}$$

Since f is continuous at c and $\delta_1 > 0$, there exists a $\delta_2 > 0$ such that

$$|x - c| < \delta_2 \Rightarrow |f(x) - f(c)| < \delta_1$$

Now let $t = f(x)$ in (1) to deduce that

$$|x - c| < \delta_2 \Rightarrow |g(f(x)) - g(f(c))| < \varepsilon$$

Hence for any $\varepsilon > 0$ there exists a $\delta = \delta_2 > 0$ such that

$$|x - c| < \delta \Rightarrow |(g \circ f)(x) - (g \circ f)(c)| < \varepsilon \qquad \square$$

So far, an arbitrary function has been denoted as $f: A \to \mathbb{R}$, where A is the domain of f. For particular functions, $f(x)$ has been additionally specified for $x \in A$. Whenever the domain of f is clear from the context, the alternative notation $x \mapsto f(x)$ will often be used.

■■   EXAMPLE 4

Given that the identity function $x \mapsto x$, the constant functions $x \mapsto k$ and the trigonometric functions sine and cosine are all continuous (on $\mathbb{R}$), the following are proved to be continuous functions.

(a) $x \mapsto \dfrac{x^2 + 2x - 3}{x^2 + x + 1}$ (b) $x \mapsto x^3 \cos x^2$

(c) $x \mapsto \begin{cases} x \sin(1/x) & \text{if } x \neq 0 \\ 0 & \text{if } x = 0 \end{cases}$

Solution

(a) The given continuous functions and 5.2.4 are used. By the product rule, $x \mapsto x^2$ and $x \mapsto 2x$ are continuous. Hence $x \mapsto x^2 + 2x - 3$ and $x \mapsto x^2 + x + 1$ are continuous by the sum rule. Now

$$x^2 + x + 1 = (x + \tfrac{1}{2})^2 + \tfrac{3}{4} > 0 \quad \text{for all } x$$

and so, by the reciprocal rule,

$$x \mapsto \frac{1}{x^2 + x + 1}$$

is continuous. Finally, the product rule shows that

$$x \mapsto \frac{x^2 + 2x - 3}{x^2 + x + 1}$$

is continuous.

(b) The functions $x \mapsto x^2$ and $x \mapsto x^3$ are continuous by the product rule. By the composite rule (5.2.6), $x \mapsto \cos x^2$ is continuous. Hence, by the product rule, $x \mapsto x^3 \cos x^2$ is continuous.

(c) Call the given function f. For $x \neq 0$, $x \mapsto 1/x$ is continuous by the reciprocal rule. So, from the composite and product rules, $x \mapsto x \sin(1/x)$ is continuous for $x \neq 0$. In other words, f is continuous for $x \neq 0$. At $x = 0$ the sandwich rule (5.2.5) must be used. Since $\lim_{x \to 0} |x| = 0$ by Example 10 of Section 5.1, the function $x \mapsto |x|$ is continuous at $x = 0$. By the product rule, $x \mapsto -|x|$ is also continuous at $x = 0$. Let $g(x) = |x|$ and $h(x) = -|x|$. Now $h(x) \leq f(x) \leq g(x)$ for all x, and h and g are continuous at 0. Hence, by the sandwich rule, f is also continuous at $x = 0$. ∎

Remark

A **rational function** is one of the form $x \mapsto P(x)/Q(x)$ where $P(x)$ and $Q(x)$ are both polynomials in x. Such a function is only defined for those x for which the denominator $Q(x) \neq 0$. Now P and Q are polynomials that are everywhere continuous, from liberal use of the sum and product rules. Also, $1/Q(x)$ is continuous for those x for which $Q(x) \neq 0$, by the reciprocal rule. Hence, by the product rule, every rational function is continuous on its domain. Example 4(a) is just a special case.

Recall that a function $f: A \to B$ where A and B are intervals is a **bijection** if, for every $b \in B$, there exists a *unique* $a \in A$ with $f(a) = b$. In this case the **inverse function** $f^{-1}: B \to A$, given by $f^{-1}(b) = a$, is defined. If $f: A \to B$ is either **strictly increasing**, in which case

$$\text{for all } a_1, a_2 \in A, \quad a_1 > a_2 \Rightarrow f(a_1) > f(a_2)$$

or **strictly decreasing**, in which case

$$\text{for all } a_1, a_2 \in A, \quad a_1 > a_2 \Rightarrow f(a_1) < f(a_2)$$

and B is the image of f, then f will be a bijection. (Why?) The next rule enables us to increase our fund of known continuous functions.

5.2.7 Inverse rule

Suppose that $f: A \to B$ is a bijection where A and B are intervals. If f is continuous on A then f^{-1} is continuous on B.

This result can be proved at the end of Section 5.3, since for the proof the interval theorem (5.3.3) must be invoked.

The next example assumes that the sine and exponential functions are continuous. By suitably restricting the domains and codomains of these functions, continuous bijections can be derived whose inverses are thus continuous.

■■   EXAMPLE 5

Each of the functions in Table 5.1 is a continuous bijection, and so by the inverse rule has the specified continuous inverse. Note that h is clearly strictly increasing on its domain. Example 2 of Section 6.2 shows that g is strictly increasing on $\mathbb{R}$, and the Appendix shows that f is strictly increasing on $[-\frac{1}{2}\pi, \frac{1}{2}\pi]$.   ■

It is now possible to prove the composite rule for sequences (3.1.4) and the composite rule for limits (5.1.5).

Proof of 5.1.5

Suppose that $\lim_{x \to a} f(x) = L$ and g is a function that is continuous in some neighbourhood of $x = L$. Since g is continuous at L, for

Table 5.1

Function	Inverse
$f: [-\frac{1}{2}\pi, \frac{1}{2}\pi] \to [-1, 1]$	$f^{-1}: [-1, 1] \to [-\frac{1}{2}\pi, \frac{1}{2}\pi]$
$f(x) = \sin x$	$f^{-1}(x) = \sin^{-1} x$
$g: \mathbb{R} \to (0, \infty)$	$g^{-1}: (0, \infty) \to \mathbb{R}$
$g(x) = e^x$	$g^{-1}(x) = \log_e x$
$h: [0, \infty) \to [0, \infty)$	$h^{-1}: [0, \infty) \to [0, \infty)$
$h(x) = x^2$	$h^{-1}(x) = \sqrt{x}$

each $\varepsilon > 0$ there exists a $\delta_1 > 0$ such that

$$|t - L| < \delta_1 \Rightarrow |g(t) - g(L)| < \varepsilon$$

For this $\delta_1 > 0$ there exists a $\delta_2 > 0$ such that

$$0 < |x - a| < \delta_2 \Rightarrow |f(x) - L| < \delta_1$$

since $\lim_{x \to a} f(x) = L$. Now put $t = f(x)$ to deduce that for $\varepsilon > 0$ there exists a $\delta = \delta_2 > 0$ such that

$$0 < |x - a| < \delta \Rightarrow |g(f(x)) - g(L)| < \varepsilon \qquad \square$$

Proof of 3.1.4

Suppose that $a_n \to L$ and that f is continuous at L. For any $\varepsilon > 0$ there exists a $\delta > 0$ such that

$$|x - L| < \delta \Rightarrow |f(x) - f(L)| < \varepsilon$$

For this $\delta > 0$ there is a natural number N such that, for all $n > N$, $|a_n - L| < \delta$. Immediately then

$$|f(a_n) - f(L)| < \varepsilon \quad \text{for all} \quad n > N \qquad \square$$

Exercises 5.2

1.   Each of the following expressions defines a function whose domain is the whole of $\mathbb{R}$ except for finitely many exceptions. In each case find these exceptional points and discuss the type of singularity exhibited.

(a) $f_1(x) = \dfrac{1}{x^2 - 1}$ (b) $f_2(x) = \dfrac{x - 1}{x^2 - 1}$ (c) $f_3(x) = \dfrac{x^2 - 1}{|x^2 - 1|}$

2.   Discuss the continuity of the following functions:

(a) $f(x) = \begin{cases} 1 & \text{if } x \text{ is rational} \\ 0 & \text{if } x \text{ is irrational} \end{cases}$

(b) $g(x) = \begin{cases} x & \text{if } x \text{ is rational} \\ 1 - x & \text{if } x \text{ is irrational} \end{cases}$

3.   Suppose that $f: \mathbb{R} \to \mathbb{R}$ satisfies $|f(x)| \leqslant M|x|$ for all $x \in \mathbb{R}$, where M is a fixed positive real number. Prove that f is continuous at 0.

4.   Given that the identity function, constant functions, exponential function and the sine and cosine functions are continuous on $\mathbb{R}$, use the rules for continuous functions (5.2.4–5.2.7) to prove that each of the following are continuous at the points indicated:

(a) $x \mapsto \dfrac{x^2}{x - 2}$ for $x \neq 2$

(b) $x \mapsto x^3 + e^{2 \sin x}$ for all $x \in \mathbb{R}$

(c) $x \mapsto \begin{cases} x^2 \cos(1/x) & \text{if } x \neq 0 \\ 0 & \text{if } x = 0 \end{cases}$ for all $x \in \mathbb{R}$

(d) $x \mapsto \sqrt{x}$ for $x \geqslant 0$

(e) $x \mapsto \cosh^{-1} x$ for $x \geqslant 1$

(*Hint*: The bijection function $\cosh: [0, \infty) \to [1, \infty)$ is defined by $\cosh x = \frac{1}{2}(e^x + e^{-x})$.)

5.   Suppose that $f: A \to B$ is strictly increasing, where A and B are intervals and $B = f(A)$. Show that $f^{-1}: B \to A$ is strictly increasing. Hence prove that the function $x \mapsto x^{1/n}$, $x > 0$ and $n \in \mathbb{N}$, is continuous.

5.3 Theorems

The important results in this section show that the algebraic definition of continuity leads naturally to the intuitive geometric interpretation that is used when sketching the graphs of continuous functions. The results are also of independent interest, since some of them form the theoretical basis of useful techniques in numerical analysis. The statement of each theorem will be accompanied by a graph and a brief commentary intended to give a 'feel' for the result. The proofs are naturally rigorous and bring into play the underlying properties of the real number system $\mathbb{R}$. Of particular importance will be the completeness axiom (2.2.3).

5.3.1   The boundedness property

Let f be continuous on the interval $[a, b]$. Then

(1) f is bounded on $[a, b]$,
(2) f attains both a maximum value and a minimum value somewhere on $[a, b]$.

Commentary

Part (1) of this result claims that there exist real numbers m and M such that $m \leqslant f(x) \leqslant M$ for all $x \in [a, b]$. Part (2) makes the stronger claim that f attains both a maximum value and a minimum value on $[a, b]$; in other words, that there exist c and d in $[a, b]$ such that $f(c) \leqslant f(x) \leqslant f(d)$ for all $x \in [a, b]$. See Figure 5.9.

Proof of the boundedness property

The proof is in three parts.

Preliminary observation

The proof of (1) requires the key observation that if a function f is continuous at c then f is bounded on an interval of the form $(c - \delta, c + \delta)$ for some $\delta > 0$. This can be deduced from the ε–δ definition of continuity (5.2.2), which states that for every $\varepsilon > 0$ there exists a $\delta > 0$ such that $|x - c| < \delta \Rightarrow |f(x) - f(c)| < \varepsilon$, by letting $\varepsilon = 1$ (other values of ε could equally well be chosen). For

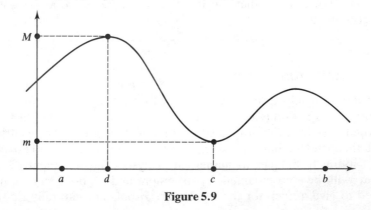

Figure 5.9

this value of ε there is a $\delta > 0$ such that

$$|x - c| < \delta \Rightarrow |f(x) - f(c)| < 1$$
$$\Rightarrow |f(x)| \leq 1 + |f(c)|$$

Hence f is bounded on the interval $(c - \delta, c + \delta)$. This argument can easily be adapted to show that, since f is right-continuous at a, f is bounded on an interval of the form $[a, a + \delta')$ for some $\delta' > 0$.

Proof of (1)

Let $B = \{x : x \in [a, b]$ and f is bounded on $[a, x]\}$, with the aim of proving that $B = [a, b]$. First, $a \in B$, and so B is a non-empty set of real numbers; secondly, B is bounded above by b. So, by the completeness axiom (2.2.3), B possesses a least upper bound. Let $c = \sup B$ be this least upper bound. All that is now required is to show that $c = b$.

Since f is right-continuous at a, f is bounded on $[a, a + \delta')$ for some $\delta' > 0$, and so f is bounded on $[a, a + \frac{1}{2}\delta']$. Hence $c \geq a + \frac{1}{2}\delta' > a$. Suppose by way of contradiction that $c < b$. Since $c > a$, f is continuous at c, and so f is bounded on $(c - \delta, c + \delta)$ for some $\delta > 0$. From the definition of c, f is bounded on $[a, c - \delta]$. Hence f is in fact bounded on $[a, c + \delta)$ and, since $\delta > 0$, f is bounded on $[a, c + \frac{1}{2}\delta]$. Hence $c + \frac{1}{2}\delta \in B$, contradicting the fact that c is the supremum of B. Therefore $c = b$.

All that is now required is to show that $b \in B$. Since f is left-continuous at b, f is bounded on $(b - \delta^*, b]$ for some δ^*, where $0 < \delta^* < b - a$. But $b = \sup B$, and so f is bounded on $[a, d]$ for any d, $b - \delta^* < d < b$. For such a d, f is also bounded on $[d, b]$, and so f is bounded on $[a, b]$, as required. This establishes (1).

Proof of (2)

Let $A = f([a, b]) = \{f(x) : a \leq x \leq b\}$. Since f is bounded on $[a, b]$, A is bounded both above and below. Let $m = \inf A$ and $M = \sup A$. What is now required is to show that m and M are both values of the function f. To this end, suppose by way of contradiction that there is no value of x for which $f(x) = M$. Define a new function g by

$$g(x) = \frac{1}{M - f(x)}$$

for $x \in [a, b]$; g is defined on $[a, b]$ and continuous on $[a, b]$ by the sum, product and reciprocal rules (5.2.4). Hence, by part (1), g is also bounded on $[a, b]$. In particular, there exists a real number

$K > 0$ such that $g(x) \leqslant K$ for all $x \in [a, b]$. But

$$g(x) \leqslant K \Rightarrow \frac{1}{M - f(x)} \leqslant K$$

$$\Rightarrow \frac{1}{K} \leqslant M - f(x)$$

$$\Rightarrow f(x) \leqslant M - \frac{1}{K}$$

This contradicts the fact that M is the least upper bound for f on $[a, b]$, so the assumption that M is not a value of f is false. Hence f attains a maximum value somewhere on $[a, b]$.

A similar argument shows that f attains a minimum value somewhere on $[a, b]$. $\square$

5.3.2 The intermediate value property

Let f be continuous on the interval $[a, b]$ and suppose that $f(a) = \alpha$ and $f(b) = \beta$. For every real number γ between α and β there exists a number c, $a < c < b$, with $f(c) = \gamma$.

Commentary

This result claims that if a continuous function f takes the values α and β at the endpoints of some interval $[a, b]$ then f must take all possible values between α and β. See Figure 5.10, where it has been assumed that $\alpha < \beta$, and so $\alpha < \gamma < \beta$.

Proof of the intermediate value property

The result will be proved when $a < b$ and $\alpha < \gamma < \beta$, the corresponding proofs in other cases being similar. The essence of the

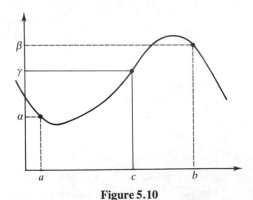

Figure 5.10

proof is to consider the set of xs for which $f(x)$ is strictly less than γ and show that if c is the supremum of this set then $f(c) = \gamma$.

So, let

$$S = \{x : x \in [a, b] \text{ and } f(x) < \gamma\}, \quad \text{where} \quad \alpha < \gamma < \beta$$

The set S is non-empty, since it contains a, and it is bounded above by b. Hence, by the completeness axiom (2.2.3), S possesses a least upper bound. Let $c = \sup S$. Then $a \leqslant c \leqslant b$.

In fact, the condition $a < c < b$ holds true. To see why this is the case, note that, since f is left-continuous at a, for $\varepsilon = \gamma - \alpha > 0$ there exists a $\delta > 0$ such that

$$0 \leqslant x - a < \delta \Rightarrow |f(x) - f(a)| < \varepsilon$$

In particular, $|f(a + \tfrac{1}{2}\delta) - f(a)| < \varepsilon$, and so $f(a + \tfrac{1}{2}\delta) - f(a) < \gamma - \alpha$. Since $f(a) = \alpha$, this gives $f(a + \tfrac{1}{2}\delta) < \gamma$, and so $a + \tfrac{1}{2}\delta \in S$. This means that $a < c$.

To establish that $c < b$, let $\varepsilon = \beta - \gamma > 0$. Then the right-continuity of f at b guarantees that there exists a $\delta > 0$ such that

$$-\delta < x - b \leqslant 0 \Rightarrow |f(x) - f(b)| < \varepsilon$$

In particular, $|f(b - \tfrac{1}{2}\delta) - f(b)| < \varepsilon$, and so $f(b) - f(b - \tfrac{1}{2}\delta) < \beta - \gamma$. Since $f(b) = \beta$, this gives $f(b - \tfrac{1}{2}\delta) > \gamma$, and so $b - \tfrac{1}{2}\delta \notin S$. This means that $c < b$.

Since $a < c < b$, f is continuous at c, a fact that is now used to show that the possibilities

(i) $f(c) < \gamma$ and
(ii) $f(c) > \gamma$

are both impossible.

Case (i)

If $f(c) < \gamma$ then $\varepsilon = \gamma - f(c) > 0$. Since f is continuous at c, there exists a $\delta > 0$ such that

$$|x - c| < \delta \Rightarrow |f(x) - f(c)| < \varepsilon$$

Choose δ such that $\delta < \max(c - a, b - c)$, so that the condition $|x - c| < \delta$ guarantees that $x \in [a, b]$, the interval under consideration. Then, in particular,

$$|f(c + \tfrac{1}{2}\delta) - f(c)| < \varepsilon$$

and so

$$f(c + \tfrac{1}{2}\delta) - f(c) < \gamma - f(c)$$

But then $f(c + \frac{1}{2}\delta) < \gamma$, and hence $c + \frac{1}{2}\delta \in S$, which contradicts the fact that c is the supremum of S. Hence $f(c) < \gamma$ is impossible.

Case (ii)

If $f(c) > \gamma$ then $\varepsilon = f(c) - \gamma > 0$. Since f is continuous at c, there exists a $\delta > 0$ such that

$$|x - c| < \delta \Rightarrow |f(x) - f(c)| < \varepsilon$$

Again, choose δ such that $\delta < \max(c - a, b - c)$. Therefore, for any x satisfying $c - \delta < x \leq c$,

$$|f(x) - f(c)| < f(c) - \gamma$$

and so $f(x) > \gamma$. Hence, *none* of the xs in the range $c - \delta < x \leq c$ lie in S, which contradicts the fact that c is the supremum of S. Hence $f(c) > \gamma$ is impossible.

Therefore the only conclusion that can be drawn is that $f(c) = \gamma$, and so γ is a value of f, as required. □

The intermediate value property has many applications, and Example 1 illustrates one of these. However, one immediate consequence follows if the result is applied to the continuous function $x \mapsto x^2$ on the interval $[1, 2]$. Since $x \mapsto x^2$ takes the value 1 at the left-hand endpoint of $[1, 2]$ and the value 4 at the right-hand endpoint, and $1 < 2 < 4$, there exists a real number between 1 and 2 such that $x^2 = 2$. This establishes the existence of $\sqrt{2}$ in a far less painful manner than that presented in Section 2.3. By considering the continuous function $x \mapsto x^n$ where n is a positive integer, the intermediate value property can be used, in a similar vein, to deduce that *every positive real number a has a positive nth root*.

■■   EXAMPLE 1

Any polynomial of odd degree has at least one real root.

Solution

Let $P(x) = a_0 + a_1 x + \ldots + a_n x^n$, where n is odd, and without loss of generality let $a_n = 1$. We know that P is continuous. Define $r(x)$ by

$$r(x) = \frac{P(x)}{x^n} - 1 \quad (x \neq 0)$$

Now

$$|r(x)| = \left| \frac{P(x)}{x^n} - 1 \right|$$

$$= \left| \frac{a_{n-1}}{x} + \ldots + \frac{a_1}{x^{n-1}} + \frac{a_0}{x^n} \right|$$

$$\leqslant \left| \frac{a_{n-1}}{x} \right| + \ldots + \left| \frac{a_1}{x^{n-1}} \right| + \left| \frac{a_0}{x^n} \right|$$

by the triangle inequality (see Section 2.2, Example 8). If M is the maximum of $|a_{n-1}|, \ldots, |a_0|$ then

$$|r(x)| \leqslant M \sum_{r=1}^{n} \frac{1}{|x|^r} < M \sum_{r=1}^{\infty} \frac{1}{|x|^r}$$

$$= \frac{M/|x|}{1 - 1/|x|} = \frac{M}{|x| - 1} \quad \text{for } |x| > 1$$

Hence $|r(x)| < 1$ for $|x| > 1 + M$. In particular, $1 + r(x) > 0$ for $|x| > 1 + M$. Hence $P(x) = x^n[1 + r(x)]$ has the same sign as x^n for $|x| > 1 + M$. Since n is odd, there exist $\alpha, \beta \in \mathbb{R}$ with $P(\alpha) > 0$ (choose $\alpha > 1 + M$) and $P(\beta) < 0$ (choose $\beta < -(1 + M)$). By 5.3.2, $P(\gamma) = 0$ for some γ, $|\gamma| < 1 + M$. Incidentally, this shows that P has a zero in the interval $(-(1 + M), 1 + M)$. In fact, all the real zeros of P lie in this interval. ∎

An immediate consequence of 5.3.1 and 5.3.2 is the following result.

5.3.3   The interval theorem

Let f be continuous on $J = [a, b]$. Then $f(J)$ is a closed bounded interval.

Commentary

The claim here is that continuous functions map *intervals onto intervals*. This means that the intuitive picture of a continuous function as one having a continuous graph is sound. See Figure 5.11.

Proof of the interval theorem

By the boundedness property (5.3.1), there exist numbers c and d in J such that $f(c) = m$ and $f(d) = M$ and $m \leqslant f(x) \leqslant M$ for all

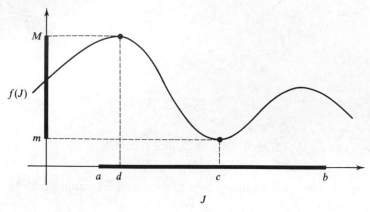

Figure 5.11

$x \in J$. Suppose for simplicity that $c \leqslant d$. Apply the intermediate value property (5.3.2) to f on the subinterval $[c, d]$ to deduce that f takes all possible values between $f(c) = m$ and $f(d) = M$. In other words, $f(J) = [m, M]$. □

5.3.4   A fixed point theorem

Let $f: [0, 1] \to [0, 1]$ be a continuous function. Then there is at least one number c that is fixed by f. That is, $f(c) = c$.

Commentary

This result says that on proceeding continuously from $(0, f(0))$ to $(1, f(1))$ then the line $y = x$ must be crossed. See Figure 5.12.

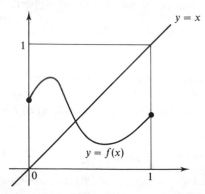

Figure 5.12

Proof of the fixed point theorem

Let $g: [0, 1] \to \mathbb{R}$ be defined by $g(x) = f(x) - x$. Since the identity function $x \mapsto x$ is continuous, the sum and product rules (5.2.4) can be applied to deduce that g is continuous on $[0, 1]$. If $f(0) = 0$ or $f(1) = 1$ then there is nothing to prove. So it is assumed that $f(0) \neq 0$ and $f(1) \neq 1$. Since f maps onto $[0, 1]$, $g(0) > 0$ and $g(1) < 0$. The intermediate value property (5.3.2) applied to g on the interval $[0, 1]$ implies that $g(c) = 0$ for some c, $0 < c < 1$. Hence $f(c) = c$. □

Before proving the inverse rule (5.2.7) whose proof has been delayed since the interval theorem is required, some examples will be given to demonstrate that if the hypotheses of the theorems in this section are violated, the conclusions will not hold.

■■ EXAMPLE 2

(1) The function $f(x) = 1/x$ is continuous on the *open* interval $(0, 1)$, but it is not bounded on $(0, 1)$. Hence the boundedness property holds only for functions continuous on closed intervals.

(2) The function $f(x) = x - [x]$ is not continuous on the interval $[0, 2]$. It is bounded on $[0, 2]$, but it does not possess a maximum value on $[0, 2]$.

(3) Consider the function

$$f(x) = \begin{cases} -2 & \text{if } x < 0 \\ 2 & \text{if } x \geq 0 \end{cases}$$

Now f is not continuous on the interval $[-1, 1]$. Although $f(-1) = -2$ and $f(1) = 2$, f does not take any of the values between -2 and 2 on the interval $[-1, 1]$. ■

Proof of 5.2.7

Consider the continuous bijection $f: A \to B$ where A and B are intervals. First it is shown that f is either strictly increasing or strictly decreasing. If f is neither strictly increasing nor decreasing then, without loss of generality, there are numbers a_1, a_2 and a_3 such that $a_1 < a_2 < a_3$ and $f(a_1) < f(a_3) < f(a_2)$. Apply the intermediate value theorem (5.3.2) to f on the interval $[a_1, a_2]$ to deduce that $f(c) = f(a_3)$ for some $c \in (a_1, a_2)$. This contradicts the fact that f is a bijection. For the rest of the proof it is assumed that f is strictly increasing, the proof for the strictly decreasing case being similar.

By Question 5 of Exercises 5.2, f^{-1} is strictly increasing. Let $b \in B$, where b is not equal to either of the endpoints of the interval B, and suppose that $f^{-1}(b) = a$, so that $f(a) = b$. For every $\varepsilon > 0$ such that $a \pm \varepsilon$ lies in the interval A, f maps the interval $J = [a - \varepsilon, a + \varepsilon]$ onto some interval $f(J) = [m, M]$ by the interval theorem (5.3.3). Since f is strictly increasing, $m < b < M$, so let δ be the minimum of $b - m$ and $M - b$. Clearly $\delta > 0$.

Now $[b - \delta, b + \delta]$ is a subset of $[m, M] = f(J)$, and so f^{-1} maps $[b - \delta, b + \delta]$ into $J = [a - \varepsilon, a + \varepsilon]$. Thus, given any $\varepsilon > 0$, there exists a $\delta > 0$ such that

$$|x - b| < \delta \Rightarrow |f^{-1}(x) - f^{-1}(b)| < \varepsilon$$

This shows that at all points of B, except the endpoints, $f^{-1}: B \to A$ is continuous. Minor changes are required in the above proof to establish the one-sided continuity claimed for the endpoints of B.

Hence $f^{-1}: B \to A$ is continuous on B, as required. $\qquad\square$

Exercises 5.3

---

1. Let $P(x) = x^3 - 3x + 1$. Calculate $P(0)$ and $P(1)$. Which theorem now guarantees that P has a root in $[0, 1]$? Calculate $P(0 \cdot 5)$, and hence describe a method whereby the root of P lying in $[0, 1]$ can be calculated to any degree of accuracy required.

2. For each of the following continuous functions f and closed intervals J calculate $f(J)$. The interval theorem is useful.

    (a) $f(x) = x^2$, $\quad J = [-1, 1]$
    (b) $f(x) = 3 \sin x$, $\quad J = [0, \frac{1}{6}\pi]$
    (c) $f(x) = x + |x|$, $\quad J = [-1, 2]$
    (d) $f(x) = x^2 - x^4$, $\quad J = [0, 1]$

3. Let $P(x) = a_0 + a_1 x + \ldots + a_n x^n$, where n is even and $a_n = 1$. If $a_0 < 0$, use the method in Example 1 to prove that P has at least two real roots.

4. Let $f: J \to J$ be a continuous function where $J = [0, 1]$. Show that the function g, given by $g(x) = [f(x)]^2$, is a continuous function from J into J. Hence use the fixed point theorem to show that there is a number c, $0 \le c \le 1$, with $f(c) = \sqrt{c}$.

Problems 5

1. Give $\varepsilon-\delta$ proofs of the following.

(a) $\lim_{x \to 0} (x + 1)^3 = 1$

(b) $\lim_{x \to 1} \left(\dfrac{x - 2}{x + 3} \right) = -\dfrac{1}{4}$

(*Hint*: for (b), if $|x - 1| < 1$ then $3 < |x + 3| < 5$; draw the corresponding intervals to verify this fact!)

2. Let

$$f(x) = \frac{x}{x + [x]} \quad (x \neq 0)$$

Determine the following:

(a) $\lim_{x \to n+} f(x)$ (*n* an integer)   (b) $\lim_{x \to n-} f(x)$ (*n* an integer)

(c) $\lim_{x \to \infty} f(x)$ (d) $\lim_{x \to -\infty} f(x)$

Sketch the graph of $y = f(x)$.

3. Evaluate the following limits.

(a) $\lim_{x \to 1} \left(\dfrac{1 - x}{1 - \sqrt{x}} \right)$ (b) $\lim_{x \to 0} \left(\dfrac{|1 - x| - |1 + x|}{x} \right)$

(c) $\lim_{x \to 2} \left(\dfrac{1}{x - 2} - \dfrac{4}{x^2 - 4} \right)$ (d) $\lim_{x \to 0} \left[\sin x \sin \left(\dfrac{1}{x} \right) \right]$

(e) $\lim_{x \to \pi} \left(\dfrac{\cos 2x - \cos x}{\cos 2x - 1} \right)$ (*Hint*: $\cos 2x = 2\cos^2 x - 1$)

4. (a) Prove that if $f = g + h$, g is continuous at c and f is not continuous at c then h is not continuous at c.

(b) Find three functions f, g and h such that $f = gh$ and f and g are continuous at c, but h is not continuous at c.

(c) If, in (b), $g(c) \neq 0$ then show that no such functions f, g and h can be found.

5. Given that the identity function, constant functions, exponential function and sine function are continuous on $\mathbb{R}$, use 5.2.4–5.2.7 to

prove that the following functions are continuous on $\mathbb{R}$:

(a) $x \mapsto |x| + 2\sin x$ (b) $x \mapsto \dfrac{1}{1 + \sin^2 x}$

(c) $x \mapsto e^{\sqrt{x^2+1}} + x^5$ (d) $x \mapsto \begin{cases} e^{-1/x^2} & \text{if } x \neq 0 \\ 0 & \text{if } x = 0 \end{cases}$

(e) $x \mapsto \begin{cases} x^2 - 1 & \text{if } x > 2 \\ x + |x - 1| & \text{if } 0 \leqslant x \leqslant 2 \\ x + 1 & \text{if } x < 0 \end{cases}$

6. The **sign function**, sgn: $\mathbb{R} \to \mathbb{R}$, is defined by

$$\operatorname{sgn}(x) = \begin{cases} 1 & \text{if } x > 0 \\ 0 & \text{if } x = 0 \\ -1 & \text{if } x < 0 \end{cases}$$

Show that $\lim_{x \to 0} \operatorname{sgn}(x)$ does not exist.
Discuss the continuity of the following functions.
(a) $x \mapsto \operatorname{sgn}(\sin x)$ (b) $x \mapsto \cos[\operatorname{sgn}(x)]$

(c) $x \mapsto x \operatorname{sgn}(x)$ (d) $x \mapsto \begin{cases} \operatorname{sgn}[\sin(1/x)] & \text{if } x \neq 0 \\ 0 & \text{if } x = 0 \end{cases}$

7. Let f be continuous on $[a, b]$ and suppose that $f(x) = 0$ whenever x is rational. Prove that $f(x) = 0$ for all $x \in [a, b]$.

8. (a) Prove that if $P(x)$ is a polynomial of odd degree then the equation $P(x) = \sin x$ has at least one solution.
(b) Find a polynomial $Q(x)$ of odd degree such that $Q(x) = \sin x$ has only one solution. (*Hint*: $|\sin x| < |x|$ for $x \neq 0$.)
(c) Find a polynomial $R(x)$ of even degree such that $R(x) = \sin x$ has no solutions.

9. For each of the following continuous functions f and closed intervals J calculate $f(J)$:

(a) $f(x) = \cos x + \sin x$, $J = [-\tfrac{1}{2}\pi, 0]$
(b) $f(x) = x^2 + \sin^2 x$, $J = [-1, 1]$

(c) $f(x) = \dfrac{x}{x + 1}$, $J = [0, 3]$

(d) $f(x) = |x - 1| + |x + 1|$, $J = [-2, 2]$

10. Suppose that f is continuous on $[a, b]$ and let $K = f(x_1) + f(x_2)$, where $a < x_1 < x_2 < b$. Prove that there exists a c, $a < c < b$, such that $f(c) = \tfrac{1}{2}K$.

CHAPTER SIX

6

Differentiation

6.1 Differentiable functions
6.2 Theorems
6.3 Taylor polynomials
6.4 Alternative forms of Taylor's theorem

Differentiation originated in the problem of drawing tangents to (smooth) curves and in finding the maximum and minimum values of functions. Although the Greeks were concerned with such problems, and indeed invented some very cunning methods for their solution, it was not until the seventeenth century that new methods were developed that, in turn, led to the discovery of differential calculus. The English scientist and mathematician Sir Isaac Newton (1642–1727), one of the greatest mathematicians the world has yet produced, invented the differential calculus in 1666 and recorded it in his *Method of Fluxions* in 1671. However, this work was not published until 1736, by which time the German mathematician Gottfried Wilhelm Leibniz (1646–1716) had discovered differential calculus independently and published his findings in 1684. Their discoveries had a profound effect on the mathematical community, and by 1700 calculus formed part of the undergraduate curriculum. Applications of calculus were pursued with, as has been previously mentioned, little regard for its logical foundations. This came later and is, as is the case in this chapter, based firmly on the theory of limits.

The account begins in Section 6.1, where the geometric concept of a tangent line to a graph is formalized as a limit whose value (if it exists) gives the slope of the tangent line. The standard rules for differentiation are established, which, together with the limit definition of differentiability, show that the class of differentiable functions includes polynomials, trigonometric functions, the exponential function and combinations of these elementary functions.

Section 6.2 discusses the properties of functions differentiable on an interval and proves some powerful results about derivatives, many of which can be described easily in geometric terms. Some applications of these theorems are given, including the use of calculus in establishing inequalities between functions and in evaluating limits of the form $\lim_{x \to a} [f(x)/g(x)]$ in the awkward case where $f(a) = g(a) = 0$.

Power series expansions of functions form the substance of Section 6.3, where it is shown that, under suitable restrictions, certain polynomials, known as Taylor polynomials, can be used to approximate the values of a given function. Moreover, a precise estimate is given for the accuracy of such approximations, a subject pursued further in Section 6.4.

The application of differential calculus to the problem of finding the maximum and minimum values of a function is a recurring theme throughout this chapter. An early result, in Section 6.1, establishes the fact that local maxima and minima of differentiable functions occur only at points where the derivative is zero. The increasing–decreasing theorem in Section 6.2 shows how the sign of the derivative aids the problem of determining local extrema. Finally, in Section 6.3, Taylor series expansions are used to obtain a classification theorem for local extrema.

6.1   Differentiable functions

Intuitively, a function $f: A \to \mathbb{R}$ is differentiable at $c \in A$ if a tangent can be drawn to the curve at the point $P = (c, f(c))$. The slope of the chord PQ in Figure 6.1 is

$$\frac{f(x) - f(c)}{x - c}$$

and, as Q moves closer to P, it is required that the slope of PQ approach the slope of the tangent line at P. This geometric idea motivates the following formal definition.

6.1.1   Definition

A function $f: A \to \mathbb{R}$ is **differentiable** at c if and only if

$$\lim_{x \to c} \frac{f(x) - f(c)}{x - c}$$

exists. The value of this limit, called the **derivative** of f at c, is denoted by $f'(c)$.

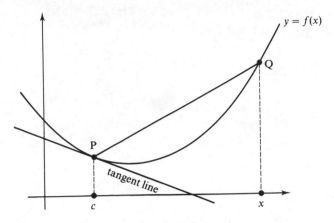

Figure 6.1

An alternative form of the limit in 6.1.1 is obtained by setting $x = c + h$. Then

$$\lim_{x \to c} \frac{f(x) - f(c)}{x - c} = \lim_{h \to 0} \frac{f(c + h) - f(c)}{h}$$

Either form may be used.

■■   EXAMPLE 1

The function $f(x) = x^2$ is differentiable for all x.

Solution

Consider

$$\frac{f(x) - f(c)}{x - c} \quad \text{for any } x \neq c$$

where c is fixed. Now

$$\lim_{x \to c} \frac{f(x) - f(c)}{x - c} = \lim_{x \to c} \frac{x^2 - c^2}{x - c} = \lim_{x \to c} (x + c) = 2c$$

Hence f is differentiable at c and $f'(c) = 2c$. Since c was arbitrary, the **derived function** f' can be defined by $f'(x) = 2x$.   ■

■■   EXAMPLE 2

The function $f(x) = |x|$ is not differentiable at $c = 0$.

Solution

Consider

$$\frac{f(x) - f(0)}{x - 0} = \frac{|x|}{x} = \begin{cases} 1 & \text{if } x > 0 \\ -1 & \text{if } x < 0 \end{cases}$$

Hence

$$\lim_{x \to 0+} \frac{f(x) - f(0)}{x - 0} = 1$$

but

$$\lim_{x \to 0-} \frac{f(x) - f(0)}{x - 0} = -1$$

and, since these right- and left-hand limits differ, f is not differentiable at 0. It is easy to show that f is differentiable for all $x \neq 0$. The graphs of f and its derived function are shown in Figure 6.2, where it is geometrically obvious that a tangent line cannot be drawn at the origin. ■

Points on the graph of a function f where f is not differentiable can often be detected by examining the left- and right-hand limits of

$$\frac{f(x) - f(c)}{x - c}$$

as $x \to c$. The limit

$$\lim_{x \to c-} \frac{f(x) - f(c)}{x - c}$$

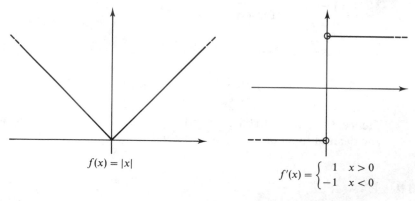

$$f(x) = |x|$$

$$f'(x) = \begin{cases} 1 & x > 0 \\ -1 & x < 0 \end{cases}$$

Figure 6.2

is called the **left-hand derivative** of f at c and is denoted by $f'_-(c)$. Similarly,

$$\lim_{x \to c+} \frac{f(x) - f(c)}{x - c}$$

is called the **right-hand derivative** of f at c and is denoted by $f'_+(c)$. Clearly $f'(c)$ exists if and only if $f'_-(c)$ and $f'_+(c)$ both exist and are equal. The next result establishes the fact that only continuous functions can be differentiable.

6.1.2   Theorem

> If f is differentiable at c then f is continuous at c.

Proof

Define the function F_c by

$$F_c(x) = \begin{cases} \dfrac{f(x) - f(c)}{x - c} & \text{if } x \ne c \\[2ex] f'(c) & \text{if } x = c \end{cases}$$

Therefore the function F_c gives the slope of the chord PQ in Figure 6.1 for $P \ne Q$ and the slope of the tangent line when $Q = P$. Since f is differentiable at c,

$$\lim_{x \to c} \frac{f(x) - f(c)}{x - c} = f'(c)$$

Hence F_c is continuous at c. Now

$$f(x) = f(c) + F_c(x)(x - c) \quad \text{for all } x$$

Since F_c and the identity and constant functions are all continuous at c, f is continuous at c, using the sum and product rules for continuous functions (5.2.4).   □

Note that Example 2 shows that there are continuous functions that are not differentiable. Table 6.1 gives certain elementary functions and their derivatives. The first two entries can be obtained directly from the definition of differentiability (6.1.1). The remaining three entries can be

Table 6.1

f	f'
$f(x) = k$, a constant	$f'(x) = 0$
$f(x) = x^n$, $n \in \mathbb{N}$	$f'(x) = nx^{n-1}$
$f(x) = \sin x$	$f'(x) = \cos x$
$f(x) = \cos x$	$f'(x) = -\sin x$
$f(x) = e^x$	$f'(x) = e^x$

obtained from the power series definitions 4.3.3–4.3.5 and depend on the fact that the sum of a power series is a differentiable function whose derivative is the power series derived from the power series for f by term-by-term differentiation. This fact is proved in the Appendix (see A.2).

■■ **EXAMPLE 3**

Let $f(x) = x^n$, where $n \in \mathbb{N}$. Consider

$$\frac{f(x) - f(c)}{x - c} = \frac{x^n - c^n}{x - c}$$

$$= x^{n-1} + x^{n-2}c + \ldots + xc^{n-2} + c^{n-1} \quad (x \neq c)$$

Hence

$$\lim_{x \to c} \frac{f(x) - f(c)}{x - c} = \lim_{x \to c} (x^{n-1} + x^{n-2}c + \ldots + xc^{n-2} + c^{n-1})$$

$$= nc^{n-1}$$

Thus f is differentiable at c for any c, and so the derivative is given by $f'(x) = nx^{n-1}$. ■

To extend the list in Table 6.1 and to develop some of the well-known techniques of calculus, it is established, from the definition of differentiability, that the following rules hold. Application of these rules leads to the results in Table 6.2.

6.1.3 Rules

Sum rule

Let f and g be functions differentiable at c. Then their sum $f + g$ is differentiable at c and

$$(f + g)'(c) = f'(c) + g'(c)$$

Product rule

Let f and g be functions differentiable at c. Then their product $f \cdot g$ is differentiable at c and

$$(f \cdot g)'(c) = f(c)g'(c) + f'(c)g(c)$$

Reciprocal rule

Let f be a function that is non-zero and differentiable at c. Then $1/f$ is differentiable at c and

$$\left(\frac{1}{f}\right)'(c) = \frac{-f'(c)}{[f(c)]^2}$$

Proof of the product rule

For $x \neq c$

$$\frac{(f \cdot g)(x) - (f \cdot g)(c)}{x - c}$$

$$= \frac{f(x)g(x) - f(c)g(c)}{x - c}$$

$$= \frac{f(x)g(x) - f(x)g(c) + f(x)g(c) - f(c)g(c)}{x - c}$$

$$= f(x)\frac{g(x) - g(c)}{x - c} + \frac{f(x) - f(c)}{x - c}g(c)$$

As $x \to c$,

$$\frac{f(x) - f(c)}{x - c} \to f'(c) \quad \text{and} \quad \frac{g(x) - g(c)}{x - c} \to g'(c)$$

since f and g are differentiable at c. Since f is differentiable at c, f is continuous at c by 6.1.2, and so $f(x) \to f(c)$ as $x \to c$. Now $g(c)$ is a constant, and so, by the sum and product rules for limits of functions (5.1.3),

$$\frac{(f \cdot g)(x) - (f \cdot g)(c)}{x - c} \rightarrow f'(c)g(c) + f(c)g'(c)$$

Hence $f \cdot g$ is differentiable at c, and the product rule for differentiation is established. □

The product and reciprocal rules can be combined as follows, to give the quotient rule.

6.1.4 Quotient rule

If f and g are differentiable at c and $g(c) \neq 0$ then f/g is differentiable at c and

$$\left(\frac{f}{g}\right)'(c) = \frac{f'(c)g(c) - f(c)g'(c)}{[g(c)]^2}$$

■■ EXAMPLE 4

Use the above rules to prove that each of the following functions is differentiable at the points indicated:

(a) $f(x) = x^2 + \sin x$ $(x \in \mathbb{R})$
(b) $f(x) = x^2 \sin x$ $(x \in \mathbb{R})$
(c) $f(x) = \tan x$ $(x \in \mathbb{R}$ and $x \neq n + \frac{1}{2}\pi$, n an integer$)$

Solution

For (a) and (b) note that $x \mapsto x^2$ and $x \mapsto \sin x$ are differentiable at any x, by Table 6.1. By the sum rule, $x \mapsto x^2 + \sin x$ is differentiable, and, by the product rule, $x \mapsto x^2 \sin x$ is differentiable. For (c) $x \mapsto \cos x$ is differentiable and non-zero for $x \neq n + \frac{1}{2}\pi$, n an integer. Since $x \mapsto \sin x$ is differentiable for all x, $x \mapsto \tan x$ is differentiable for the specified x by the quotient rule. The derivatives of the three functions are given by $x \mapsto 2x + \cos x$, $x \mapsto 2x \sin x + x^2 \cos x$ and $x \mapsto \sec^2 x$ respectively. ■

6.1.5 Sandwich rule

Let f, g and h be three functions such that $g(x) \leq f(x) \leq h(x)$ for all x in some neighbourhood of c and such that $g(c) = f(c) = h(c)$. If g and h are differentiable at c then so is f, and $f'(c) = g'(c) = h'(c)$.

Proof

The given inequalities imply that

$$\frac{g(x) - g(c)}{x - c} \leqslant \frac{f(x) - f(c)}{x - c} \leqslant \frac{h(x) - h(c)}{x - c} \quad \text{for all } x > c$$

and the inequality signs are reversed for $x < c$. Thus

$$g'(c) = \lim_{x \to c} \frac{g(x) - g(c)}{x - c}$$

$$= \lim_{x \to c+} \frac{g(x) - g(c)}{x - c}$$

$$\leqslant \lim_{x \to c+} \frac{h(x) - h(c)}{x - c}$$

$$= \lim_{x \to c} \frac{h(x) - h(c)}{x - c} = h'(c)$$

and

$$g'(c) = \lim_{x \to c} \frac{g(x) - g(c)}{x - c}$$

$$= \lim_{x \to c-} \frac{g(x) - g(c)}{x - c}$$

$$\geqslant \lim_{x \to c-} \frac{h(x) - h(c)}{x - c}$$

$$= \lim_{x \to c} \frac{h(x) - h(c)}{x - c} = h'(c)$$

Hence $g'(c) = h'(c)$. The result now follows by the sandwich rule for limits of functions (5.2.5). □

Table 6.2

f	f'
$f(x) = x^n,\ n \in \mathbb{Z},\ n < 0$	$f'(x) = nx^{n-1},\ x \neq 0$
$f(x) = \sec x$	$f'(x) = \sec x \tan x,\ x \neq n + \frac{1}{2}\pi$
$f(x) = \operatorname{cosec} x$	$f'(x) = \operatorname{cosec} x \cot x,\ x \neq n\pi$
$f(x) = \tan x$	$f'(x) = \sec^2 x,\ x \neq n + \frac{1}{2}\pi$
$f(x) = \cot x$	$f'(x) = -\operatorname{cosec}^2 x,\ x \neq n\pi$

■■ **EXAMPLE 5**

The function

$$f(x) = \begin{cases} x^2 \sin\left(\dfrac{1}{x}\right) & \text{if } x \neq 0 \\ 0 & \text{if } x = 0 \end{cases}$$

can be sandwiched between $h(x) = -x^2$ and $g(x) = x^2$ at $x = 0$. Since g and h are differentiable at 0 with common derivative of value 0, the sandwich rule gives that f is differentiable at $x = 0$. By the other rules, including the forthcoming composite rule (6.1.6), f is also differentiable for $x \neq 0$. Moreover,

$$f'(x) = \begin{cases} 2x \sin\left(\dfrac{1}{x}\right) - \cos\left(\dfrac{1}{x}\right) & \text{if } x \neq 0 \\ 0 & \text{if } x = 0 \end{cases}$$

Note that $\lim_{x \to 0} f'(x)$ does not exist, so that $f'(0)$ exists but f' is not continuous at 0. ■

6.1.6 Composite rule

> Let f be differentiable at c and let g be differentiable at $b = f(c)$. Then $g \circ f$ is differentiable at c and
>
> $$(g \circ f)'(c) = g'(f(c))f'(c)$$

Proof

Let

$$F_c(x) = \begin{cases} \dfrac{f(x) - f(c)}{x - c} & \text{if } x \neq c \\ f'(c) & \text{if } x = c \end{cases}$$

Then F_c is continuous at $x = c$, and, for *all* x,

$$f(x) = f(c) + (x - c)F_c(x)$$

Let

$$G_b(y) = \begin{cases} \dfrac{g(y) - g(b)}{y - b} & \text{if } y \neq b \\ g'(b) & \text{if } y = b \end{cases}$$

Then G_b is continuous at $y = b$, and, for *all* y,

$$g(y) = g(b) + (y - b)G_b(y)$$

Now

$$\begin{aligned}
(g \circ f)(x) &= g(f(x)) \\
&= g(y) \\
&= g(b) + (y - b)G_b(y) \\
&= g(f(c)) + [f(x) - f(c)]G_b(f(x)) \\
&= g(f(c)) + (x - c)F_c(x)G_b(f(x))
\end{aligned}$$

so

$$\frac{(g \circ f)(x) - (g \circ f)(c)}{x - c} = F_c(x)G_b(f(x)) \tag{1}$$

Since f is differentiable at c, it is continuous at c by 6.1.2, and, since G_b is continuous at $f(c)$, $(G_b \circ f)$ is continuous at c by 5.2.6. But F_c is also continuous at c, and so, by 5.2.4, the function on the right-hand side of (1) is continuous at $x = c$. Hence

$$\lim_{x \to c} \frac{(g \circ f)(x) - (g \circ f)(c)}{x - c} = F_c(c)G_b(f(c))$$

$$= F_c(c)G_b(b) = f'(c)g'(f(c))$$

as required.   □

The composite rule is often called the **chain rule**, and the formula it gives for the derivative of a composite is more suggestive in Leibniz notation. Let $h = \Delta x$ and $f(x + h) - f(x) = \Delta y$; then

$$f'(x) = \lim_{h \to 0} \frac{f(x + h) - f(x)}{h} = \lim_{\Delta x \to 0} \frac{\Delta y}{\Delta x}$$

The Leibniz notation for this limit is dy/dx. Write $y = g(u)$, where $u = f(x)$. Then $f'(x) = du/dx$ and $g'(f(x)) = dy/du$ and $(g \circ f)'(x) = dy/dx$. The chain rule can now be written as

$$\frac{dy}{dx} = \frac{dy}{du}\frac{du}{dx}$$

■■   EXAMPLE 6

Show that $h(x) = \sin x^2$ is differentiable.

Solution

Let $g(x) = \sin x$ and $f(x) = x^2$; then $h = g \circ f$. Since f and g are everywhere differentiable, the composite rule gives that h is differentiable. Moreover,

$$h'(x) = g'(f(x))f'(x) = 2x \cos x^2$$

■

Gottfried Wilhelm Leibniz (1646–1716) was born and grew up in Leipzig, where he studied theology, law, philosophy and mathematics at the university. He entered the diplomatic service, to which he devoted 40 years of his life. This proved to be a lucrative employment that left Leibniz plenty of time to pursue his interests; these were diverse, and he produced scholarly work in Sanskrit, philosophy and mathematics. Leibniz had a keen appreciation of the importance of logical reasoning and of the need for an appropriate language and notation in any new subject. It was he who, on 29 October 1675, first used the sign $\int$ for integration; derived from the first letter of the Latin word *summa*, meaning sum. By the time his paper on differential calculus appeared in 1684, Leibniz was writing differentials as we do today. Many of the standard rules for differentation were obtained by Leibniz, and the notation $\mathrm{d}y/\mathrm{d}x$ is still associated with his name, as is Leibniz' formula (see 6.2.6) for finding the nth derivative of a product of two functions. The last seven years of his life were spent in bitter controversy with Newton over who had priority in the discovery of calculus. When, in 1714, Leibniz' employer became the first Hanoverian king of England, Leibniz was left in Germany to spend his declining years as a neglected scholar. Newton, on the other hand, when he died in 1727, was buried with pomp and circumstance in Westminster Abbey. However, the priority dispute itself alienated British mathematicians from their continental counterparts for much of the eighteenth century, much to the detriment of British mathematics.

6.1.7 Inverse rule

Suppose that $f: A \to B$ is a bijection where A and B are intervals. If f is differentiable at $a \in A$ and $f'(a) \neq 0$ then f^{-1} is differentiable at $b = f(a)$ and

$$(f^{-1})'(b) = \frac{1}{f'(a)}$$

Proof

For each $a \in A$ let

$$F_a(x) = \begin{cases} \dfrac{f(x) - f(a)}{x - a} & \text{if } x \neq a \\ f'(a) & \text{if } x = a \end{cases}$$

Then F_a is continuous at $x = a$ and, for *all* $x \in A$,

$$f(x) = f(a) + (x - a)F_a(x)$$

Given $f(a) = b$ and letting $f(x) = y$ for arbitrary $x \in A$,

$$F_a(x) = \frac{y - b}{x - a} \quad \text{for } x \neq a$$

Consider

$$G_b(y) = \frac{x - a}{y - b} \quad \text{for } y \neq b$$

Then

$$G_b(y) = \frac{x - a}{f(x) - f(a)} = \frac{1}{F_a(x)} = \frac{1}{(F_a \circ f^{-1})(y)} \quad (y \neq b)$$

Since f is differentiable, it is continuous (6.1.2), and so f^{-1} is continuous (5.2.7). Also, $f^{-1}(b) = a$, and F_a is continuous at $x = a$. Hence $F_a \circ f^{-1}$ is continuous at $y = b$ (5.2.6).
    Now

$$(F_a \circ f^{-1})(b) = F_a(f^{-1}(b))$$
$$= F_a(a)$$
$$= f'(a)$$
$$\neq 0$$

by hypothesis, and so

$$\frac{x - a}{y - b} = G_b(y)$$
$$= \frac{1}{(F_a \circ f^{-1})(y)} \rightarrow \frac{1}{(F_a \circ f^{-1})(b)} \Big/ \text{as } y \rightarrow b$$

In other words,

$$\lim_{y \to b} \frac{f^{-1}(y) - f^{-1}(b)}{y - b} = \frac{1}{f'(a)} \qquad \qquad \square$$

This rule can be written in Leibniz notation as follows. Let $y = f(x)$; then $x = f^{-1}(y)$, and so

$$(f^{-1})'(y) = \frac{dx}{dy} \quad \text{and} \quad f'(x) = \frac{dy}{dx}$$

Thus

$$\frac{dx}{dy} = \frac{1}{dy/dx}$$

a highly memorable form.

■■ EXAMPLE 7

The function $f: (0, \infty) \to (0, \infty)$ given by $f(x) = x^2$ is a bijection. Its inverse function is given by $f^{-1}(y) = \sqrt{y}$.

Now f is differentiable for $x > 0$ and $f'(x) = 2x \neq 0$ for $x > 0$. By the inverse rule, f^{-1} is differentiable for $y > 0$, and

$$(f^{-1})'(y) = \frac{1}{f'(x)} = \frac{1}{2x} = \frac{1}{2\sqrt{y}}$$

■

■■ EXAMPLE 8

The function $g: (-\frac{1}{2}\pi, \frac{1}{2}\pi) \to (-1, 1)$ given by $g(x) = \sin x$ is a bijection with inverse given by $g^{-1}(y) = \sin^{-1} y$. Now g is differentiable and $g'(x) = \cos x \neq 0$ for $|x| < \frac{1}{2}\pi$. Hence, by the inverse rule, g^{-1} is differentiable for $|y| < 1$, and

$$(g^{-1})'(y) = \frac{1}{\cos x} = \frac{1}{\sqrt{1 - \sin^2 x}}$$

$$\text{(since } \cos x > 0 \text{ for } |x| < \tfrac{1}{2}\pi)$$

$$= \frac{1}{\sqrt{1 - y^2}}$$

■

■■ EXAMPLE 9

The function $h: \mathbb{R} \to (0, \infty)$ given by $h(x) = e^x$ is a bijection with inverse given by $h^{-1}(y) = \log_e y$. Now h is differentiable and $h'(x) = e^x \neq 0$ for all x. Hence h^{-1} is differentiable for $y > 0$ and

$$(h^{-1})'(y) = \frac{1}{e^x} = \frac{1}{y}$$

■

■■   **EXAMPLE 10**

By contrast, the function $k: \mathbb{R} \to \mathbb{R}$ given by $k(x) = x^3$ is a differentiable bijection, but its inverse $k^{-1}(y) = y^{1/3}$ is not differentiable at 0. (Why not?)   ■

Chapter 5 showed that the limit definition of a continuous function could be used to establish theorems that supported the intuitive notion that the graph of a continuous function is a 'continuous curve'. The graph of a differentiable function can be dubbed a 'smooth' continuous curve. The theorems to support this idea are developed in the next section, but this section concludes with a result that helps to locate the local maxima and minima of a differentiable function.

6.1.8 Definition

> A function f has a **local maximum value** at c if c is contained in some open interval J for which $f(x) \leqslant f(c)$ for each $x \in J$.
>
>   If $f(x) \geqslant f(c)$ for each $x \in J$ then f has a **local minimum value** at c.

See Figure 6.3 for a diagrammatic representation of these two types of local extrema.

6.1.9 Local extremum theorem

> If f is differentiable at c and possesses a local maximum or a local minimum at c then $f'(c) = 0$.

Proof

Consider the case of a local minimum at $x = c$. There is an open interval J with centre c such that $f(x) - f(c) \geqslant 0$ for all $x \in J$. If

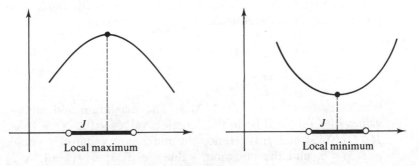

Local maximum   Local minimum

Figure 6.3

$x > c$ then

$$\frac{f(x) - f(c)}{x - c} \geq 0$$

and if $x < c$ then

$$\frac{f(x) - f(c)}{x - c} \leq 0$$

Thus $f'_+(c) \geq 0$ and $f'_-(c) \leq 0$. But $f'(c)$ exists, and so $f'_+(c) = f'_-(c)$. Thus $f'(c) = 0$. □

Note that, although f' must vanish at a local extremum, this is not sufficient for such a point. For example, consider the behaviour of $f(x) = x^3$ at $x = 0$. Here $f'(0) = 0$, but $(0, 0)$ is neither a local maximum nor a local minimum. The next example shows that the local extremum theorem together with the boundedness property for continuous functions (see 5.3.1) enables the global maxima and minima of a differentiable function to be determined, since on a closed bounded interval these occur either at points where $f' = 0$ or at the endpoints of the interval.

■■ EXAMPLE 11

Determine the maximum and minimum values of the function

$$f(x) = x(x - 1)(x - 2)$$

on the interval $[0, 3]$.

Solution

First investigate the local extrema of f on $[0, 3]$. Since f is a polynomial, it is differentiable on $[0, 3]$. Moreover,

$$\begin{aligned} f'(x) &= x(2x - 3) + (x - 1)(x - 2) \\ &= 3x^2 - 6x + 2 \end{aligned}$$

which has roots at $x = 1 \pm 1/\sqrt{3}$. The maximum and minimum values of f on $[0, 3]$ lie in the function values $f(0)$, $f(1 - 1/\sqrt{3})$, $f(1 + 1/\sqrt{3})$ and $f(3)$. Hence the maximum value of f on $[0, 3]$ is $f(3) = 6$, and the minimum value on $[0, 3]$ is $f(1 + 1/\sqrt{3}) = -2/3\sqrt{3}$. ■

Exercises 6.1

1. Use the limit definition of differentiability to find the derivatives of the following functions:

    (a) $f(x) = x^4 + 3x$ (b) $f(x) = 1/x$

2. Prove the sum and reciprocal rules (see 6.1.3) for differentiable functions.

3. Find the left- and right-hand derivatives of each of the following at the point indicated:

    (a) $f(x) = x - [x]$ at $x = 2$

    (b) $f(x) = \begin{cases} \dfrac{x}{1 + e^{1/x}} & \text{if } x \neq 0 \\ 0 & \text{if } x = 0 \end{cases}$ at $x = 0$

4. For each of the following functions use the rules for differentiability to prove that f is differentiable. You may assume that the identity, constant, exponential and sine and cosine functions are differentiable on $\mathbb{R}$:

    (a) $f(x) = e^x \cos x + 1$ (b) $f(x) = \dfrac{1}{1 + x^4}$

    (c) $f(x) = \dfrac{1 + x^2}{1 + x^4}$ (d) $f(x) = \tan^3 x$ $(x \neq n + \tfrac{1}{2}\pi)$

    (e) $f(x) = \begin{cases} x^2 \cos\left(\dfrac{1}{x}\right) & \text{if } x \neq 0 \\ 0 & \text{if } x = 0 \end{cases}$ (f) $f(x) = x^{1/5}$ $(x > 0)$

    (g) $f(x) = \sinh^{-1} x$ (*Hint*: The bijective function $\sinh: \mathbb{R} \to \mathbb{R}$ is defined by $\sinh x = \tfrac{1}{2}(e^x - e^{-x})$.)

5. Calculate f' for each of the functions in Question 4.

6.2 Theorems

This section establishes some basic properties of differentiable functions. The main result is the mean value theorem, which will be proved by first establishing a special case of it known as Rolle's theorem.

6.2.1 Rolle's theorem

> Let f be differentiable on (a, b) and continuous on $[a, b]$. If
> $f(a) = f(b)$ then there exists a c, $a < c < b$, such that
> $f'(c) = 0$. (See Figure 6.4.)

Proof

Since f is continuous on $[a, b]$, it attains a maximum value $f(c_1)$
and a minimum value $f(c_2)$ on $[a, b]$ by the boundedness property
(5.3.1). If $f(c_1) = f(c_2)$ then f is constant for all $x \in [a, b]$, whence
$f'(x) = 0$ for all $x \in [a, b]$, and the result follows. If $f(c_1) \neq f(c_2)$
then at least one of c_1 and c_2 is not at a or b. Hence f has a local
maximum or minimum (or both) inside the interval $[a, b]$. By the
local extremum theorem (6.1.9), f' is zero at at least one point
inside $[a, b]$. □

Although Figure 6.4 seems to imply that Rolle's theorem is obvious, it
is important to note that the result is true for any differentiable function
and not just the idealized one depicted in Figure 6.4.

Michel Rolle (1652–1729) was a French mathematician foremost
among those who questioned the validity of many of the new infinitesimal methods derived from the calculus of Newton and Leibniz.
Indeed, Rolle attacked the calculus itself as a collection of ingenious
fallacies. It was attacks such as these by respected mathematicians
that led to attempts to reconcile the new calculus with the more
ancient geometry of Euclid. This reconciliation concentrated on
those (limiting) processes underlying the calculus and developed
into the subject now known as mathematical analysis. Although
Rolle exposed several absurdities that arose from the indiscriminate

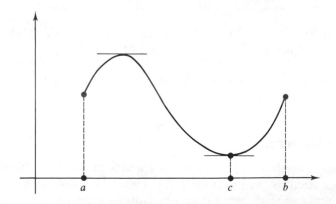

Figure 6.4

use of calculus, he was eventually convinced of the soundness of the developing analytical approach to the subject. Among several contributions to analysis is the theorem that bears his name, published in an obscure book in 1691. Its fundamental importance was not recognized at the time, since it arose as incidental to methods of finding approximate solutions of equations.

6.2.2 Mean value theorem

Let f be differentiable on (a, b) and continuous on $[a, b]$. Then there exists a c, $a < c < b$, such that

$$f'(c) = \frac{f(b) - f(a)}{b - a}$$

Proof

The result is illustrated in Figure 6.5.

For the proof let $g(x) = f(x) - \lambda x$, where λ is chosen so that $g(a) = g(b)$; that is,

$$\lambda = \frac{f(b) - f(a)}{b - a}$$

the slope of the chord joining $(a, f(a))$ to $(b, f(b))$. Now f is differentiable on (a, b), as is the linear function $x \mapsto -\lambda x$. By the sum rule (6.1.3), g is also differentiable on (a, b). Similarly, g is continuous on $[a, b]$. The choice of λ means that, in addition, $g(a) = g(b)$. Applying Rolle's theorem to g yields a c, $a < c < b$, such that $g'(c) = 0$. Hence $f'(c) - \lambda = 0$, so that

$$f'(c) = \frac{f(b) - f(a)}{b - a} \qquad\qquad \square$$

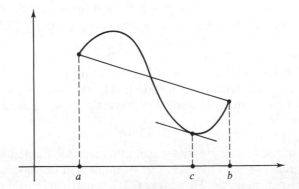

Figure 6.5

An immediate consequence of the mean value theorem is the following result, which in turn gives a technique for analysing the graphs of differentiable functions.

6.2.3 The increasing–decreasing theorem

If f is differentiable on (a, b) and continuous on $[a, b]$ then

(1) $f'(x) > 0$ for all $x \in (a, b) \Rightarrow f$ is strictly increasing on $[a, b]$.
(2) $f'(x) < 0$ for all $x \in (a, b) \Rightarrow f$ is strictly decreasing on $[a, b]$.
(3) $f'(x) = 0$ for all $x \in (a, b) \Rightarrow f$ is constant on $[a, b]$.

Proof

(1) Let $x_1, x_2 \in [a, b]$ with $x_1 < x_2$. Since f satisfies the hypotheses of the mean value theorem on the interval $[x_1, x_2]$,

$$\frac{f(x_2) - f(x_1)}{x_2 - x_1} = f'(c)$$

for some c, $x_1 < c < x_2$. But $f'(c) > 0$, and so $f(x_2) > f(x_1)$. In other words, f is strictly increasing on $[a, b]$.
   The proofs of (2) and (3) are similar. □

■■ EXAMPLE 1

Find and describe the local extrema of $f(x) = x^2 e^{-x}$.

Solution

Now f is everywhere differentiable and continuous and

$$f'(x) = 2xe^{-x} - x^2 e^{-x} = e^{-x}(2 - x)x$$

Local extrema occur only when $f'(x) = 0$, and so $x = 0$ or $x = 2$.
   Since $e^{-x}e^x = e^{x-x} = e^0 = 1$, $e^x \neq 0$ for all $x \in \mathbb{R}$. Then $e^x > 0$ for all x, otherwise the intermediate value property (3.3.2) implies that e^x vanishes somewhere.
   Hence if $x < 0$ then $f'(x) < 0$; if $0 < x < 2$ then $f'(x) > 0$; and if $x > 2$ then $f'(x) < 0$. Thus f is decreasing on $(-\infty, 0]$, increasing on $[0, 2]$ and decreasing again on $[2, \infty)$. This means that $x = 0$ gives a local minimum, and $x = 2$ gives a local maximum of f. See Figure 6.6. ■

The use of the increasing–decreasing theorem for finding and classifying local extrema seen in the previous example will be superseded in the sequel by crisper methods. However, 6.2.3 is extremely useful for establishing inequalities between functions.

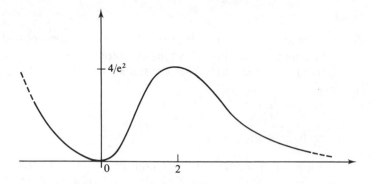

Figure 6.6

■■ EXAMPLE 2

Prove that $e^x \geq 1 + x$ for all x.

Solution

From the definition of e^x (4.3.3), $e^x > 1$ for all $x > 0$. Hence if $x_1 < x_2$ then $e^{x_2} = e^{x_2 - x_1}e^{x_1} > e^{x_1}$. Thus e^x is strictly increasing on $\mathbb{R}$. Let $f(x) = e^x - 1 - x$. Then f is differentiable and $f'(x) = e^x - 1$. Hence $f'(x) > 0$ for $x > 0$, and so f is strictly increasing on $(0, \infty)$. Since $e^{-x} = (e^x)^{-1}$, $e^{-x} < 1$ for $x > 0$, and so $f'(x) < 0$ for $x < 0$. Hence f is strictly decreasing on $(-\infty, 0)$. Since $f(0) = 0$, $f(x) \geq 0$ for all x, and so $e^x \geq 1 + x$ for all x.   ■

The next result, which is difficult to interpret geometrically, will be needed to prove L'Hôpital's rule. This is a rule that is well suited to the evaluation of limits of the form

$$\lim_{x \to x_0} \frac{f(x)}{g(x)}$$

where $f(x_0) = g(x_0) = 0$.

6.2.4 Cauchy's mean value theorem

Let f and g be differentiable on (a, b) and continuous on $[a, b]$. Then there exists a c, $a < c < b$, such that

$$\frac{f'(c)}{g'(c)} = \frac{f(b) - f(a)}{g(b) - g(a)}$$

provided that $g'(x) \neq 0$ for all $x \in (a, b)$.

Proof

First note that $g(a) \neq g(b)$, otherwise Rolle's theorem applied to g on $[a, b]$ would mean that g' vanished somewhere on (a, b). Let $h(x) = f(x) - \lambda g(x)$, where λ is chosen so that $h(a) = h(b)$; that is,

$$\lambda = \frac{f(b) - f(a)}{g(b) - g(a)}$$

By the sum and product rules for continuity and differentiability and our choice of λ, h satisfies all the hypotheses of Rolle's theorem. Hence there is a c, $a < c < b$, such that $h'(c) = 0$. This gives $f'(c) = \lambda g'(c)$, and the result now follows. $\qquad\square$

6.2.5 L'Hôpital's rule (version A)

Let f and g satisfy the hypotheses of Cauchy's mean value theorem and let x_0 satisfy $a < x_0 < b$. If $f(x_0) = g(x_0) = 0$ then

$$\lim_{x \to x_0} \frac{f(x)}{g(x)} = \lim_{x \to x_0} \frac{f'(x)}{g'(x)}$$

provided that the latter limit exists.

Proof

Apply Cauchy's mean value theorem to f and g on the interval $[x_0, x]$, where $x_0 < x \leq b$. Hence there exists a c, $x_0 < c < x$, such that

$$\frac{f'(c)}{g'(c)} = \frac{f(x) - f(x_0)}{g(x) - g(x_0)} = \frac{f(x)}{g(x)}$$

Now

$$\lim_{x \to x_0+} \frac{f(x)}{g(x)} = \lim_{c \to x_0+} \frac{f'(c)}{g'(c)} = \lim_{x \to x_0+} \frac{f'(x)}{g'(x)}$$

provided that the latter exists. A similar argument applied on the interval $[x, x_0]$, where $a \leq x < x_0$ gives

$$\lim_{x \to x_0-} \frac{f(x)}{g(x)} = \lim_{x \to x_0-} \frac{f'(x)}{g'(x)}$$

again provided that the latter exists. The rule now follows. $\qquad\square$

The **Marquis de L'Hôpital** (1661–1704) wrote the first textbook on calculus, *Analyse des infiniment petits*, which appeared in Paris in 1696. This text contains the rule with which his name is associated. However, the rule was in fact discovered by Jean Bernoulli (1667–1748), one of the dozen or so eminent Swiss mathematicians spawned by six successive generations of the Bernoulli family during the seventeenth, eighteenth and nineteenth centuries. Bernoulli instructed the young Marquis in 1692 and, in return for a regular salary, entered into an agreement whereby L'Hôpital received Bernoulli's mathematical discoveries and had permission to do with them as he pleased. The *Analyse* takes a Leibnizian approach to calculus, covers the basic rules of differentiation and makes use of them in a wide variety of applications. The author pays due credit to both Leibniz and Bernoulli in his preface, and there is no doubt that some of the text was developed independently by L'Hôpital, a fine mathematician in his own right.

■■   EXAMPLE 3

$$\lim_{x \to 0} \frac{\sin x}{x} = 1$$

Solution

The functions $f(x) = \sin x$ and $g(x) = x$ satisfy the hypotheses of L'Hôpital's rule. Moreover

$$\lim_{x \to 0} \frac{f'(x)}{g'(x)} = \lim_{x \to 0} \frac{\cos x}{1} = 1$$ ■

■■   EXAMPLE 4

$$\lim_{x \to 0} \frac{\sinh 2x}{\log_e (1 + x)} = 2$$

Solution

Let $f(x) = \sinh 2x$ and $g(x) = \log_e (1 + x)$. Then $f'(x) = 2 \cosh 2x$ and $g'(x) = 1/(1 + x)$. By L'Hôpital's rule,

$$\lim_{x \to 0} \frac{f(x)}{g(x)} = \lim_{x \to 0} 2(1 + x) \cosh 2x = 2$$ ■

■■   EXAMPLE 5

$$\lim_{x \to 0} (1 + x)^{1/x} = e$$

Solution

Now

$$\log_e \left[\lim_{x \to 0} (1 + x)^{1/x}\right] = \lim_{x \to 0} \left[\log_e (1 + x)^{1/x}\right]$$

by the composite rule for limits of functions (5.1.5). By L'Hôpital's rule,

$$\lim_{x \to 0} \frac{\log_e (1 + x)}{x} = \lim_{x \to 0} \frac{1}{1 + x} = 1$$

Hence

$$\lim_{x \to 0} (1 + x)^{1/x} = \exp \left\{\lim_{x \to 0} \left[\log_e (1 + x)^{1/x}\right]\right\} = e$$

by the continuity of log and exp. ∎

Note

The sequence $((1 + 1/n)^n)$ was shown to be convergent in Example 3 of Section 3.4. Setting $x = 1/n$ in Example 5 above shows that

$$\lim_{x \to 0} (1 + x)^{1/x} = \lim_{x \to 0} \left(1 + \frac{1}{n}\right)^n$$

$$= \lim_{n \to \infty} \left(1 + \frac{1}{n}\right)^n$$

and so e is the limit of the convergent sequence $((1 + 1/n)^n)$.

L'Hôpital's rule can be used to evaluate many seemingly indeterminate limits once they have been expressed as the limit of a quotient of differentiable functions; provided of course that

$$\lim_{x \to x_0} \frac{f'(x)}{g'(x)}$$

can be evaluated. Often this final limit is itself indeterminate (in other words $f'(x_0) = g'(x_0) = 0$), and it may be tempting to apply L'Hôpital's rule again. But this requires that f' and g' themselves be differentiable. If the derived function f' of a given differentiable function is itself differentiable, it is said that f is **twice differentiable**, and f'' or $f^{(2)}$ is written for its second derivative. In general, f is n-times differentiable if f is $(n - 1)$-times differentiable and its $(n - 1)$th derivative is differentiable. The **nth derivative** is denoted by $f^{(n)}$. If, moreover, $f^{(n)}$ is a continuous function then f is said to be **n-times continuously differentiable**.

■■   **EXAMPLE 6**

Consider

$$f(x) = \begin{cases} x^2 & \text{if } x \geq 0 \\ x^3 & \text{if } x < 0 \end{cases}$$

For $x > 0$, f is differentiable and $f'(x) = 2x$. Similarly, for $x < 0$, f is differentiable and $f'(x) = 3x^2$. At $x = 0$ the calculation of left- and right-hand derivatives has to be resorted to. Hence

$$\lim_{x \to 0+} \frac{f(x) - f(0)}{x - 0} = \lim_{x \to 0+} \frac{x^2}{x} = \lim_{x \to 0+} x = 0$$

and

$$\lim_{x \to 0-} \frac{f(x) - f(0)}{x - 0} = \lim_{x \to 0-} \frac{x^3}{x} = \lim_{x \to 0-} x^2 = 0$$

Hence f is also differentiable at $x = 0$. The derived function f' is given by

$$f'(x) = \begin{cases} 2x & \text{if } x \geq 0 \\ 3x^2 & \text{if } x < 0 \end{cases}$$

It is straightforward to show that f' is everywhere continuous, and so f is continuously differentiable. However, f' is not differentiable at $x = 0$ (in fact, $f_+^{(2)}(0) = 2$ but $f_-^{(2)}(0) = 0$, and so f is not twice differentiable at $x = 0$.   ■

   In the next example the functions considered are all infinitely differentiable (in other words, n-times continuously differentiable for all $n \in \mathbb{N}$). Repeated differentiation enables the general formula for the nth derivative to be obtained. This can of course be established by induction on n.

■■   **EXAMPLE 7**

   (a) If $f(x) = x^m$, $m \in \mathbb{N}$, then

$$f^{(n)}(x) = \begin{cases} \dfrac{m!}{(m - n)!} x^{m-n} & \text{for } n \leq m \\ 0 & \text{for } n > m \end{cases}$$

   (b) If $f(x) = \sin x$ then

$$f^{(n)}(x) = \sin(x + \tfrac{1}{2}n\pi) \quad \text{for } n \geq 0$$

(c) If $f(x) = \log_e x$ then

$$f^{(n)}(x) = \frac{(-1)^{n-1}(n-1)!}{x^n} \quad \text{for } n \geq 1 \qquad \blacksquare$$

Now the successive differentiation of products of simple functions is investigated. Recall that

$$(f \cdot g)' = f \cdot g' + f' \cdot g$$

Hence

$$(f \cdot g)^{(2)} = (f \cdot g')' + (f' \cdot g)'$$
$$= f \cdot g^{(2)} + 2f' \cdot g' + f^{(2)} \cdot g$$

Similarly,

$$(f \cdot g)^{(3)} = f \cdot g^{(3)} + 3f' \cdot g^{(2)} + 3f^{(2)} \cdot g' + f^{(3)} \cdot g'$$

Generally the following formula applies. This can be proved by induction on n.

6.2.6 Leibniz's formula

Let f and g be n-times continuously differentiable at c. Then $h = f \cdot g$ is also n-times continuously differentiable at c, and

$$h^{(n)}(c) = \sum_{r=0}^{n} \binom{n}{r} f^{(r)}(c) g^{(n-r)}(c)$$

■■ EXAMPLE 8

Calculate $h^{(6)}(x)$ when $h(x) = x^3 \sin x$.

Solution

If $f(x) = x^3$ then $f'(x) = 3x^2$, $f^{(2)}(x) = 6x$, $f^{(3)}(x) = 6$ and $f^{(r)}(x) = 0$ for $r \geq 4$. If $g(x) = \sin x$ then $g^{(r)}(x) = \sin(x + \frac{1}{2}r\pi)$ for $r \geq 0$.
So

$$h^{(6)}(c) = \sum_{r=0}^{6} \binom{6}{r} f^{(r)}(c) g^{(6-r)}(c)$$

$$= \binom{6}{0} x^3 \sin\left(x + \frac{6\pi}{2}\right) + \binom{6}{1} 3x^2 \sin\left(x + \frac{5\pi}{2}\right)$$

$$+ \binom{6}{2} 6x \sin\left(x + \frac{4\pi}{2}\right) + \binom{6}{3} 6 \sin\left(x + \frac{3\pi}{2}\right)$$

$$= x^3(-\sin x) + 18x^2(\cos x) + 90x(\sin x)$$

$$+ 120(-\cos x)$$

$$= x(90 - x^2)\sin x + 6(3x^2 - 20)\cos x \qquad \blacksquare$$

This section concludes with a generalization of L'Hôpital's rule that is merely the repeated application of the earlier version A.

6.2.7 L'Hôpital's rule (version B)

Let f and g be n-times continuously differentiable on the interval $[a, b]$ and let x_0 satisfy $a < x_0 < b$. If

$$f^{(r)}(x_0) = g^{(r)}(x_0) = 0 \quad \text{for } 0 \leqslant r \leqslant n - 1$$

and

$$g^{(n)}(x_0) \neq 0$$

then

$$\lim_{x \to x_0} \frac{f(x)}{g(x)} = \lim_{x \to x_0} \frac{f^{(n)}(x)}{g^{(n)}(x)}$$

provided the latter exists.

■■ EXAMPLE 9

$$\lim_{x \to 0} \frac{1 - \cos x}{x^2} = \tfrac{1}{2}$$

Solution

If $f(x) = 1 - \cos x$ and $g(x) = x^2$ then

$$f(0) = g(0) = 0$$

Now $f'(x) = \sin x$ and $g'(x) = 2x$, and so

$$f'(0) = g'(0) = 0$$

Now $f''(x) = \cos x$ and $g''(x) = 2$, and so

$$g''(0) \neq 0$$

Finally, then,

$$\lim_{x \to 0} \frac{1 - \cos x}{x^2} = \lim_{x \to 0} \frac{\cos x}{2} = \tfrac{1}{2} \qquad \blacksquare$$

■■ EXAMPLE 10

$$\lim_{x \to 0} (x^{-1} - \cot x) = 0$$

Solution

First note that

$$(x^{-1} - \cot x) = \frac{\sin x - x \cos x}{x \sin x}$$

If $f(x) = \sin x - x \cos x$ and $g(x) = x \sin x$ then

$$f(0) = g(0) = 0$$

Now $f'(x) = x \sin x$ and $g'(x) = \sin x + x \cos x$, and so

$$f'(0) = g'(0) = 0$$

Thus $f''(x) = g'(x)$ and $g''(x) = 2 \cos x - x \sin x$, and $g''(0) \neq 0$. Finally, then,

$$\lim_{x \to 0} (x^{-1} - \cot x) = \lim_{x \to 0} \frac{\sin x + x \cos x}{2 \cos x - x \sin x} = \frac{0}{2} = 0 \qquad \blacksquare$$

Exercises 6.2

---

1.   Use Rolle's theorem to show that

$$P(x) = x^3 + ax + b \quad (a > 0)$$

has precisely one real root.

2.   Suppose that $f: [0, 1] \to [0, 1]$ is differentiable on $(0, 1)$ and continuous on $[0, 1]$. By 5.3.4, it is known that there is at least one $c \in [0, 1]$ for which $f(c) = c$. If $f'(x) \neq 1$ for $x \in (0, 1)$, prove that there is exactly one such point c. (*Hint*: consider the function

$h(x) = f(x) - x$; suppose that f has two fixed points, and obtain a contradiction.)

3.   Suppose that f is a twice differentiable function and that

$$f(a_1) = f(a_2) = f(a_3) \quad (a_1 < a_2 < a_3)$$

Prove that $f''(c) = 0$ for some c between a_1 and a_3.

4.   Use the mean value theorem to prove the following:

(a) $|\sin a - \sin b| \leqslant |a - b|$ for all $a, b \in \mathbb{R}$

(b) $\frac{1}{10} < \sqrt{83} - 9 < \frac{1}{9}$

5.   Use the local extremum theorem and the increasing–decreasing theorem to find all the local maxima and minima of f when

(a) $f(x) = x + \dfrac{1}{x}$ (b) $f(x) = e^{-x^2}$ (c) $f(x) = \dfrac{2}{x} + \log_e x$

6.   Establish the following inequalities:

(a) $x - \frac{1}{2}x^2 < \log_e(1 + x)$ for $x > 0$

(b) $\tan^{-1} x > \dfrac{x}{1 + \frac{1}{3}x^2}$ for $x > 0$

(c) $\sin x < x < \tan x$ for $0 < x < \frac{1}{2}\pi$

7.   Use L'Hôpital's rule to evaluate the following:

(a) $\displaystyle\lim_{x \to 1} \left(\frac{3x^2}{x^3 - 1} - \frac{1}{x - 1} \right)$ (b) $\displaystyle\lim_{x \to 0} (\operatorname{cosec} x - \cot x)$

(c) $\displaystyle\lim_{x \to \pi/2} (\sin x)^{\operatorname{cosec} 2x}$ (d) $\displaystyle\lim_{x \to 0} \frac{x^2}{\cosh x - 1}$

(e) $\displaystyle\lim_{x \to 0} \frac{\sqrt{1 + x} - 1}{\sqrt{1 - x} - 1}$ (f) $\displaystyle\lim_{x \to 0} (\cosh x)^{1/x^2}$

8.   Use Leibniz's formula to find

(a) the fourth derivative of $x^2 e^{-x}$

(b) the sixth derivative of $x^3 \log_e x$

6.3   Taylor polynomials

Determining the value of a polynomial expression, such as $1 + x + \frac{1}{2}x^2$, for any specified value of x involves only simply addition and multiplication.

Moreover, given x to any specified degree of accuracy, it is straightforward to evaluate such polynomial expressions to a comparable degree of accuracy. When faced with the problem of evaluating $\sin x$ or e^x to a high degree of accuracy, especially when more decimal places are required than an electronic calculator can provide, the situation is not so clear-cut. For the elementary functions, one approach would be to use an nth partial sum of the power series defining the function, for a suitably large value of n. This would be a polynomial of degree $n - 1$ whose value at any specified x provides an approximation to the value at x of the elementary function under consideration. However, there is no obvious measure of how accurate such an approximation might be. For other functions there is also the question of whether or not they may be represented by power series. This section provides answers to such questions. In order to determine the form of those polynomials that may be used to approximate the values of a given function, and ultimately to derive a power series expansion of the function, it is first necessary to examine the coefficients of powers of x in a general power series.

Suppose that $\sum_{n=0}^{\infty} a_n x^n$ is a power series with radius of convergence R. Let $f(x)$ be the sum of this series for $|x| < R$. It is proved in the Appendix that f is differentiable and that

$$f'(x) = \sum_{n=1}^{\infty} na_n x^{n-1} \quad \text{for } |x| < R$$

Continually differentiating in this manner leads to

$$f^{(r)}(x) = \sum_{n=r}^{\infty} n(n-1)(n-2) \cdots (n-r+1)a_n x^{n-r} \quad \text{for } |x| < R$$

If $x = 0$ then

$$a_r = \frac{f^{(r)}(0)}{r!} \quad \text{for } r = 1, 2, \ldots$$

Thus the coefficient of x^n in any power series is $f^{(n)}(0)/n!$, where $f(x)$ is the sum of the given power series. Hence

$$f(x) = \sum_{n=0}^{\infty} \frac{f^{(n)}(0)}{n!} x^n \quad \text{for } |x| < R$$

For small values of x the sum function $f(x)$ can be approximated by the polynomial

$$f(0) + \frac{f'(0)}{1!} x + \frac{f''(0)}{2!} x^2 + \ldots + \frac{f^{(N)}(0)}{N!} x^N$$

for any value of N.

This section investigates how good an approximation this polynomial is when $f(x)$ is an *arbitrary* function of x and not necessarily the sum of a given power series. It shows that certain functions can then be expressed as the sum of suitable power series. This investigation begins with a definition.

6.3.1   Definition

> Let f be an n-times continuously differentiable function at 0.
> The **Taylor polynomial of degree n for f at 0** is defined by
>
> $$T_n f(x) = f(0) + \frac{f'(0)}{1!} x + \frac{f''(0)}{2!} x^2 + \ldots + \frac{f^{(n)}(0)}{n!} x^n$$

■■   EXAMPLE 1

Let $f(x) = e^x$. Then $f^{(r)}(x) = e^x$ for $r = 1, 2, \ldots$, and so $f^{(r)}(0) = 1$. Thus

$$T_0 f(x) = 1$$
$$T_1 f(x) = 1 + x$$
$$T_2 f(x) = 1 + x + \tfrac{1}{2} x^2$$

and so on. As can be seen from Figure 6.7, as n increases, the Taylor polynomials become progressively better approximations to $f(x) = e^x$ *near $x = 0$.*   ■

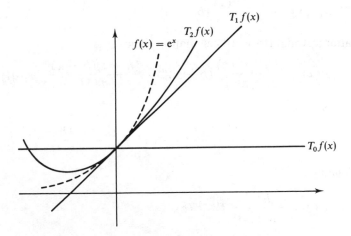

Figure 6.7

The first result in this section provides an estimate of the difference between $f(b)$, the value of a given function at $x = b$, and $T_n f(b)$, the value of its Taylor polynomial of degree n at $x = b$.

6.3.2 The first remainder theorem

Let f be $(n + 1)$-times continuously differentiable on an open interval containing the points 0 and b. Then the difference between f and $T_n f$ at $x = b$ is given by

$$f(b) - T_n f(b) = \frac{b^{n+1}}{(n + 1)!} f^{(n+1)}(c)$$

for some c between 0 and b.

Proof

For simplicity, assume that $b > 0$. Let

$$h_n(x) = f(b) - \sum_{r=0}^{n} \left[\frac{f^{(r)}(x)}{r!} (b - x)^r \right] \quad (x \in [0, b])$$

Then $h_n(b) = 0$ and $h_n(0) = f(b) - T_n f(b)$. Let

$$g(x) = h_n(x) - \left(\frac{b - x}{b} \right)^{n+1} h_n(0) \quad (x \in [0, b])$$

Then g is continuous on $[0, b]$ and differentiable on $(0, b)$, and $g(0) = g(b)$. Hence, by Rolle's theorem (6.2.1), $g'(c) = 0$ for some c between 0 and b.

Now

$$h'_n(x) = - \frac{f^{(n+1)}(x)}{n!} (b - x)^n$$

after a straightforward calculation. Thus

$$g'(x) = - \frac{f^{(n+1)}(x)}{n!} (b - x)^n + \frac{(n + 1)(b - x)^n}{b^{n+1}} h_n(0)$$

and so

$$0 = g'(c) = - \frac{f^{(n+1)}(c)}{n!} (b - c)^n + \frac{(n + 1)(b - c)^n}{b^{n+1}} h_n(0)$$

leading to

$$h_n(0) = \frac{b^{n+1}}{(n + 1)!} f^{(n+1)}(c)$$

as required. □

Denote $f(b) - T_nf(b)$ by $R_nf(b)$ and call it the **remainder term** at $x = b$. Thus $f(b) = T_nf(b) + R_nf(b)$, and so the error in approximating $f(b)$ by $T_nf(b)$ is given by the remainder term $R_nf(b)$. Since $f^{(n+1)}$ is continuous on an interval containing 0 and b, it is bounded on that interval. Therefore there exists a number M such that $|f^{(n+1)}(c)| \leq M$, and so

$$|R_nf(b)| \leq \left| \frac{b^{n+1}}{(n+1)!} \right| M$$

Thus, for fixed n, the remainder term will be small for b close to zero. In other words, Taylor polynomials provide good approximations to functions near $x = 0$. The next example illustrates this.

■■   EXAMPLE 2

When $f(x) = \sin x$, calculate
(a) $T_7f(x)$ (b) $R_7f(x)$
(c) $\sin 0.1$ accurate to 12 decimal places

Solution

First, $f^{(r)}(x) = \sin(x + \tfrac{1}{2}r\pi)$ for $r \geq 0$.

(a) $T_7f(x) = \sum_{r=0}^{7} \frac{f^{(r)}(0)}{r!} x^r = \sum_{r=0}^{7} \frac{x^r}{r!} \sin\left(\frac{r\pi}{2}\right)$

Thus

$$T_7f(x) = x - \frac{x^3}{3!} + \frac{x^5}{5!} - \frac{x^7}{7!}$$

(b) $R_7f(x) = (x^8/8!)\sin c$ for some c between 0 and x.
(c) By the first remainder theorem,

$$|f(0.1) - T_7f(0.1)| = |R_7f(0.1)| = \frac{(0.1)^8}{8!} |\sin c| \leq \frac{(0.1)^8}{8!}$$

Now

$$T_7f(0.1) = (0.1) - \frac{(0.1)^3}{3!} + \frac{(0.1)^5}{5!} - \frac{(0.1)^7}{7!}$$

$$= 0.099\,833\,416\,646\,826\ldots$$

and

$$|R_7 f(0.1)| \leqslant \frac{(0.1)^8}{8!} = 2.48 \times 10^{-13}$$

Thus $|\sin 0.1 - 0.099\,833\,416\,646\,826| \leqslant 2.48 \times 10^{-13}$

Therefore $\sin 0.1 = 0.099\,833\,416\,647$ correct to 12 decimal places. ∎

It can now be shown how Taylor polynomials can be used to generate power series expansions for functions f that are infinitely differentiable on an open interval containing 0 and x. For each real number x

$$f(x) = T_n f(x) + R_n f(x)$$

Now

$$\lim_{n \to \infty} T_n f(x) = \sum_{r=0}^{\infty} \frac{f^{(r)}(0)}{r!} x^r \quad \text{for } |x| < R$$

where R is the radius of convergence of the resulting power series. If it can be shown that $\lim_{n \to \infty} R_n f(x) = 0$ for $|x| < R' < R$ for some R' then

$$f(x) = \sum_{r=0}^{\infty} \frac{f^{(r)}(0)}{r!} x^r \quad \text{for } |x| < R'$$

This power series is called the **Maclaurin series** for $f(x)$.

Colin Maclaurin (1698–1746) was a Scotsman who ranks as one of the finest British mathematicians in the generation after Newton. Although principally interested in geometric curves, he is best known for the series that bears his name. These appeared in his *Treatise of Fluxions* in 1742, and had been anticipated by some half dozen earlier workers. In fact, this series is only a special case of the more general Taylor series published by **Brook Taylor** (1685–1731) in his *Methodus incrementorum directa et inversa*. Taylor, a Cambridge graduate, used his series to solve equations numerically by obtaining successively closer and closer approximations. In fact, the general Taylor series expression was known to Jean Bernoulli years before it appeared in print in Taylor's *Methodus incrementorum*. History, as usual, is fickle in the matter of attaching names to theorems.

▪▪ EXAMPLE 3

Derive the Maclaurin series for $f(x) = e^x$.

Solution

First,

$$T_n f(x) = 1 + x + \frac{x^2}{2!} + \ldots + \frac{x^n}{n!}$$

and

$$R_n f(x) = \frac{x^{n+1}}{(n+1)!} e^c$$

for some c between 0 and x. Now $\sum_{r=0}^{\infty} x^r / r!$ is absolutely convergent for all x by the ratio test (4.2.3), so for any fixed real number x

$$\lim_{n \to \infty} T_n f(x) = \sum_{r=0}^{\infty} \frac{x^r}{r!}$$

By the vanishing condition (4.1.2), $x^n / n! \to 0$ as $n \to 0$. Thus

$$|R_n f(x)| = \left| \frac{x^{n+1}}{(n+1)!} e^c \right| \to 0 \quad \text{as } n \to \infty$$

for fixed x. Hence, for any x, $\lim_{n \to \infty} R_n f(x) = 0$. Hence

$$e^x = \sum_{r=0}^{\infty} \frac{x^r}{r!} = 1 + x + \frac{x^2}{2!} + \ldots + \frac{x^n}{n!} + \ldots, \quad \text{valid for all } x$$

Notice that the Maclaurin series for e^x is quite naturally the power series used to define e^x. ▪

Power series can be generated in a similar manner for all the standard functions. In each case the power series is just $\lim_{n \to \infty} T_n f(x)$, and the range of validity is precisely those x for which:

(a) the resulting power series converges, and
(b) the remainder $R_n f(x) \to 0$ as $n \to \infty$.

In deriving the following list (6.3.3) the trickiest part is establishing (b). The form of the remainder found in the first remainder theorem is called the **Lagrange form**. Alternative expressions are sometimes needed. See Section 6.4 for further details.

6.3.3 Standard series

(1) $e^x = 1 + x + \dfrac{x^2}{2!} + \ldots = \displaystyle\sum_{r=0}^{\infty} \dfrac{x^r}{r!},$ valid for all x

(2) $\sin x = x - \dfrac{x^3}{3!} + \ldots = \displaystyle\sum_{r=0}^{\infty} \dfrac{(-1)^r x^{2r+1}}{(2r+1)!},$ valid for all x

(3) $\cos x = 1 - \dfrac{x^2}{2!} + \ldots = \displaystyle\sum_{r=0}^{\infty} \dfrac{(-1)^r x^{2r}}{(2r)!},$ valid for all x

(4) $(1+x)^t = 1 + tx + \dfrac{t(t-1)}{2!} x^2 + \ldots,$ where $t \notin \mathbb{N}$

$$= 1 + \sum_{r=1}^{\infty} \frac{t(t-1)\ldots(t-r+1)x^r}{r!}$$

valid for $-1 < x < 1$

(5) $\log_e (1+x) = x - \tfrac{1}{2}x^2 + \ldots$

$$= \sum_{r=1}^{\infty} \frac{(-1)^{r-1}x^r}{r},$$ valid for $-1 < x \leqslant 1$

Algebraic and calculus techniques can now be used to generate from these standard series the series for other functions such as $\sin^{-1} x$, $\cosh x$, $\sec x$ and so on. See Exercises 6.3.

The first few terms of the Maclaurin series for a given function f provide a good approximation to $f(x)$ for x close to 0, but what if approximations for x close to some other real number a are required? Polynomials must then be considered not in powers of x but in powers of $x - a$.

6.3.4 Definition

Let f be n-times continuously differentiable at a, a fixed real number. Define the **Taylor polynomial of degree n for f at a** by

$$T_{n,a}f(x) = f(a) + f'(a)(x-a) + \frac{f''(a)}{2!}(x-a)^2 + \ldots$$
$$+ \frac{f^{(n)}(a)}{n!}(x-a)^n$$

The first remainder theorem can now be generalized.

6.3.5   Taylor's theorem

> Let f be $(n + 1)$-times continuously differentiable on an open interval containing the points a and b. Then the difference between f and $T_{n,a}f$ at b is given by
>
> $$f(b) - T_{n,a}f(b) = \frac{(b - a)^{n+1}}{(n + 1)!}\, f^{(n+1)}(c)$$
>
> for some c between a and b.

Proof

This result is restated as 6.4.1(2) and is proved in Section 6.4.   □

From 6.3.5, the error in approximating $f(x)$ by the polynomial $T_{n,a}f(x)$ is just the **remainder term**.

$$R_{n,a}f(x) = \frac{(x - a)^{n+1}}{(n + 1)!}\, f^{(n+1)}(c)$$

where c lies between a and x. The approximation is good for x close to a.

Just as before, power series can be generated in powers of $x - a$ called **Taylor series**, for suitable functions of x. The range of validity is again those x for which

(a) the resulting power series converges, and

(b) $R_{n,a}f(x) \to 0$ as $n \to \infty$.

A form of Taylor's theorem much used in numerical analysis is derived below and used to round off the investigation of local extrema of functions begun in Section 6.1.

From Taylor's theorem,

$$f(x) = T_{n,a}f(x) + R_{n,a}f(x)$$

$$= f(a) + (x - a)f'(a) + \ldots + \frac{(x - a)^n}{n!}\, f^{(n)}(a)$$

$$+ \frac{(x - a)^{n+1}}{(n + 1)!}\, f^{(n+1)}(c)$$

for some c between a and x.

Let $x - a = h$. Then c lies between a and $a + h$. Thus, whether h is positive or negative, $c = a + \theta h$ for some $0 < \theta < 1$, which immediately gives the following.

6.3.6 Result

$$f(a + h) = f(a) + hf'(a) + \ldots + \frac{h^n}{n!} f^{(n)}(a)$$

$$+ \frac{h^{n+1}}{(n + 1)!} f^{(n+1)}(a + \theta h)$$

for some $\theta, 0 < \theta < 1$.

This expression emphasizes that the value of f at $a + h$ is determined by the values of f and its derivatives at a, with θ measuring the degree of indeterminacy. Note that θ depends on h; it is not a constant that works for all values of h.

When the nature of a function f that has a stationary point at $x = a$ (that is, $f'(a) = 0$) is investigated, it is required that the sign of $f(a + h) - f(a)$ for all small h be examined. The expression in 6.3.6 relates $f(a + h) - f(a)$ to the derivatives of f at a, thus enabling the following to be proved.

6.3.7 Classification theorem for local extrema

If f is $(n + 1)$-times continuously differentiable on a neighbourhood of a and $f^{(r)}(a) = 0$ for $r = 1, 2, \ldots, n$ (in particular, $f'(a) = 0$ and so $x = a$ is a stationary point of f) and $f^{(n+1)}(a) \neq 0$ then

(1) $n + 1$ even and $f^{(n+1)}(a) > 0 \Rightarrow f$ has a local minimum at $x = a$

(2) $n + 1$ even and $f^{(n+1)}(a) < 0 \Rightarrow f$ has a local maximum at $x = a$

(3) $n + 1$ odd $\Rightarrow f$ has neither a local maximum nor a local minimum at $x = a$

Proof

Since $f^{(r)}(a) = 0$ for $r = 1, 2, \ldots, n$,

$$f(a + h) = f(a) + hf'(a) + \ldots + \frac{h^n}{n!} f^{(n)}(a)$$

$$+ \frac{h^{n+1}}{(n + 1)!} f^{(n+1)}(a + \theta h)$$

$$= f(a) + \frac{h^{n+1}}{(n + 1)!} f^{(n+1)}(a + \theta h)$$

for some $\theta, 0 < \theta < 1$

Hence

$$f(a + h) - f(a) = \frac{h^{n+1}}{(n + 1)!} f^{(n+1)}(a + \theta h)$$

where $0 < \theta < 1$. Since $f^{(n+1)}(a) \neq 0$ and $f^{(n+1)}$ is continuous, there is a $\delta > 0$ such that $f^{(n+1)}(x) \neq 0$ for $|x - a| < \delta$. Thus, for all h satisfying $|h| < \delta$, $f^{(n+1)}(a + \theta h)$ has the same sign as $f^{(n+1)}(a)$. So $f(a + h) - f(a)$ has the same sign as $h^{n+1} f^{(n+1)}(a)$ for all h, $|h| < \delta$.

(1) If $n + 1$ is even and $f^{(n+1)}(a) > 0$ then $f(a + h) - f(a) > 0$ on the open interval $(a - \delta, a + \delta)$. Hence $x = a$ gives a local minimum of f.

(2) If $n + 1$ is even and $f^{(n+1)}(a) < 0$ then $f(a + h) - f(a) < 0$ on the open interval $(a - \delta, a + \delta)$. Hence $x = a$ gives a local maximum of f.

(3) If $n + 1$ is odd, the sign of $f(a + h) - f(a)$ changes with the sign of h. It is said that $x = a$ gives a **horizontal point of inflection**. $\qquad \square$

■■ EXAMPLE 4

Determine the nature of the stationary points of

$$f(x) = x^6 - 6x^4$$

Solution

Since f is a polynomial function, it is infinitely differentiable. Now

$$f'(x) = 6x^5 - 24x^3 = 6x^3(x^2 - 4)$$

Thus f has stationary points at $x = 0$ and at $x = \pm 2$. Now $f''(x) = 30x^4 - 72x^2$, and so $f''(\pm 2) = 192 > 0$. Thus f has local minima at $x = \pm 2$.

However, $f''(0) = 0$. Since $f^{(3)}(x) = 120x^3 - 144x$, $f^{(3)}(0) = 0$. Now $f^{(4)}(x) = 360x^2 - 144$, and so $f^{(4)}(0) = -144 < 0$. Thus f has a local maximum at $x = 0$. See Figure 6.8. ■

The above rather simple example can be analysed without resorting to derivatives higher than the first, but it serves to illustrate the use of the classification theorem. The theorem is not, however, a universal panacea. For example, it cannot be applied to the function $f(x) = e^{-1/x^2}$, $x \neq 0$, $f(0) = 0$, where $f^{(n)}(0) = 0$ for all n and f has a stationary point at $x = 0$. In this case the Maclaurin series does not converge to $f(x)$ for any nonzero x.

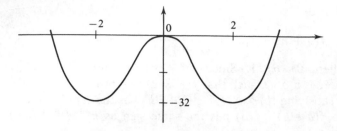

Figure 6.8

Exercises 6.3

1. Determine the Taylor polynomial $T_5f(x)$ for $f(x) = \log_e(1+x)$. Estimate $R_5f(b)$ when (a) $b = 1$ and (b) $b = 0.1$. Hence find $\log_e 1.1$ accurate to four decimal places.

2. Find the Taylor polynomial $T_5f(x)$ for $f(x) = \cos x$. Evaluate $T_5f(0.1)$ correct to nine decimal places, and hence use the first remainder theorem to determine $\cos 0.1$ accurate to eight decimal places.

3. Show that

    $$\sum_{r=n+1}^{\infty} \frac{1}{r!} < \frac{1}{n!n}$$

    Deduce that

    $$e - \sum_{r=0}^{n} \frac{1}{r!} < \frac{1}{n!n}$$

    Hence prove that e is irrational. (*Hint*: let $e = p/q$ and choose $n = q$ in the second inequality above.)

4. Using standard series (see 6.3.3), determine the Maclaurin series of $(\cosh x) \log_e(1+x^2)$ up to the term in x^6. Hence evaluate

    $$\lim_{x \to 0} \frac{(\cosh x) \log_e(1+x^2) - x^2}{x^6}$$

5. If f is twice continuously differentiable in a neighbourhood of a, use 6.3.6 to show that

    $$f''(a) = \lim_{h \to 0} \frac{f(a+h) + f(a-h) - 2f(a)}{h^2}$$

6. Find the stationary points of each of the following functions and determine their nature:

(a) $f(x) = x^4 e^{-x}$ (b) $f(x) = x^2 \log_e x$ (c) $f(x) = 2e^x - e^{2x}$

Sketch their graphs.

6.4 Alternative forms of Taylor's theorem

This section presents proofs of three different versions of Taylor's theorem, each giving rise to a different form of the remainder term. The third version requires the use of the fundamental theorem of calculus, which is one of the main results in Section 7.1. Taylor's theorem can be restated as follows.

6.4.1 Taylor's theorem

Let f be $(n + 1)$-times continuously differentiable on an open interval containing the points a and x. Let $T_{n,a}f(x)$ be the Taylor polynomial of degree n for f at a and define $R_{n,a}f(x)$ by

$$f(x) = T_{n,a}f(x) + R_{n,a}f(x)$$

Then

(1) $R_{n,a}f(x) = \dfrac{f^{(n+1)}(c)}{n!} (x - c)^n (x - a)$

for some c between a and x

(2) $R_{n,a}f(x) = \dfrac{f^{(n+1)}(c)}{(n + 1)!} (x - a)^{n+1}$

for some c between a and x

(3) $R_{n,a}f(x) = \displaystyle\int_a^x \dfrac{f^{(n+1)}(t)}{n!} (x - t)^n \, dt$

(The c in (2) may be different from the c in (1).)

Proof

For each t between a and x

$$f(x) = f(t) + f'(t)(x - t) + \ldots + \frac{f^{(n)}(t)}{n!} (x - t)^n + F(t)$$

where $F(t) = R_{n,t}f(x)$. Differentiating with respect to t gives

$$0 = f'(t) + \left[-f'(t) + \frac{f''(t)}{1!} (x - t) \right]$$

$$+ \left[-\frac{f''(t)}{1!} (x - t) + \frac{f^{(3)}(t)}{2!} (x - t)^2 \right] + \dots$$

$$+ \left[-\frac{f^{(n)}(t)}{(n-1)!} (x - t)^{n-1} + \frac{f^{(n+1)}(t)}{n!} (x - t)^n \right] + F'(t)$$

Cancellation now gives

$$F'(t) = -\frac{f^{(n+1)}(t)}{n!} (x - t)^n$$

(1) Apply the mean value theorem (6.2.2) to the function F on the interval with endpoints a and x. Hence there exists a c between a and x such that

$$\frac{F(x) - F(a)}{x - a} = F'(c) = -\frac{f^{(n+1)}(c)}{n!} (x - c)^n$$

Hence

$$\frac{R_{n,x}f(x) - R_{n,a}f(x)}{x - a} = -\frac{f^{(n+1)}(c)}{n!} (x - c)^n$$

and, since $R_{n,x}f(x) = 0$,

$$R_{n,a}f(x) = \frac{f^{(n+1)}(c)}{n!} (x - c)^n (x - a)$$

This is called **Cauchy's form of the remainder**.

(2) Apply Cauchy's mean value theorem (6.2.4) to the functions F and G on the interval with endpoints a and x, where $G(t) = (x - t)^{n+1}$. Thus there is a number c between a and x such that

$$\frac{F(x) - F(a)}{G(x) - G(a)} = \frac{F'(c)}{G'(c)} = \frac{-[f^{(n+1)}(c)/n!](x - c)^n}{-(n + 1)(x - c)^n}$$

Hence

$$\frac{R_{n,a}f(x)}{(x - a)^{n+1}} = \frac{f^{(n+1)}(c)}{(n + 1)!}$$

and so

$$R_{n,a}f(x) = \frac{f^{(n+1)}(c)}{(n + 1)!} (x - a)^{n+1}$$

This gives **Lagrange's form of the remainder**.

(3) By the fundamental theorem of calculus (see 7.1.7),

$$F(x) - F(a) = \int_a^x F'(t)\,dt = -\int_a^x \frac{f^{(n+1)}(t)}{n!} (x-t)^n\,dt$$

Hence

$$R_{n,a}f(x) = \int_a^x \frac{f^{(n+1)}(t)}{n!} (x-t)^n\,dt$$

This is called the **integral form of the remainder**.   □

Joseph Louis Lagrange (1736–1813) was born in Turin, and in mathematical stature ranks with Euler. He was the first to recognize the parlous state of the foundations of analysis and, accordingly, to attempt to provide a rigorous foundation for calculus. This attempt, *Théorie des fonctions analytiques contenant les principes du calcul differentiel*, which appeared in 1797, was far from successful. Lagrange's key idea was to represent a function $f(x)$ by its Taylor series and realize $f'(x)$, $f''(x)$ and so on as coefficients in the corresponding Taylor expansion of $f(x + h)$ in terms of h (see 6.3.6). The notation $f'(x)$ in use today is due to Lagrange. However, insufficient attention was paid in *Théorie des fonctions analytiques* to matters of convergence and divergence – matters that were successfully laid to rest by Cauchy in the early nineteenth century. Lagrange is better known for his contributions to algebra and number theory, where his concise style and attention to rigour proved much more successful.

Section 7.1 discusses the estimation of integrals, and so the integral form of the remainder proves useful in practice. The Lagrange form has already been seen in action in the derivation of the Maclaurin series for e^x in Section 6.3, Example 3.

The binomial expansion and the series for $\log_e (1 + x)$ are best derived using the Cauchy form of the remainder.

■■   EXAMPLE 1

Let $f(x) = \log_e (1 + x)$. Now

$$f^{(r)}(x) = \frac{(-1)^{r-1}(r-1)!}{(1+x)^r} \quad \text{for } r \geqslant 1$$

and so

$$T_n f(x) = x - \frac{x^2}{2} + \frac{x^3}{3} \cdots + \frac{(-1)^{n-1} x^n}{n} + R_n f(x)$$

where

$$R_n f(x) = R_{n,0} f(x) = \frac{f^{(n+1)}(c)}{n!}(x - c)^n x$$

c lying between 0 and x. Write $c = \theta x$, $0 < \theta < 1$. Thus

$$R_n f(x) = \frac{(-1)^n n!}{(1 + \theta x)^{n+1} n!}(x - \theta x)^n x = \frac{(-1)^n x^{n+1}}{(1 + \theta x)^{n+1}}(1 - \theta)^n$$

For $x > -1$, the domain of f, choose $k > 0$ such that $-1 < -k < x$. Since $0 < \theta < 1$,

$$0 < 1 - \theta < 1 + \theta x \quad \text{and} \quad 1 - k < 1 - \theta k < 1 + \theta x$$

Hence

$$|R_n f(x)| = \frac{|x|^{n+1}}{1 + \theta x}\left(\frac{1 - \theta}{1 + \theta x}\right)^n < \frac{|x|^{n+1}}{1 - k} \to 0$$

$$\text{as } n \to \infty \text{ for } |x| < 1$$

Also

$$|R_n f(1)| = \left(\frac{1 - \theta}{1 + \theta}\right)^n \frac{1}{1 + \theta} \to 0 \quad \text{as } n \to \infty$$

So, for $-1 < x \leqslant 1$, it follows that $R_n f(x) \to 0$ as $n \to \infty$. Since

$$\sum_{r=1}^{\infty}(-1)^{r-1} \frac{x^r}{r}$$

has radius of convergence 1, it can be deduced that

$$\log_e(1 + x) = x - \frac{x^2}{2} + \frac{x^3}{3} - \cdots, \quad \text{valid for } -1 < x \leqslant 1$$

Note that putting $x = 1$ in this series gives that the sum of the alternating series

$$\sum_{r=1}^{\infty} \frac{(-1)^{r-1}}{r}$$

is $\log_e 2$. ∎

Exercises 6.4

1. Determine the Taylor series for $f(x) = \sin x$ in powers of $x - \frac{1}{6}\pi$ using the Lagrange form of the remainder.

2. Use the result of Question 1 to write down the Taylor polynomial of degree four for $f(x) = \sin x$ at $\frac{1}{6}\pi$, and the corresponding Lagrange form of the remainder. Hence find a value for $\sin 32°$, and determine its accuracy.

3. (a) Use the Maclaurin series for $\log_e (1 + t)$ to find the Taylor series for $\log_e x$ in powers of $x - 1$.
   (b) Use the Maclaurin's series for $1/(1 - t)$ to find the Taylor series for $1/x$ in powers of $x + 2$. (*Hint*: try $t = \frac{1}{2}(x + 2)$.)

Problems 6

1. Calculate the left- and right-hand derivatives of each of the following functions at their points of non-differentiability:

   (a) $f(x) = |x| + |x - 2|$ (b) $f(x) = |\sin x|$

   (c) $f(x) = \dfrac{1}{|x| + 1}$ (d) $f(x) = e^{|x|}$

2. For each of the following functions use the rules for differentiability to prove that f is differentiable. You may assume that the identity, constant, exponential, sine and cosine functions are differentiable on $\mathbb{R}$.

   (a) $f(x) = e^x \cos^2 3x$ (b) $f(x) = \dfrac{x}{\cosh x}$

   (c) $f(x) = \begin{cases} \sin x \sin (1/x) & \text{if } x \neq 0 \\ 0 & \text{if } x = 0 \end{cases}$

   (d) $f(x) = \sqrt{1 + x^2}$ (e) $f(x) = x^2 \log_e (x^2 + 1)$

3. Use Rolle's theorem to show that

   $$f(x) = x^3 - \tfrac{3}{2}x^2 + \lambda, \quad \lambda \in \mathbb{R}$$

   never has two zeros in $[0, 1]$.

4. Suppose that $f: \mathbb{R} \to \mathbb{R}$ is differentiable and that $|f'(x)| \leqslant M$ for all $x \in \mathbb{R}$. Prove that $|f(x) - f(y)| \leqslant M|x - y|$ for all $x, y \in \mathbb{R}$.

5.   Suppose that f is differentiable on $[0, 1]$ and that $f(0) = 0$. Prove that if f' is increasing on $(0, 1)$ then $f(x)/x$ is increasing on $(0, 1)$.

6.   Establish the following.

(a) $f(x) = x - \sin x$ is increasing on $[\frac{1}{4}\pi, \frac{1}{2}\pi]$.
(b) $f(x) = x^n e^{-x}$ is increasing on $[\frac{1}{2}, n - 1]$, $n > 1$.
(c) $f(x) = x \log_e x$ is increasing on $[1, \infty)$.
(d) $f(x) = 1/(1 + x)$ is decreasing on $[1, \infty)$.

7.   Find the maximum and minimum values of each of the following functions on the interval $[0, 2]$ (Note that endpoints need to be considered):

(a) $f(x) = |x^2 - x|$ (b) $f(x) = xe^{-x}$
(c) $f(x) = |2x + 1| - |2x - 1| + |2x - 3|$

8.   Establish the following inequalities:

(a) $4x^{1/4} \leq x + 3$ for $0 \leq x \leq 1$
(b) $1 - 1/x \leq \log_e x$ for $x \geq 1$

9.   Use L'Hôpital's rule(s) to evaluate the following:

(a) $\displaystyle\lim_{x \to 0} \frac{\sinh(x + \sin x)}{\sin x}$

(b) $\displaystyle\lim_{x \to 1} \frac{(5x + 3)^{1/3} - (x + 1)^{1/2}}{(x - 1)}$

(c) $\displaystyle\lim_{x \to 0} \frac{\sinh x - x}{\sin x^2}$

10.   Let f be differentiable on $[a, b]$, where $0 < a < b$. Prove that there exists a number c, $a < c < b$ such that

$$f(b) - f(a) = cf'(c) \log_e \left(\frac{b}{a}\right)$$

By taking $f(x) = x^{1/n}$, n an integer, $n \geq 1$, deduce that

$$\lim_{n \to \infty} n(a^{1/n} - 1) = \log_e a$$

11.   Let $f(x) = \sin(m \sin^{-1} x)$, m an integer, $|x| < 1$. Show that

(a) $(1 - x^2)f''(x) - xf'(x) + m^2 f(x) = 0$

(b) $(1 - x^2)f^{(n+2)}(x) - (2n + 1)xf^{(n+1)}(x) + (m^2 - n^2)f^{(n)}(x) = 0,$
$$n \geqslant 0$$

(*Hint*: differentiate both sides of (a) n times)

Hence show that

$$\sin(m \sin^{-1} x) = m\left[x + \frac{1^2 - m^2}{3!}x^3\right.$$

$$\left. + \frac{(1^2 - m^2)(3^2 - m^2)}{5!}x^5 + \cdots\right]$$

and deduce that if m is odd, then $\sin(m \sin^{-1} x)$ is a polynomial of degree m.

12.   Use standard series to determine the Maclaurin series for each of the following functions up to the term in x^6:

(a) $f(x) = \dfrac{\cosh x}{1 - x}$

(b) $f(x) = \log_e(1 + x + x^2)$ (*Hint*: factorize $1 - x^3$)

13.   Determine the nature of the stationary points of each of the following functions:

(a) $f(x) = \left(\dfrac{x}{1 + x}\right)^2$ (b) $f(x) = \sin x - \tan x$

(c) $f(x) = x^2 - \sin^2 x$ (d) $f(x) = \dfrac{4}{x} - \dfrac{1}{x - 1}$

14.   Determine the Taylor polynomial $T_{4,a}f(x)$ of degree 4 for f at a for each of the following functions:

(a) $f(x) = (1 + x)^{-2},$ $a = \frac{1}{2}$
(b) $f(x) = \cos x,$ $a = \frac{1}{3}\pi$
(c) $f(x) = \log_e(1 + x),$ $a = 2$

15.   Let

$$f(x) = \frac{x}{x + 3} \quad (x \neq -3)$$

Show that the Taylor polynomial of degree 3 for f at 2 approximates $f(x)$ to within 10^{-4} for all $x \in [2, \frac{5}{2}]$.

7

Integration

7.1 The Riemann integral
7.2 Techniques
7.3 Improper integrals

Historically, the idea of integral calculus developed before that of differential calculus. Whereas differentiation was created in connection with tangents to curves and maxima and minima of functions, integration arose as a summation process for finding certain areas and volumes. Archimedes in the third century BC had devised the ingenious **method of equilibrium** for finding the area of a planar region by subdividing the region into a large number of thin strips, and then hanging these pieces at one end of a lever so that equilibrium was established with a figure whose area and centroid were known. This idea was rediscovered in Europe in the seventeenth century and formulae for the areas between many standard curves were obtained. When Newton discovered the calculus in 1666, the inverse link between finding areas under curves and constructing tangents to related curves was firmly established, and Newton appears to be the first mathematician to formally treat integration as the reverse process to differentiation.

Section 7.1 begins by defining (indefinite) integration as 'antidifferentiation'. This is followed by a rigorous definition of the Riemann integral (or definite integral), the modern analytic analogue of Archimedes' method for finding areas. The twin concepts of antiderivative (or primitive) and the Riemann integral are related by the triumphant fundamental theorem of calculus (7.1.11), first stated by Newton in somewhat more flowing language as

> (the) method of tangents ... extends itself not only to the drawing of
> tangents to any curved lines ... but also to the resolving ... of problems
> about areas, lengths, centres of gravity etc.

Principia 1687

In Section 7.2 two of the standard techniques of integration are established: integration by parts and change of variables, which are respectively the reverse of the product rule and the composite rule for differentiation.

The definition of the Riemann integral (see 7.1.1–7.1.3) was coined in
the nineteenth century, and has been followed in this century by more
general notions of integration. The Riemann integral requires that the
function in question be (at least) defined and bounded on a closed
bounded interval. In Section 7.3 the effect of relaxing one or both of
these two boundedness conditions is explored. The resulting improper
integrals, which are limits of sequences of (proper) Riemann integrals and
arise in areas including mathematical physics and probability theory, are
referred to as convergent or divergent integrals depending on whether or
not the limit in question exists.

7.1 The Riemann integral

Given a function $f: A \to \mathbb{R}$, where A is an interval, any differentiable
function $F: A \to \mathbb{R}$ such that $F' = f$ is called a **primitive** of f. If F_1 and F_2
are two, possibly different, primitives for f then $F_1'(x) = F_2'(x)$, and so
$(F_1 - F_2)'(x) = 0$. By the increasing–decreasing theorem (6.2.3),
$F_1 - F_2$ is constant on A. Hence any two primitives of f differ by a
constant. Therefore the process of finding a primitive of a given function
f, if one exists, determines a function F that, up to an additive constant,
is unique. The notation $F = \int f$ will be used whenever F is a primitive
of f.

■■ EXAMPLE 1

The knowledge of simple differentiable functions and their derivatives enables a table of elementary functions and a corresponding
primitive to be written down (Table 7.1). ■

■■ EXAMPLE 2

Show that $\sin^{-1} x + \cos^{-1} x = \frac{1}{2}\pi$, for $x \in (-1, 1)$.

Table 7.1

f	$F = \int f$
$f(x) = x^n \ (n \in \mathbb{N})$	$F(x) = \dfrac{x^{n+1}}{n+1}$
$f(x) = \cos x$	$F(x) = \sin x$
$f(x) = \sin x$	$F(x) = -\cos x$
$f(x) = e^x$	$F(x) = e^x$
$f(x) = \cosh x$	$F(x) = \sinh x$
$f(x) = \sinh x$	$F(x) = \cosh x$
$f(x) = \dfrac{1}{1 + x^2}$	$F(x) = \tan^{-1} x$
$f(x) = \dfrac{1}{\sqrt{1 + x^2}}$	$F(x) = \sinh^{-1} x$
$f(x) = x^t \ (t \in \mathbb{R}, \ t \neq -1)$	$F(x) = \dfrac{x^{t+1}}{t+1} \qquad (x \in (0, \infty))$
$f(x) = \dfrac{1}{x}$	$F(x) = \log_e |x| \qquad (x \neq 0)$
$f(x) = \dfrac{1}{\sqrt{1 - x^2}}$	$F(x) = \sin^{-1} x \qquad (x \in (-1, 1))$
$f(x) = \dfrac{1}{\sqrt{x^2 - 1}}$	$F(x) = \cosh^{-1} x \qquad (x \in (1, \infty))$

Solution

Example 8 of Section 6.1 showed that the derivative of $\sin^{-1} x$ is $1/\sqrt{1 - x^2}$ for $|x| < 1$. An analogous argument applied to the bijection from $(0, \pi)$ onto $(-1, 1)$ given by $x \mapsto \cos x$ yields that $\cos^{-1} x$ has derivative $-1/\sqrt{1 - x^2}$ for $|x| < 1$.

Thus $F_1(x) = \sin^{-1} x$ and $F_2(x) = -\cos^{-1} x$ are both primitives of

$$f(x) = \frac{1}{\sqrt{1 - x^2}} \qquad (|x| < 1)$$

Thus

$$F_1(x) - F_2(x) = \sin^{-1} x + \cos^{-1} x$$

is constant for $|x| < 1$. But $\sin^{-1} 0 + \cos^{-1} 0 = 0 + \tfrac{1}{2}\pi$. Hence

$$\sin^{-1} x + \cos^{-1} x = \tfrac{1}{2}\pi \quad \text{for} \quad x \in (-1, 1) \qquad \blacksquare$$

Two questions arise concerning primitives. First, given a function $f : A \to \mathbb{R}$, does f possess a primitive? Secondly, if f has a primitive, how

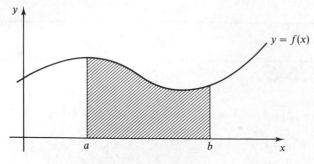

Figure 7.1

is it calculated? Techniques relevant to the second question are discussed in Section 7.2. In order to tackle the first, it is necessary to digress to a discussion of area. It seems reasonable to suppose that there is a well-defined area enclosed by the graph of a continuous function, the x-axis and the ordinates $x = a$ and $x = b$ (see Figure 7.1); this will turn out to be the case for any reasonable function. It is intended to give one form of the definition of the **Riemann integral** $\int_a^b f(x)\,dx$, which will assign a value to the aforementioned area. The definition involves the areas of rectangles and applies to a wider class of functions than continuous ones.

7.1.1 Definition

> Let $[a, b]$ be a given finite interval. A **partition** P of $[a, b]$ is a finite set of points $\{x_0, x_1, \ldots, x_n\}$ satisfying $a = x_0 < x_1 < \ldots < x_n = b$.

Suppose now that f is a function defined and bounded on $[a, b]$ (if f were continuous on $[a, b]$, this would certainly be the case by 5.3.1). Then f is bounded on each of the subintervals $[x_{i-1}, x_i]$. Hence f has a least upper bound M_i and a greatest lower bound m_i on $[x_{i-1}, x_i]$. See Figure 7.2.

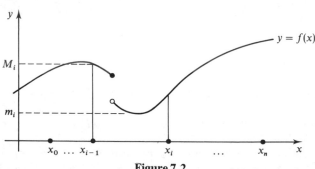

Figure 7.2

7.1.2 Definition

The **upper sum of f relative to P** is defined by

$$U(P) = \sum_{i=1}^{n} M_i(x_i - x_{i-1})$$

where

$$M_i = \sup\{f(x) : x_{i-1} \leqslant x \leqslant x_i\}$$

The **lower sum of f relative to P** is defined by

$$L(P) = \sum_{i=1}^{n} m_i(x_i - x_{i-1})$$

where

$$m_i = \inf\{f(x) : x_{i-1} \leqslant x \leqslant x_i\}$$

As can be seen from Figure 7.3, the area under the graph of f between $x = a$ and $x = b$ lies between $L(P)$ and $U(P)$. Now f is bounded above

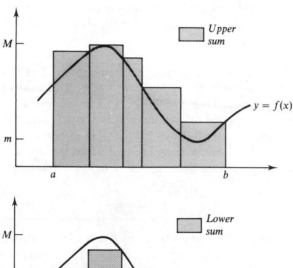

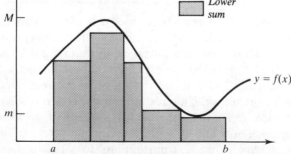

Figure 7.3

and below on the whole of $[a, b]$. So there exist numbers M and m with $m \leqslant f(x) \leqslant M$ for all $x \in [a, b]$. Thus for any partition P of $[a, b]$

$$L(P) \leqslant M \sum_{i=1}^{n}(x_i - x_{i-1}) = M(b - a)$$

and

$$U(P) \geqslant m \sum_{i=1}^{n}(x_i - x_{i-1}) = m(b - a)$$

Hence the set

$$L = \{L(P) : P \text{ is a partition of } [a, b]\}$$

is bounded above and the set

$$U = \{U(P) : P \text{ is a partition of } [a, b]\}$$

is bounded below. So $\mathcal{L} = \sup L$ and $\mathcal{U} = \inf U$ exist.

7.1.3   Definition

> A function defined and bounded on $[a, b]$ is **Riemann integrable** on $[a, b]$ if and only if $\mathcal{L} = \mathcal{U}$. This common value is denoted by $\int_a^b f(x)\, dx$.

The reason for introducing this notion of integrability is that there are functions, as will be seen later, for which $\mathcal{L} \neq \mathcal{U}$. Luckily, as will be proved in due course, there are large classes of functions for which $\mathcal{L} = \mathcal{U}$, including continuous functions and monotone functions. A function f is **monotone** on an interval $[a, b]$ if f is either increasing on the whole of $[a, b]$ or else decreasing on $[a, b]$. Before looking at some examples, the intuitively obvious fact that $\mathcal{L} \leqslant \mathcal{U}$ for a bounded function is established.

Georg Bernhard Riemann (1826–1866), a German mathematician, was one of the leaders of the nineteenth-century drive towards greater abstraction and generalization in mathematics. His doctoral thesis in 1851 led to the concept of a Riemann surface, and in a famous lecture in 1854 he laid the foundations necessary for a unified view of geometry. This resulted in generalized ideas of space, which in turn generated the modern concept of an abstract space and the subject now known as differential geometry. In his relatively short life Riemann made contributions to virtually every area of pure mathematics, notably in analysis, where he was the first to find necessary and sufficient conditions for a bounded function to be integrable; in honour of this, the Riemann integral is named after him. This notion of integrability was generalized during the twentieth century but Riemann integration still dominates the undergraduate mathematics curriculum today.

7.1.4   Theorem

If f is defined and bounded on $[a, b]$ then $\mathscr{L} \leqslant \mathscr{U}$.

Proof

Let P be a partition of $[a, b]$ and P' be the partition $P \cup \{y\}$, where $x_{r-1} < y < x_r$ for one particular r, $1 \leqslant r \leqslant n$. In other words, P' is obtained by adding one more point to P. It is now shown that $L(P) \leqslant L(P')$ and $U(P) \geqslant U(P')$. Let

$$M'_r = \sup \{f(x) : x_{r-1} \leqslant x \leqslant y\}$$

and

$$M''_r = \sup \{f(x) : y \leqslant x \leqslant x_r\}$$

Clearly $M'_r \leqslant M_r$ and $M''_r \leqslant M_r$. Hence

$$U(P') = \sum_{i=1}^{r-1} M_i(x_i - x_{i-1}) + M'_r(y - x_{r-1})$$

$$+ M''_r(x_r - y) + \sum_{i=r+1}^{n} M_i(x_i - x_{i-1})$$

$$\leqslant \sum_{i=1}^{r-1} M_i(x_i - x_{i-1}) + M_r(y - x_{r-1})$$

$$+ M_r(x_r - y) + \sum_{i=r+1}^{n} M_i(x_i - x_{i-1})$$

$$= \sum_{i=1}^{n} M_i(x_i - x_{i-1}) = U(P)$$

In a similar fashion it can be shown that $L(P) \leqslant L(P')$.

It now follows that if $P'' = P \cup \{y_1, y_2, \ldots, y_m\}$, where the y_i are distinct numbers in $[a, b]$, then $L(P) \leqslant L(P'')$ and $U(P) \geqslant U(P'')$. Now suppose that P_1 and P_2 are two partitions of $[a, b]$ and let $P_3 = P_1 \cup P_2$. Thus $L(P_1) \leqslant L(P_3)$ and $U(P_2) \geqslant U(P_3)$. Since $L(P_3) \leqslant U(P_3)$, it can be deduced that $L(P_1) \leqslant U(P_2)$. In other words, the lower sum relative to a given partition of $[a, b]$ does not exceed the upper sum relative to any partition of $[a, b]$.

Hence every lower sum is a lower bound for the set of upper sums. So $L(P) \leqslant \mathcal{U}$ for all possible partitions P. But then $\mathcal{U}$ is an upper bound for the set of lower sums. Thus $\mathcal{L} \leqslant \mathcal{U}$. ☐

■■ EXAMPLE 3

Prove that $f(x) = x$ is Riemann integrable on $[0, 1]$.

Solution

For each $n \in \mathbb{N}$ let P_n be the partition of $[0, 1]$ given by

$$P_n = \left\{0, \frac{1}{n}, \frac{2}{n}, \ldots, 1\right\}$$

Thus P_n subdivides $[0, 1]$ into n subintervals each of length $1/n$. Hence $m_i = f(x_{i-1}) = (i-1)/n$ and $M_i = f(x_i) = i/n$. Now

$$U(P_n) = \sum_{i=1}^{n} \frac{i}{n} \frac{1}{n}$$

$$= \frac{1}{n^2} \sum_{i=1}^{n} i$$

$$= \frac{n(n+1)}{2n^2}$$

$$= \frac{n+1}{2n}$$

and similarly

$$L(P_n) = \frac{(n-1)}{2n}$$

So

$$\frac{n-1}{2n} \leqslant \mathcal{L} \leqslant \mathcal{U} \leqslant \frac{n+1}{2n}$$

and, by letting $n \to \infty$, it can be deduced that $\mathcal{L} = \mathcal{U} = \frac{1}{2}$. In other words, f is Riemann integrable on $[0, 1]$ and $\int_0^1 x \, dx = \frac{1}{2}$. ■

■■ EXAMPLE 4

Show that

$$f(x) = \begin{cases} 1 & \text{if } x \text{ is rational} \\ 0 & \text{if } x \text{ is irrational} \end{cases}$$

is not Riemann integrable on any interval $[a, b]$.

Solution

For any partition P it follows that $L(P) = 0$ and $U(P) = b - a$, since any interval of real numbers contains infinitely many rationals and irrationals. Hence $\mathscr{L} = 0$ and $\mathscr{U} = b - a$, and so $\mathscr{L} \neq \mathscr{U}$. ∎

In Example 3, the difference between the upper and lower sums relative to the partition $P_n = \{0, 1/n, 2/n, \ldots, 1\}$ is given by

$$U(P_n) - L(P_n) = \frac{n+1}{2n} - \frac{n-1}{2n} = \frac{1}{n}$$

The fact that $1/n \to 0$ as $n \to \infty$ was sufficient to enable the condition $\mathscr{L} = \mathscr{U}$ to be established. This suggests that in order to establish that a bounded function is Riemann integrable on $[a, b]$, if indeed it is, it should suffice to show that for any $\varepsilon > 0$ there is a partition P of the interval $[a, b]$ such that $U(P) - L(P) < \varepsilon$. This turns out to be a necessary and sufficient condition for a bounded function to be Riemann integrable on $[a, b]$, as the following theorem demonstrates.

7.1.5 Riemann's condition

> Let f be defined and bounded on $[a, b]$. Then f is Riemann integrable on $[a, b]$ if and only if for every $\varepsilon > 0$ there exists a partition P of $[a, b]$ such that $U(P) - L(P) < \varepsilon$.

Proof

Suppose that f is bounded and Riemann integrable on $[a, b]$, and $\varepsilon > 0$ is given. Since $\mathscr{L} = \sup \{L(P) : P$ is a partition of $[a, b]\}$ exists, there is a partition P_1 of $[a, b]$ such that

$$L(P_1) > \mathscr{L} - \tfrac{1}{2}\varepsilon$$

Since $\mathscr{U} = \inf \{U(P) : P$ is a partition of $[a, b]\}$ exists, there is a partition P_2 of $[a, b]$ such that

$$U(P_2) < \mathscr{U} + \tfrac{1}{2}\varepsilon$$

Let $P = P_1 \cup P_2$, another partition of $[a, b]$. Then, arguing as in the proof of 7.1.4,

$$U(P) \leqslant U(P_2)$$

and

$$L(P) \geqslant L(P_1)$$

Hence

$$U(P) - L(P) \leqslant U(P_2) - L(P_1)$$
$$< (\mathcal{U} + \tfrac{1}{2}\varepsilon) - (\mathcal{L} - \tfrac{1}{2}\varepsilon)$$
$$= \varepsilon$$

since $\mathcal{L} = \mathcal{U}$.

For the converse, assume that for a given $\varepsilon > 0$ there is a partition P of $[a, b]$ such that $U(P) - L(P) < \varepsilon$. Then

$$L(P) \leqslant \mathcal{L} \leqslant \mathcal{U} \leqslant U(P)$$

Hence

$$0 \leqslant \mathcal{U} - \mathcal{L}$$
$$\leqslant U(P) - L(P)$$
$$< \varepsilon$$

Since ε is arbitrarily small (see Question 6, Problems 2), $\mathcal{L} = \mathcal{U}$, and so f is Riemann integrable on $[a, b]$. $\qquad\square$

Riemann's condition can now be used to prove that monotone functions are Riemann integrable.

7.1.6 Theorem

> If f is monotone on $[a, b]$ then f is Riemann integrable on $[a, b]$.

Proof

Suppose that f is increasing on $[a, b]$ and let P_n be the partition given by

$$a = x_0 < x_1 < \ldots < x_n = b$$

where the x_i are chosen so that $x_i - x_{i-1} = (b - a)/n$ for $i = 1, 2, \ldots, n$. In other words, P partitions $[a, b]$ into n subintervals of equal length. Since f is increasing on each subinterval $[x_{i-1}, x_i]$,

$$M_i = \sup \{f(x) : x_{i-1} \leqslant x \leqslant x_i\} = f(x_i)$$

and

$$m_i = \inf \{f(x) : x_{i-1} \leqslant x \leqslant x_i\} = f(x_{i-1})$$

Hence

$$U(P_n) - L(P_n) = \sum_{i=1}^{n}[f(x_i) - f(x_{i-1})](x_i - x_{i-1})$$

$$= \sum_{i=1}^{n}[f(x_i) - f(x_{i-1})]\frac{b - a}{n}$$

$$= \frac{b - a}{n}\sum_{i=1}^{n}[f(x_i) - f(x_{i-1})]$$

$$= \frac{(b - a)[f(b) - f(a)]}{n}$$

Given any $\varepsilon > 0$,

$$\frac{(b - a)[f(b) - f(a)]}{n} < \varepsilon$$

provided that

$$n > \frac{(b - a)[f(b) - f(a)]}{\varepsilon}$$

By the Archimedean postulate (2.3.1), there is an integer N exceeding $(b - a)[f(b) - f(a)]/\varepsilon$. Hence for the partition P_N

$$U(P_N) - L(P_N) < \varepsilon$$

By 7.1.5, f is Riemann integrable on $[a, b]$, as claimed. (The proof in the case that f is decreasing is similar.)   □

Another important consequence of Riemann's condition is that all continuous functions are Riemann integrable. The proof of this result will also require some deeper properties of continuous functions than those covered in Chapter 5. Recall that if a function f is continuous on an interval $[a, b]$ then it is continuous for each x', $a < x' < b$, and it is right-continuous at a and left-continuous at b. So for every x', $a < x' < b$, and for every $\varepsilon > 0$ there exists a $\delta > 0$ such that

$$|x - x'| < \delta \Rightarrow |f(x) - f(x')| < \varepsilon \tag{*}$$

If $x' = a$ or b, minor adjustments are required to (∗) in order to ensure that $x \in [a, b]$. The δ depends of course on the ε given, but it also depends on the x' under consideration; for a fixed $\varepsilon > 0$ each different x' may require a different $\delta > 0$ in order for (∗) to be satisfied. This is a direct consequence of the definition of continuity adopted in Section 5.2, where it was the concept of *continuity at a point* that was first formulated.

If a function f is such that (*) holds, where the δ can be chosen *independently* of x', then f is said to be uniformly continuous on $[a, b]$.

7.1.7 Definition

A function f defined on the interval $[a, b]$, is **uniformly continuous** if for any given $\varepsilon > 0$ there exists a $\delta > 0$ such that for all $x, y \in [a, b]$

$$|x - y| < \delta \Rightarrow |f(x) - f(y)| < \varepsilon$$

The δ in this definition depends on ε, but it does *not* depend on x or y. The concept of uniform continuity is therefore stronger than the concept of continuity in the sense that all functions that are uniformly continuous on $[a, b]$ are automatically continuous on $[a, b]$. What is surprising is that for a closed bounded interval $[a, b]$ the two concepts are in fact equivalent.

7.1.8 Uniform continuity theorem

Let f be defined and continuous on $[a, b]$. Then f is uniformly continuous on $[a, b]$.

Proof

The proof is by contradiction. Suppose that f is continuous on $[a, b]$ but not uniformly continuous on $[a, b]$. Then, the negation of Definition 7.1.7, implies that there exists an $\varepsilon > 0$ such that for every $\delta > 0$, there are $x, y \in [a, b]$, depending on δ, such that

$$|x - y| < \delta \quad \text{and} \quad |f(x) - f(y)| \geq \varepsilon$$

In particular, when $\delta = 1$, there exist $x_1, y_1 \in [a, b]$ such that

$$|x_1 - y_1| < 1 \quad \text{and} \quad |f(x_1) - f(y_1)| \geq \varepsilon$$

When $\delta = \frac{1}{2}$, there exist $x_2, y_2 \in [a, b]$ such that

$$|x_2 - y_2| < \tfrac{1}{2} \quad \text{and} \quad |f(x_2) - f(y_2)| \geq \varepsilon$$

In general, for every integer $n \geq 1$ there exist $x_n, y_n \in [a, b]$ such that

$$|x_n - y_n| < \tfrac{1}{2} \quad \text{and} \quad |f(x_n) - f(y_n)| \geq \varepsilon \tag{†}$$

Consider the sequence (x_n). Since each $x_n \in [a, b]$, the sequence (x_n) is bounded. Hence, by the Bolzano–Weierstrass theorem (3.4.2), (x_n) contains a convergent subsequence (x_{n_r}). Suppose that $x_{n_r} \to c$ as $r \to \infty$. Since $|x_{n_r} - y_{n_r}| < 1/n_r \to 0$ as $r \to \infty$, it also follows that $y_{n_r} \to c$ as $r \to \infty$. Now f is continuous at c, and so $\lim_{r \to \infty} f(x_{n_r}) = f(c)$ and $\lim_{r \to \infty} f(y_{n_r}) = f(c)$, by the composite rule for sequences (3.1.4). (If $c = a$ or b, the left-continuity of f at b or the right-continuity of f at a allow the same conclusions to be drawn.) Hence for the ε in question there exists an integer n_k such that

$$|f(x_{n_k}) - f(c)| < \tfrac{1}{2}\varepsilon \quad \text{and} \quad |f(y_{n_k}) - f(c)| < \tfrac{1}{2}\varepsilon$$

But then

$$\begin{aligned}
|f(x_{n_k}) - f(y_{n_k})| &= |[f(x_{n_k}) - f(c)] - [f(y_{n_k}) - f(c)]| \\
&\leq |f(x_{n_k}) - f(c)| + |f(y_{n_k}) - f(c)| \\
&< \tfrac{1}{2}\varepsilon + \tfrac{1}{2}\varepsilon = \varepsilon
\end{aligned}$$

But this contradicts the condition (†), and so the original assumption that f is not uniformly continuous is false. Hence continuity on $[a, b]$ implies uniform continuity on $[a, b]$.   □

It is worth mentioning at this point that if the closed bounded interval $[a, b]$ in 7.1.8 is replaced by an infinite interval such as $[0, \infty)$ then the result is false. For example, it can be shown that the function $x \mapsto x^2$, which is certainly continuous on $[0, \infty)$, is not uniformly continuous on $[0, \infty)$.

The uniform continuity theorem and Riemann's condition play a key role in the next result, which establishes the fact that continuous functions are Riemann integrable.

7.1.9   Theorem

> If f is continuous on $[a, b]$ then f is Riemann integrable on $[a, b]$.

Proof

Let $\varepsilon > 0$ be given. Then, for $a < b$, $\varepsilon/(b - a) > 0$. By 7.1.8, f is uniformly continuous on $[a, b]$, and so there exists a $\delta > 0$ such that for all $x, y \in [a, b]$

$$|x - y| < \delta \Rightarrow |f(x) - f(y)| < \frac{\varepsilon}{b - a}$$

Choose a partition P of $[a, b]$ in which $x_i - x_{i-1} < \delta$ for $i = 1, 2, \ldots, n$. Then

$$U(P) - L(P) = \sum_{i=1}^{n} (M_i - m_i)(x_i - x_{i-1})$$

where

$$M_i = \sup \{f(x) : x_{i-1} \leqslant x \leqslant x_i\}$$

and

$$m_i = \inf \{f(x) : x_{i-1} \leqslant x \leqslant x_i\}$$

Since $x_i - x_{i-1} < \delta$,

$$|f(x) - f(y)| < \frac{\varepsilon}{b - a} \quad \text{for all } x, y \in [x_{i-1}, x_i]$$

Hence

$$M_i - m_i < \frac{\varepsilon}{b - a} \quad \text{for } i = 1, 2, \ldots, n$$

Therefore

$$U(P) - L(P) < \frac{\varepsilon}{b - a} \sum_{i=1}^{n} (x_i - x_{i-1})$$

$$< \frac{\varepsilon}{b - a} (b - a) = \varepsilon$$

By Riemann's condition (7.1.5), f is Riemann integrable on $[a, b]$.

$\square$

7.1.10 Properties of the Riemann integral

If f and g are Riemann integrable on $[a, b]$ then all the following integrals exist and have the indicated properties:

(1) $\displaystyle\int_a^b [\alpha(f(x) + \beta g(x)] \, dx = \alpha \int_a^b f(x) \, dx + \beta \int_a^b g(x) \, dx$

$$(\alpha, \beta \in \mathbb{R})$$

(2) $\displaystyle\int_a^b f(x) \, dx = \int_a^c f(x) \, dx + \int_c^b f(x) \, dx \quad (a \leqslant c \leqslant b)$

(3) If $f(x) \leqslant g(x)$ on $[a, b]$ then $\displaystyle\int_a^b f(x) \, dx \leqslant \int_a^b g(x) \, dx$

(4) $\displaystyle\left| \int_a^b f(x) \, dx \right| \leqslant \int_a^b |f(x)| \, dx$

Note The Riemann integral $\int_a^b f(x)\,dx$ is only one way of assigning areas to bounded regions. There are others, notably the Lebesgue integral; all, however, give the same 'answer' for areas under the graphs of continuous functions. The properties detailed in 7.1.10 are essentially properties of areas. Property (1) is described as the **linearity of the integral** and (2) is called the **additive property**.

Proof of (2)

Let P_1 and P_2 be partitions of $[a, c]$ and $[c, b]$ respectively. Then $P = P_1 \cup P_2$ is a partition of $[a, b]$. Clearly, $L(P) = L(P_1) + L(P_2)$. Let

$$\mathcal{L}_1 = \sup\{L(P_1) : P_1 \text{ is a partition of } [a, c]\}$$

and

$$\mathcal{L}_2 = \sup\{L(P_2) : P_2 \text{ is a partition of } [c, b]\}$$

By definition,

$$L(P_1) + L(P_2) \leqslant \int_a^b f(x)\,dx$$

Hence

$$L(P_1) \leqslant \int_a^b f(x)\,dx - L(P_2)$$

and so

$$\mathcal{L}_1 \leqslant \int_a^b f(x)\,dx - L(P_2)$$

Hence

$$L(P_2) \leqslant \int_a^b f(x)\,dx - \mathcal{L}_1$$

and so

$$\mathcal{L}_2 \leqslant \int_a^b f(x)\,dx - \mathcal{L}_1, \quad \text{or} \quad \mathcal{L}_1 + \mathcal{L}_2 \leqslant \int_a^b f(x)\,dx$$

Now consider upper sums and observe that $U(P) = U(P_1) + U(P_2)$. Let

$$\mathcal{U}_1 = \inf\{U(P_1) : P_1 \text{ is a partition of } [a, c]\}$$

and

$$\mathcal{U}_2 = \inf\{U(P_2) : P_2 \text{ is a partition of } [c, b]\}$$

By definition,

$$U(P_1) + U(P_2) \geqslant \int_a^b f(x)\,dx$$

Hence

$$U(P_1) \geqslant \int_a^b f(x)\,dx - U(P_2)$$

and so

$$\mathcal{U}_1 \geqslant \int_a^b f(x)\,dx - U(P_2)$$

Hence

$$U(P_2) \geqslant \int_a^b f(x)\,dx - \mathcal{U}_1$$

and so

$$\mathcal{U}_2 \geqslant \int_a^b f(x)\,dx - \mathcal{U}_1, \quad \text{or} \quad \mathcal{U}_1 + \mathcal{U}_2 \geqslant \int_a^b f(x)\,dx$$

Thus

$$\mathcal{L}_1 + \mathcal{L}_2 \leqslant \int_a^b f(x)\,dx \leqslant \mathcal{U}_1 + \mathcal{U}_2$$

Since f is Riemann integrable on $[a, b]$, Riemann's condition (7.1.5) gives that, for any $\varepsilon > 0$, P can be chosen such that $U(P) - L(P) < \varepsilon$. By adjoining c if necessary, it can be assumed that P contains c. Then $P_1 = P \cap [a, c]$ and $P_2 = P \cap [c, b]$ are partitions of $[a, c]$ and $[c, b]$ respectively. Now

$$[U(P_1) - L(P_1)] + [U(P_2) - L(P_2)] = [U(P_1) + U(P_2)]$$
$$- [L(P_1) + L(P_2)]$$
$$= U(P) - L(P) < \varepsilon$$

Hence $0 \leqslant U(P_1) - L(P_1) < \varepsilon$ and $0 \leqslant U(P_2) - L(P_2) < \varepsilon$. By 7.1.5, f is Riemann integrable on both $[a, c]$ and $[c, b]$. Hence

$$\mathscr{L}_1 = \mathscr{U}_1 = \int_a^c f(x) \, dx \quad \text{and} \quad \mathscr{L}_2 = \mathscr{U}_2 = \int_c^b f(x) \, dx$$

and the additive property now follows from the inequalities

$$\mathscr{L}_1 + \mathscr{L}_2 = \int_a^b f(x) \, dx \leqslant \mathscr{U}_1 + \mathscr{U}_2$$

The proofs of (1), (3) and (4) are similar. See Exercises 7.1. $\square$

The concepts of antiderivative (or primitive) and the Riemann integral may now be formally related in the central result of this section, the fundamental theorem of calculus.

7.1.11   The fundamental theorem of calculus

If f is Riemann integrable on $[a, b]$ and $F(x) = \int_a^x f(t) \, dt$ then F is continuous on $[a, b]$. Furthermore, if f is continuous on $[a, b]$ then F is differentiable on $[a, b]$ and $F' = f$.

Proof

Since f is integrable, it is bounded on $[a, b]$. So there exists some number M with $|f(t)| \leqslant M$ for all $t \in [a, b]$. For fixed $c \in [a, b]$

$$|F(x) - F(c)| = \left| \int_a^x f(t) \, dt - \int_a^c f(t) \, dt \right|$$

$$= \left| \int_c^x f(t) \, dt \right| \quad \text{by 7.1.10(2)}$$

$$\leqslant \left| \int_c^x |f(t)| \, dt \right| \quad \text{by 7.1.10(4)}$$

$$\leqslant \left| \int_c^x M \, dt \right| \quad \text{by 7.1.10(3)}$$

Since the constant function $x \mapsto M$ has

$$U(P) = M|x - c|$$

for the trivial partition P of the interval with endpoints x and c,

$$|F(x) - F(c)| \leqslant M|x - c|$$

Now, given $\varepsilon > 0$, choose $\delta = \varepsilon/M$. Hence

$$|x - c| < \delta \Rightarrow |F(x) - F(c)| < \varepsilon$$

In other words, F is continuous at c, and, since c was arbitrary, F is continuous on $[a, b]$.

Now assume further that f is continuous on $[a, b]$. Let $c \in [a, b]$ and consider $x \in [a, b]$. Then

$$\left| \frac{F(x) - F(c)}{x - c} - f(c) \right| = \left| \frac{\int_a^x f(t)\,dt - \int_a^c f(t)\,dt}{x - c} - f(c) \right|$$

$$\leq \left| \frac{\int_c^x f(t)\,dt}{x - c} - f(c) \right| \qquad \text{by 7.1.10(2)}$$

$$= \left| \frac{\int_c^x [f(t) - f(c)]\,dt}{x - c} \right|$$

$$\text{since } f(c) \text{ is a constant}$$

$$\leq \left| \frac{\int_c^x |f(t) - f(c)|\,dt}{x - c} \right| \qquad \text{by 7.1.10(4)}$$

Given $\varepsilon > 0$, there exists a $\delta > 0$ such that $|f(t) - f(c)| < \varepsilon$ for $|t - c| < \delta$. Hence for $0 < |x - c| < \delta$

$$\left| \frac{F(x) - F(c)}{x - c} - f(c) \right| \leq \left| \frac{\int_c^x |f(t) - f(c)|\,dt}{x - c} \right| \leq \left| \frac{\int_c^x \varepsilon\,dt}{x - c} \right|$$

$$< \varepsilon$$

In other words, $F'(c) = f(c)$. (The derivative is one-sided if $c = a$ or b.) Hence f is differentiable and $F' = f$. $\qquad\square$

This is a theorem *par excellence*, since it not only states that every continuous function possesses a primitive, answering one of the questions posed earlier, but it gives a method for evaluating the area $\int_a^b f(x)\,dx$. Suppose then that it is required to evaluate $\int_{x_1}^{x_2} f(t)\,dt$, where x_1, $x_2 \in [a, b]$ and f is continuous on $[a, b]$. Now

$$\int_{x_1}^{x_2} f(t)\,dt = \int_a^{x_2} f(t)\,dt - \int_a^{x_1} f(t)\,dt$$

using the additive property (7.1.10(2)). By the fundamental theorem,

$$\int_{x_1}^{x_2} f(t) \, dt = F(x_2) - F(x_1) = [F(x)]_{x_1}^{x_2}$$

where $F(x) = \int_a^x f(t) \, dt$. Since any two primitives of f differ by a constant, F may be chosen, in fact, to be *any* primitive of f and not just the particular one arising in 7.1.11. Note that this method of evaluating definite integrals hinges on the ability to determine a primitive of f. Unfortunately, most functions do not possess primitives expressible in terms of the elementary functions alone. In such cases it is necessary to settle for numerical estimates.

■■   EXAMPLE 5

Determine the area bounded by the function $f(x) = x^3 + 2$, the x-axis and the ordinates $x = 0$ and $x = 1$. See Figure 7.4(a).

Solution

$$\text{Area} = \int_0^1 f(x) \, dx = \int_0^1 (x^3 + 2) \, dx$$

$$= [\tfrac{1}{4}x^4 + 2x]_0^1 = \tfrac{9}{4} - 0 = \tfrac{9}{4} \qquad ■$$

A note of caution is in order here. The definition of the Riemann integral involved rectangles whose 'heights' were given by certain values of the function f. If any of these values are negative, the rectangle involved contributes negatively to the upper or lower sum being calculated. The result is that the Riemann integral counts areas below the x-axis negatively.

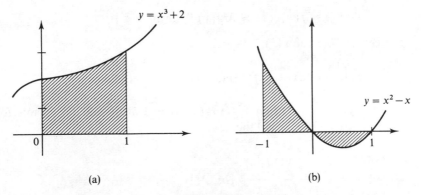

(a)   (b)

Figure 7.4

■■ EXAMPLE 6

Determine the area between the graph of $f(x) = x^2 - x$, the x-axis and the ordinates $x = -1$ and $x = 1$. See Figure 7.4(b).

Solution

First,

$$\int_{-1}^{0} f(x) \, dx = [\tfrac{1}{3}x^3 - \tfrac{1}{2}x^2]_{-1}^{0} = \frac{5}{6}$$

and

$$\int_{0}^{1} f(x) \, dx = [\tfrac{1}{3}x^3 - \tfrac{1}{2}x^2]_{0}^{1} = -\frac{1}{6}$$

and so the required area is 1. ■

The next result is useful for estimating the value of an integral when the analytic techniques surveyed in the next section fail.

7.1.12 The integral mean value theorem

Let f and g be continuous on $[a, b]$ with $g(x) \geq 0$ for $x \in [a, b]$. Then there exists a c between a and b with

$$\int_{a}^{b} f(x)g(x) \, dx = f(c) \int_{a}^{b} g(x) \, dx$$

Proof

By the interval theorem (5.3.3) applied to f on the interval $[a, b]$, $m \leq f(x) \leq M$ for all $x \in [a, b]$, where m is the infimum and M the supremum of f on $[a, b]$. Since $g(x) \geq 0$,

$$mg(x) \leq f(x)g(x) \leq Mg(x) \quad \text{for all } x \in [a, b]$$

By 7.1.10(1) and (3),

$$m \int_{a}^{b} g(x) \, dx \leq \int_{a}^{b} f(x)g(x) \, dx \leq M \int_{a}^{b} g(x) \, dx$$

If $\int_{a}^{b} g(x) \, dx = 0$ then $\int_{a}^{b} f(x)g(x) \, dx = 0$ also, and the result follows. If $\int_{a}^{b} g(x) \, dx \neq 0$ then

$$K = \frac{\displaystyle\int_{a}^{b} f(x)g(x) \, dx}{\displaystyle\int_{a}^{b} g(x) \, dx} \in [m, M]$$

Now, by the intermediate value property (5.3.2) applied to f on the interval $[a, b]$, there exists a c between a and b with $f(c) = K$. Hence

$$f(c) = \frac{\displaystyle\int_a^b f(x)g(x)\,dx}{\displaystyle\int_a^b g(x)\,dx}$$

and the result now follows.   □

7.1.13 Corollary

If f is continuous on $[a, b]$ then there exists a $c \in [a, b]$ such that

$$\int_a^b f(x)\,dx = f(c)(b - a)$$

Proof

Put $g(x) = 1$ in the integral mean value theorem.   □

This section concludes by providing the proof of the integral test for series (4.2.5).

Proof of 4.2.5

Suppose that $f: \mathbb{R}^+ \to \mathbb{R}^+$ is a decreasing function. Then $f(r + 1) \leqslant f(x) \leqslant f(r)$ for all $x \in [r, r + 1]$, r an integer. Since f is monotone, f is Riemann integrable on $[r, r + 1]$, by 7.1.6. By 7.1.10(3),

$$\int_r^{r+1} f(r + 1)\,dx \leqslant \int_r^{r+1} f(x)\,dx \leqslant \int_r^{r+1} f(r)\,dx$$

Hence

$$f(r + 1) \leqslant \int_r^{r+1} f(x)\,dx \leqslant f(r)$$

and so

$$\sum_{r=2}^{n+1} f(r) = \sum_{r=1}^{n} f(r + 1) \leqslant \int_1^{n+1} f(x)\,dx \leqslant \sum_{r=1}^{n} f(r)$$

Now let $a_n = f(n)$ and $j_n = \int_1^n f(x)\,dx$. Then

$$\sum_{r=2}^{n} a_r \leqslant j_n \leqslant \sum_{r=1}^{n-1} a_r$$

If (j_n) converges then the nth partial sums of $\sum_{r=1}^{\infty} a_r$ are increasing and bounded above. By 3.4.1, $\sum_{r=1}^{\infty} a_r$ is a convergent series. Conversely, if $\sum_{r=1}^{\infty} a_r$ is a convergent series then (j_n) is increasing and bounded above, and hence is a convergent sequence. $\quad\square$

Exercises 7.1

1. Find primitives of the following functions:

   (a) $f(x) = x^2 + 3x - 2$ (b) $f(x) = 1 + \cos 3x$

   (c) $f(x) = e^x \cosh 2x$ (d) $f(x) = \dfrac{1}{\sqrt{9 - x^2}}$

   (e) $f(x) = |x|$

2. Let $f(x) = x^2$ and P_n be the partition $\{0, 1/n, 2/n, \ldots, 1\}$ of $[0, 1]$. Calculate $L(P_n)$ and $U(P_n)$. Hence prove that f is Riemann integrable on $[0, 1]$.

3. Given that $f(x) = 1/x$ is Riemann integrable on $[1, 2]$, calculate $L(P_n)$ and $U(P_n)$, where

   $$P_n = \left\{1, 1 + \frac{1}{n}, 1 + \frac{2}{n}, \ldots, 2\right\}$$

   Deduce that

   $$\frac{1}{n+1} + \frac{1}{n+2} + \ldots + \frac{1}{2n} \to \log_e 2 \quad \text{as } n \to \infty$$

4. Prove Properties (3) and (4) of 7.1.10.

5. Let

   $$J_n = \frac{1}{n} \int_0^1 \frac{\sin nx}{1 + x^2} \, dx$$

   Use the properties of the Riemann integral to show that $|J_n| \le \pi/4n$. Hence evaluate $\lim_{n \to \infty} J_n$.

6. What is wrong with the following argument? By the fundamental theorem of calculus,

   $$\int_0^2 \frac{1}{(x-1)^2} \, dx = \left[\frac{1}{1-x}\right]_0^2 = -2$$

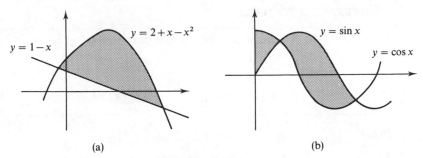

$y = 1 - x$ $y = 2 + x - x^2$ $y = \sin x$ $y = \cos x$

(a)   (b)

Figure 7.5

7. If $f(x) \geq g(x)$ on the interval $[a, b]$, show that the area between the graphs of f and g and the ordinates $x = a$ and $x = b$ is given by $\int_a^b [f(x) - g(x)]\, dx$. Hence calculate the areas shown in Figure 7.5.

8. Let f be continuous on $[0, 1]$ and

$$J_1 = \int_0^{1/\sqrt{n}} \frac{nf(x)}{1 + n^2 x^2}\, dx$$

$$J_2 = \int_{1/\sqrt{n}}^{1} \frac{nf(x)}{1 + n^2 x^2}\, dx$$

Use 7.1.12 to prove that

$$J_1 = f(c_n) \tan^{-1} \sqrt{n}$$

$$J_2 = f(d_n)(\tan^{-1} n - \tan^{-1} \sqrt{n})$$

where $0 < c_n < 1/\sqrt{n} < d_n < 1$. Deduce that

$$\int_0^1 \frac{nf(x)}{1 + n^2 x^2}\, dx \to \tfrac{1}{2}\pi f(0) \quad \text{as } n \to \infty$$

7.2   Techniques

In applications of integral calculus it is necessary to find primitives (when such primitives exist and are expressible in terms of simple functions). Various techniques exist for determining primitives, and this short section looks at two of the most important: integration by parts and change of variables.

7.2.1   Integration by parts

Let f and g be functions continuously differentiable on $[a, b]$. Then

$$\int_a^b f(x)g'(x)\,dx = [f(x)g(x)]_a^b - \int_a^b f'(x)g(x)\,dx$$

Proof

The function $h = f \cdot g$ is differentiable and its derivative h' is continuous on $[a, b]$. The fundamental theorem of calculus (7.1.11) applied to h' on the interval $[a, b]$ gives

$$\int_a^b h'(x)\,dx = [h(x)]_a^b = [f(x)g(x)]_a^b$$

By the product rule for differentiation (6.1.3),

$$h'(x) = f(x)g'(x) + f'(x)g(x)$$

Hence

$$\int_a^b [f(x)g'(x) + f'(x)g(x)]\,dx = [f(x)g(x)]_a^b$$

From the linearity of the integral (7.1.10(1)), it follows that

$$\int_a^b f(x)g'(x)\,dx = [f(x)g(x)]_a^b - \int_a^b f'(x)g(x)\,dx \qquad \square$$

The value of this formula lies in the hope that the integral on the right-hand side is easier to evaluate than the original integral.

■■ EXAMPLE 1

Show that $\int_1^2 xe^{2x}\,dx = \frac{1}{4}e^2(3e^2 - 1)$.

Solution

Let $f(x) = x$ and $g(x) = \frac{1}{2}e^{2x}$ (so $g'(x) = e^{2x}$), and integrate by parts.

$$
\begin{aligned}
\int_1^2 xe^{2x}\,dx &= [x(\tfrac{1}{2}e^{2x})]_1^2 - \int_1^2 1 \cdot \tfrac{1}{2}e^{2x}\,dx \\
&= (e^4 - \tfrac{1}{2}e^2) - [\tfrac{1}{4}e^{2x}]_1^2 \\
&= (e^4 - \tfrac{1}{2}e^2) - (\tfrac{1}{4}e^4 - \tfrac{1}{4}e^2) \\
&= \tfrac{1}{4}e^2(3e^2 - 1)
\end{aligned}
$$

■

■■   **EXAMPLE 2**

Many so-called reduction formulae can be established by repeated integration by parts. For example, let

$$J_n = \int \cos^n x \, dx$$

Then

$$J_n = \int (\cos^{n-1} x \cos x) \, dx$$

$$= (\cos^{n-1} x \sin x) + (n-1) \int (\cos^{n-2} x \sin x \sin x) \, dx$$

$$= (\cos^{n-1} x \sin x) + (n-1) \int (\cos^{n-2} x \, (1 - \cos^2 x) \, dx$$

$$= (\cos^{n-1} x \sin x) + (n-1)J_{n-2} - (n-1)J_n$$

Hence

$$nJ_n = \cos^{n-1} x \sin x + (n-1)J_{n-2} \quad \text{for } n \geq 2$$

This formula, together with the fact that $J_0 = x$ and $J_1 = \sin x$, leads to the evaluation of $\int \cos^n x \, dx$ for any $n \in \mathbb{N}$.   ■

7.2.2 Change of variables

Let f be continuous and g be continuously differentiable. Then

$$\int_a^b f(x) \, dx = \int_\alpha^\beta (f \circ g)(t)g'(t) \, dt$$

where $g(\alpha) = a$ and $g(\beta) = b$.

Proof

Let $F(x) = \int_a^x f(t) \, dt$ and $G(t) = (F \circ g)(t)$. By the composite rule for differentiation (6.1.6), $G'(t) = F'(g(t))g'(t)$. But, by the fundamental theorem of calculus (7.1.11), $F'(x) = f(x)$. Thus

$$G'(t) = F'(g(t))g'(t)$$

$$= f(g(t))g'(t)$$

$$= (f \circ g)(t)g'(t)$$

Hence

$$\int_\alpha^\beta (f \circ g)(t) g'(t) \, dt = \int_\alpha^\beta G'(t) \, dt$$

$$= [G(t)]_\alpha^\beta$$

$$= G(\beta) - G(\alpha)$$

$$= (F \circ g)(\beta) - (F \circ g)(\alpha)$$

$$= F(g(\beta)) - F(g(\alpha))$$

$$= F(b) - F(a)$$

$$= [F(x)]_a^b = \int_a^b f(x) \, dx \qquad \square$$

■■ **EXAMPLE 3**

Evaluate $\int_1^2 t^2 \sqrt{t^3 - 1} \, dt$.

Solution

Let $f(x) = \sqrt{x}$ and $g(t) = t^3 - 1$. Then $g'(t) = 3t^2$, and therefore, by 7.2.2,

$$\int_1^2 (\sqrt{t^3 - 1}) \, 3t^2 \, dt = \int_{g(1)}^{g(2)} \sqrt{x} \, dx$$

Hence

$$\int_1^2 t^2 \sqrt{t^3 - 1} \, dt = \tfrac{1}{3} \int_0^7 \sqrt{x} \, dx$$

$$= \tfrac{1}{3} [\tfrac{2}{3} x^{3/2}]_0^7$$

$$\approx 4.116 \qquad \blacksquare$$

■■ **EXAMPLE 4**

Evaluate

$$\int_2^4 \frac{1}{x \log_e x} \, dx$$

Solution

Let $f(x) = 1/(x \log_e x)$ and $g(t) = e^t$. Then $g'(t) = e^t$, and so, by 7.2.2,

$$\int_2^4 \frac{1}{x \log_e x}\, dx = \int_{\log_e 2}^{\log_e 4} \frac{1}{e^t t} e^t\, dt$$

$$= \int_{\log_e 2}^{\log_e 4} \frac{1}{t}\, dt$$

$$= [\log_e t]_{\log_e 2}^{\log_e 4}$$

$$= \log_e \left(\frac{\log_e 4}{\log_e 2} \right) \qquad \blacksquare$$

The change of variables formula is often called **integration by substitution**, where $x = g(t)$ gives the substitution to be used. It is extensively used in elementary calculus books, where trial substitutions that depend on the form of the integrand are suggested.

■■ **EXAMPLE 5**

Show that

$$\int \frac{1}{5 + 3 \cos x}\, dx = \tfrac{1}{2} \tan^{-1} \left(\tfrac{1}{2} \tan \tfrac{1}{2} x \right)$$

Solution

Let

$$f(x) = \frac{1}{5 + 3 \cos x}$$

and

$$x = g(t) = 2 \tan^{-1} t$$

Hence

$$g'(t) = \frac{2}{1 + t^2}$$

and, after some trigonometric manipulations,

$$\cos x = \frac{1 - t^2}{1 + t^2}$$

Therefore

$$\int \frac{1}{5 + 3\cos x}\,dx = \int \frac{1}{5 + 3\left(\dfrac{1 - t^2}{1 + t^2}\right)} \frac{2}{1 + t^2}\,dt$$

$$= \int \frac{1}{4 + t^2}\,dt$$

$$= \tfrac{1}{2}\tan^{-1}\tfrac{1}{2}t$$

$$= \tfrac{1}{2}\tan^{-1}\left(\tfrac{1}{2}\tan\tfrac{1}{2}x\right) \qquad \blacksquare$$

The last integral above requires substitution for the variable $\tfrac{1}{2}t$.

The final two examples detail some of the algebraic techniques that are often needed when evaluating integrals.

■■ **EXAMPLE 6**

Show that

$$\int \frac{1}{\sqrt{3 + 2x - x^2}}\,dx = \sin^{-1}\tfrac{1}{2}(x - 1)$$

Solution

Completing the square of $3 + 2x - x^2$ gives $4 - (x - 1)^2$. Hence

$$\int \frac{1}{\sqrt{3 + 2x - x^2}}\,dx = \int \frac{1}{\sqrt{4 - (x - 1)^2}}\,dx$$

$$= \int \frac{1}{\sqrt{4 - u^2}}\,du$$

via the substitution $x = u + 1$

$$= \sin^{-1}\tfrac{1}{2}u$$

via the substitution $u = 2\sin t$

$$= \sin^{-1}\tfrac{1}{2}(x - 1) \qquad \blacksquare$$

■■ **EXAMPLE 7**

Show that

$$\int \frac{x^4 + x - 1}{x^3 + x}\,dx = \tfrac{1}{2}x^2 - \log_e |x| + \tan^{-1}x$$

Solution

By division,

$$\frac{x^4 + x - 1}{x^3 + x} = x - \frac{x^2 - x + 1}{x^3 + x}$$

By partial fractions

$$\frac{x^2 - x + 1}{x^3 + x} = \frac{A}{x} + \frac{Bx + C}{x^2 + 1}$$

leading to $A = 1$, $B = 0$ and $C = -1$. Hence

$$\int \frac{x^4 + x - 1}{x^3 + x}\, dx = \int \left(x - \frac{1}{x} + \frac{1}{x^2 + 1}\right) dx$$

$$= \tfrac{1}{2}x^2 - \log_e |x| + \tan^{-1} x \qquad \blacksquare$$

Exercises 7.2

1.   Evaluate the following:

(a) $\int_1^2 (x^3 - x + 3)\, dx$ (b) $\int_0^{\pi/2} \sin 3x\, dx$

(c) $\int_0^1 \frac{3}{(1 + x)(2 - x)}\, dx$ (d) $\int_{-4}^4 \frac{1}{16 + x^2}\, dx$

(e) $\int_0^4 |2x - 3|\, dx$ (f) $\int_0^\pi x \cos x\, dx$

2.   Prove that if f is continuously differentiable on $[a, b]$ then

$$\int_a^b f(x) \cos nx\, dx \to 0 \quad \text{as } n \to \infty$$

(*Hint*: use integration by parts)

3.   Let f, f' and g be continuous on $[a, b]$, with $f' \neq 0$. Show that

$$\int_a^b f(x)g(x)\, dx = f(b)G(b) - \int_a^b f'(x)G(x)\, dx$$

where $G(x) = \int_a^x g(t)\, dt$. Apply 7.1.12 to the integral on the right-hand side of the previous equation to deduce that

$$\int_a^b f(x)g(x)\,\mathrm{d}x = f(a)\int_a^c g(x)\,\mathrm{d}x + f(b)\int_c^b g(x)\,\mathrm{d}x$$

for some $c \in (a, b)$. This result is called the **second mean value theorem for integrals**.

4.   Prove that if $f(x) = 1/x$ and $g(x) = \sin x$, and $0 < a < b$, then

$$\left| \int_a^b \frac{\sin x}{x}\,\mathrm{d}x \right| < \frac{4}{a}$$

(*Hint*: use Question 3.) Find the value of

$$\lim_{n \to \infty} \int_n^{n+1} \frac{\sin x}{x}\,\mathrm{d}x$$

7.3   Improper integrals

The definition of the Riemann integral applies only to bounded functions defined on bounded intervals. This section relaxes these conditions and, in so doing, defines **improper integrals**. First, however, the integration of a wider class of functions than continuous ones is examined.

7.3.1   Definition

> A function f is called **piecewise-continuous** on $[a, b]$ if there exists a partition $P = \{x_0, x_1, \ldots, x_n\}$ of $[a, b]$ and continuous functions f_i defined on $[x_{i-1}, x_i]$, such that $f(x) = f_i(x)$ for $x_{i-1} < x < x_i$, $i = 1, 2, \ldots, n$.

A partition Q can be chosen that contains the points $x_0, x_1, \ldots, x_n$. Then

$$Q = P_1 \cup P_2 \cup \ldots \cup P_n$$

where P_i is a partition of $[x_{i-1}, x_i]$. Hence

$$L(Q) = \sum_{i=1}^n L(P_i)$$

where $L(Q)$ is the lower sum of f relative to Q and, for each i, $L(P_i)$ is the lower sum of f_i relative to P_i. Thus $\sum_{i=1}^n L(P_i) \le \mathcal{L}$, where $\mathcal{L}$ is the supremum of all the lower sums for f on $[a, b]$. Since f_i is continuous on $[x_{i-1}, x_i]$, f_i is Riemann integrable on $[x_{i-1}, x_i]$. Hence

$$\sum_{i=1}^n \left(\int_{x_{i-1}}^{x_i} f_i(x)\,\mathrm{d}x \right) \le \mathcal{L}$$

In a similar fashion

$$\mathcal{U} \leq \sum_{i=1}^{n}\left(\int_{x_{i-1}}^{x_i} f_i(x)\,dx\right)$$

where $\mathcal{U}$ is the infimum of all the upper sums for f on $[a, b]$. Hence, using 7.1.4, $\mathcal{L} = \mathcal{U}$.

In other words, a piecewise-continuous function is Riemann integrable and

$$\int_a^b f(x)\,dx = \int_{x_0}^{x_1} f_1(x)\,dx + \int_{x_1}^{x_2} f_2(x)\,dx + \ldots + \int_{x_{n-1}}^{x_n} f_n(x)\,dx$$

Moreover, there exists a kind of primitive for f, in the sense that there is a function F defined and continuous on $[a, b]$, differentiable for $x \neq x_i$ with $F' = f$ for $x \neq x_i$.

■■ EXAMPLE 1

The function given by $f(x) = x - [x]$ is piecewise-continuous on $[0, 3]$. This can be seen by observing that $f_1(x) = x$ on $[0, 1)$, $f_2(x) = x - 1$ on $[1, 2)$ and $f_3(x) = x - 2$ on $[2, 3)$. See Figure 7.6. ■

■■ EXAMPLE 2

Calculate $\int_0^2 f(x)\,dx$ for the function f of Example 1.

Solution

Now

$$\int_0^2 (x - [x])\,dx = \int_0^1 x\,dx + \int_1^2 (x - 1)\,dx$$
$$= [\tfrac{1}{2}x^2]_0^1 + [\tfrac{1}{2}x^2 - x]_1^2$$
$$= \tfrac{1}{2} + \tfrac{1}{2} = 1$$

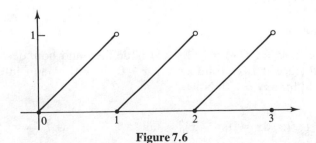

Figure 7.6

A suitable 'primitive' of f on $[0, 2]$ is given by

$$F(x) = \begin{cases} \frac{1}{2}x^2 & \text{for } 0 \leq x \leq 1 \\ \frac{1}{2}x^2 - x + 1 & \text{for } 1 \leq x \leq 2 \end{cases}$$

Note that $\int_0^2 (x - [x])\, dx = [F(x)]_0^2$. ∎

The 'integral' of a function defined and bounded on an interval that is not bounded can now be defined. This is called an **improper integral of the first kind.**

7.3.2 Definition

> Let f be a function bounded on $[a, \infty)$ and Riemann integrable on $[a, b]$ for every $b > a$. If $\lim_{b \to \infty} [\int_a^b f(x)\, dx]$ exists, it is said that $\int_a^\infty f(x)\, dx$ **converges**. Otherwise, $\int_a^\infty f(x)\, dx$ **diverges**.

A completely analogous definition holds for integrals of the form $\int_{-\infty}^b f(x)\, dx$. Since it is necessary to preserve the additivity of the integral for improper integrals, the following is defined:

$$\int_{-\infty}^\infty f(x)\, dx = \int_{-\infty}^0 f(x)\, dx + \int_0^\infty f(x)\, dx$$

provided that both the improper integrals on the right-hand side converge.

■■ EXAMPLE 3

The integral

$$\int_0^\infty \frac{1}{1 + x^2}\, dx$$

converges to $\frac{1}{2}\pi$.

Solution

The function $f(x) = 1/(1 + x^2)$ is defined and bounded on $[0, \infty)$, and since it is continuous for $x \geq 0$, it is Riemann integrable on $[0, b]$ for any $b > 0$. Now

$$\int_0^b f(x)\, dx = [\tan^{-1} x]_0^b = \tan^{-1} b \to \tfrac{1}{2}\pi \quad \text{as } b \to \infty$$

Hence

$$\int_0^\infty \frac{1}{1+x^2}\, dx$$

converges to $\frac{1}{2}\pi$.   ∎

■■ EXAMPLE 4

The integral

$$\int_1^\infty \frac{1}{x^2}\, dx$$

converges to 1.

Solution

The function $f(x) = 1/x^2$ is defined and bounded on $[1, \infty)$. Since it is continuous for $x \ge 1$, it is Riemann integrable on $[1, b]$ for any $b > 1$. Now

$$\int_1^b f(x)\, dx = \left[\frac{-1}{x}\right]_1^b = 1 - \frac{1}{b} \to 1 \quad \text{as } b \to \infty$$

Hence $\int_1^\infty (1/x^2)\, dx$ converges to 1.   ∎

■■ EXAMPLE 5

The integral $\int_1^\infty (1/\sqrt{x})\, dx$ diverges.

Solution

Since $f(x) = 1/\sqrt{x}$ is continuous on $[1, \infty)$, it follows that for $b > 1$

$$\int_1^b \frac{1}{\sqrt{x}}\, dx = [2\sqrt{x}]_1^b = 2(\sqrt{b} - 1)$$

This has no finite limit as $b \to \infty$. Hence $\int_1^\infty (1/\sqrt{x})\, dx$ diverges. ∎

■■ EXAMPLE 6

The integral $\int_{-\infty}^\infty \sin x\, dx$ diverges.

Solution

Now

$$\int_0^b \sin x \, dx = [-\cos x]_0^b = 1 - \cos b$$

which has no limit as $b \to \infty$. In fact, $\int_{-\infty}^0 \sin x \, dx$ also diverges. From the definition of $\int_{-\infty}^\infty \sin x \, dx$, it follows that $\int_{-\infty}^\infty \sin x \, dx$ diverges. However,

$$\lim_{b \to \infty} \left(\int_{-b}^b \sin x \, dx \right) = 0$$

In such cases $\lim_{b \to \infty} (\int_{-b}^b f(x) \, dx)$ is called the **Cauchy principal value** of the integral. ∎

The 'integral' of a function over a bounded interval where the function is not bounded can now be defined. This is called an **improper integral of the second kind**.

7.3.3 Definition

Let f be a function defined on $(a, b]$ and Riemann integrable on $[a + \varepsilon, b]$ for $0 < \varepsilon < b - a$. If

$$\lim_{\varepsilon \to 0+} \int_{a+\varepsilon}^b f(x) \, dx$$

exists then it is said that $\int_a^b f(x) \, dx$ **converges**.

■■ EXAMPLE 7

The integral $\int_0^1 (1/\sqrt{x}) \, dx$ converges to 2.

Solution

The function $f(x) = 1/\sqrt{x}$ is defined for $x > 0$ and is continuous on $[\varepsilon, 1]$ for $0 < \varepsilon < 1$. However, f is not bounded on $(0, 1]$. Now

$$\int_\varepsilon^1 \frac{1}{\sqrt{x}} \, dx = [2\sqrt{x}]_\varepsilon^1 = 2(1 - \sqrt{\varepsilon}) \to 2 \quad \text{as} \quad \varepsilon \to 0+$$

Hence $\int_0^1 (1/\sqrt{x}) \, dx$ converges to 2. ∎

■■   **EXAMPLE 8**

The integral

$$\int_0^1 \frac{1}{\sqrt{1 - x^2}}\, dx$$

converges to $\frac{1}{2}\pi$.

Solution

In this case the bad behaviour of

$$f(x) = \frac{1}{\sqrt{1 - x^2}}$$

occurs at the right-hand endpoint of the interval $[0, 1]$. So to establish convergence of the given integral, it is observed that f is defined on $[0, 1)$ and is Riemann integrable on $[0, 1 - \varepsilon]$ for $0 < \varepsilon < 1$ and that

$$\int_0^{1-\varepsilon} \frac{1}{\sqrt{1 - x^2}}\, dx = [\sin^{-1} x]_0^{1-\varepsilon}$$

$$= \sin^{-1}(1 - \varepsilon) \to \tfrac{1}{2}\pi \quad \text{as } \varepsilon \to 0+ \qquad ■$$

■■   **EXAMPLE 9**

The integral $\int_{-1}^1 (1/x)\, dx$ diverges.

Solution

Write

$$\int_{-1}^1 \frac{1}{x}\, dx = \int_{-1}^0 \frac{1}{x}\, dx + \int_0^1 \frac{1}{x}\, dx$$

Now $\int_0^1 (1/x)\, dx$ is divergent since

$$\int_\varepsilon^1 \frac{1}{x}\, dx = [\log_e |x|]_\varepsilon^1 = -\log_e \varepsilon$$

which has no limit as $\varepsilon \to 0+$. Hence $\int_{-1}^1 (1/x)\, dx$ also diverges. ■

The next example shows how the additivity of the integral is used to handle integrals of mixed type.

■■   EXAMPLE 10

The integral $\int_0^\infty (1/\sqrt{x})\,dx$ diverges.

Solution

Write

$$\int_0^\infty \frac{1}{\sqrt{x}}\,dx = \int_0^1 \frac{1}{\sqrt{x}}\,dx + \int_1^\infty \frac{1}{\sqrt{x}}\,dx$$

and observe that, although $\int_0^1 (1/\sqrt{x})\,dx = 2$ by Example 7, $\int_1^\infty (1/\sqrt{x})\,dx$ diverges by Example 5.   ■

As with infinite series, improper integrals with infinite integrands and/ or infinite intervals of integration can be investigated using various convergence tests. One such test is proved below as an illustration.

7.3.4   Comparison test for integrals

Let f and g be defined on $[a, \infty)$ and Riemann integrable on $[a, b]$ for every $b > a$. Suppose that

(a) $0 \leqslant f(x) \leqslant g(x)$ for all $x \geqslant a$, and
(b) $\int_a^\infty g(x)\,dx$ converges.

Then $\int_a^\infty f(x)\,dx$ also converges.

Proof

Now

$$0 \leqslant \int_a^n f(x)\,dx \leqslant \int_a^n g(x)\,dx \quad \text{by 7.1.10(3)}$$

Since $0 \leqslant f(x) \leqslant g(x)$, $\int_a^n g(x)\,dx$ increases to its limiting value as $n \to \infty$. Hence $\int_a^n f(x)\,dx$ is increasing and bounded above. By 3.4.1, $\int_a^\infty f(x)\,dx$ is a convergent integral.   □

A comparison test for improper integrals of the second kind is easily formulated and proved in a similar manner to 7.3.4. See Question 5 of Exercises 7.3.

■■ EXAMPLE 11

The integral

$$\int_0^\infty \frac{e^{-x}}{1+x^2}\, dx$$

converges.

Solution

Let

$$f(x) = \frac{e^{-x}}{1+x^2} \quad \text{and} \quad g(x) = \frac{1}{1+x^2}$$

Now $0 \le f(x) \le g(x)$ for all $x \ge 0$. By Example 3, $\int_0^\infty g(x)\,dx$ converges to $\frac{1}{2}\pi$. Hence, by the comparison test, $\int_0^\infty f(x)\,dx$ converges.

■

Many of the important functions of mathematical physics are defined using improper integrals. This chapter finishes by briefly examining one of them, the **gamma function** $\Gamma(x)$.

7.3.5 Definition

For $x > 0$ define

$$\Gamma(x) = \int_0^\infty t^{x-1}\, e^{-t}\, dt$$

Now

$$\int_0^\infty t^{x-1}\, e^{-t}\, dt = \int_0^1 t^{x-1}\, e^{-t}\, dt + \int_1^\infty t^{x-1}\, e^{-t}\, dt$$

a sum of integrals of the second and first kinds respectively. Consider

$$\int_1^\infty t^{x-1}\, e^{-t}\, dt$$

Now

$$e^t > \frac{t^n}{n!} \quad \text{for } t > 0$$

and hence

$$0 < e^{-t} < \frac{n!}{t^n} \quad \text{for } t > 0$$

If $b, t > 0$ then

$$0 < e^{-bt} < \frac{n!}{b^n t^n}$$

and so

$$0 < t^a e^{-bt} < \frac{n!}{b^n t^{n-a}}$$

for all a. Choose $n > a$ and let $t \to \infty$ to deduce that $t^a e^{-bt} \to 0$ for all a and all $b > 0$. Hence, for fixed x and large t,

$$t^{x-1} e^{-t} = (t^{x-1} e^{-t/2}) e^{-t/2} < e^{-t/2}$$

It is easy to show that $\int_1^\infty e^{-t/2} \, dt$ converges (to 2), and so, by the comparison test for integrals (7.3.4), $\int_1^\infty t^{x-1} e^{-t} \, dt$ converges for all x.
Now consider

$$\int_0^1 t^{x-1} e^{-t} \, dt$$

Since $t^{x-1} e^{-t} \leqslant t^{x-1}$ for $0 \leqslant t \leqslant 1$ and $\int_0^1 t^{x-1} \, dt$ converges for $x > 0$ (see Question 4 of Exercises 7.3), the comparison test for integrals of the second kind (see Question 5 of Exercises 7.3) gives that $\int_0^1 t^{x-1} e^{-t} \, dt$ converges for $x > 0$.
Hence $\Gamma(x) = \int_0^\infty t^{x-1} e^{-t} \, dt$ is well defined for $x > 0$. The properties of the gamma function can be deduced from its integral definition.

■■ EXAMPLE 12

Prove that $\Gamma(n + 1) = n!$ for $n = 0, 1, 2, \ldots$

Solution

First, for $x \geqslant 0$,

$$\int_0^b t^x e^{-t} \, dt = [-t^x e^{-t}]_0^b + x \int_0^b t^{x-1} e^{-t} \, dt$$

by integration by parts. Since $\lim_{t \to \infty} t^x e^{-t} = 0$, it follows that $\Gamma(x + 1) = x\Gamma(x)$. By induction on n, $\Gamma(n + 1) = n!\Gamma(1)$. But $\Gamma(1) = \int_0^\infty e^{-t} \, dt = 1$. Hence $\Gamma(n + 1) = n!$ ■

Exercises 7.3

1. Find $\int_{-2}^{2} f(x)\, dx$ in the following cases:

    (a) $f(x) = x + [x]$

    (b) $f(x) = \begin{cases} -2x & \text{if } -2 \leqslant x < 0 \\ x & \text{if } 0 \leqslant x \leqslant 2 \end{cases}$

2. Decide which of the following improper integrals converge and evaluate those which do:

    (a) $\displaystyle\int_{1}^{\infty} \frac{1}{x^3}\, dx$ (b) $\displaystyle\int_{0}^{\infty} \sin^2 x\, dx$

    (c) $\displaystyle\int_{0}^{\infty} e^{-2x}\, dx$ (d) $\displaystyle\int_{-\infty}^{\infty} \operatorname{sech} x\, dx$ (*Hint*: substitute $u = e^x$)

3. Decide which of the following improper integrals converge and evaluate those that do:

    (a) $\displaystyle\int_{0}^{1} \log_e x\, dx$ (b) $\displaystyle\int_{1}^{2} \frac{1}{x-1}\, dx$

    (c) $\displaystyle\int_{0}^{\pi/2} (\sec x - \tan x)\, dx$

4. Prove that $\int_{0}^{1} t^{x-1}\, dt$ converges for $x > 0$.

5. Let f and g be defined on $(a, b]$ and Riemann integrable on $[a + \varepsilon, b]$ for $0 < \varepsilon < b - a$. Suppose that

    (a) $0 \leqslant f(x) \leqslant g(x)$ for all $x \in (a, b]$
    (b) $\int_{a}^{b} g(x)\, dx$ converges.

    Prove that $\int_{a}^{b} f(x)\, dx$ converges.

6. Use the comparison test for integrals (7.3.4) to show that the following integrals converge:

    (a) $\displaystyle\int_{1}^{\infty} \left(\frac{\sin x}{x}\right)^2 dx$ (b) $\displaystyle\int_{1}^{\infty} \frac{x}{1 + x^3}\, dx$

    (c) $\displaystyle\int_{1}^{\infty} e^{-x^2}\, dx$

7. Use Question 5 to discuss the convergence of the following integrals:

(a) $\displaystyle\int_0^{\pi/2} \frac{1}{\sin x}\, dx$ (b) $\displaystyle\int_0^1 e^{1/x}\, dx$

(c) $\displaystyle\int_0^1 \frac{1}{\sqrt{\tan x}}\, dx$

(*Hint*: $e^x \geq 1 + x$ for all x and $\sin x < x < \tan x$ for $0 < x < \frac{1}{2}\pi$.)

8. Given that $\Gamma(\frac{1}{2}) = \sqrt{\pi}$, show by induction on n that

$$\Gamma(n + \tfrac{1}{2}) = \frac{(2n)!\sqrt{\pi}}{4^n n!} \quad \text{for } n \geq 0$$

Problems 7

1. Show that all the following functions are primitives of $f(x) = \operatorname{sech} x$, $|x| < \frac{1}{2}\pi$:

(a) $F_1(x) = \tan^{-1}(\sinh x)$ (b) $F_2(x) = 2\tan^{-1} e^x$

(c) $F_3(x) = \sin^{-1}(\tanh x)$ (d) $F_4(x) = 2\tan^{-1}(\tanh \frac{1}{2}x)$

2. Let

$$f(x) = \begin{cases} 1 - x & \text{if } 0 \leq x < 1 \\ 2 & \text{if } x = 1 \end{cases}$$

and let P_n be the partition $\{0, 1/n, 2/n, \ldots, 1\}$ of $[0, 1]$. Calculate $L(P_n)$ and $U(P_n)$. Hence prove that f is Riemann integrable on $[0, 1]$ and evaluate $\int_0^1 f(x)\, dx$.

3. Prove that

$$f(x) = \begin{cases} 1 + x & \text{if } 0 \leq x \leq 1 \text{ and } x \text{ is rational} \\ 1 - x & \text{if } 0 \leq x \leq 1 \text{ and } x \text{ is irrational} \end{cases}$$

is not Riemann integrable on $[0, 1]$.

4. Suppose that f and g are Riemann integrable on $[a, b]$. Prove that $h(x) = \max(f(x), g(x))$ is also Riemann integrable on $[a, b]$. (*Hint*: $\max(f(x), g(x)) = \frac{1}{2}[f(x) + g(x) + |f(x) - g(x)|]$.)

5. The function $f(x) = 1/(1 + x^2)^3$ is Riemann integrable on $[0, 1]$. Calculate $L(P_n)$ and $U(P_n)$, where P_n is the partition $\{0, 1/n, 2/n, \ldots, 1\}$ of $[0, 1]$. Hence evaluate

$$\lim_{n \to \infty} n^5 \sum_{r=1}^{n} \frac{1}{(n^2 + r^2)^3}$$

6. Let f, f' and g be continuous on $[a, b]$ with $f > 0$ and $f' < 0$. If $m \leqslant \int_a^c g(x) \, dx \leqslant M$ for all $c \in [a, b]$, prove that

$$mf(a) \leqslant \int_a^b f(x)g(x) \, dx \leqslant M f(a)$$

Deduce that

$$\int_a^b f(x)g(x) \, dx = f(a) \int_a^d g(x) \, dx$$

for some $d \in (a, b)$. This result is called **Bonnet's form of the second mean value theorem for integrals**.

7. Let $f(x) = [x] + 1$ and $F(x) = \int_0^x f(t) \, dt$. Find explicit formulae for $F(x)$ when $0 \leqslant x \leqslant 2$, and show that $F'(1) \neq f(1)$. Explain why this does not contradict the fundamental theorem of calculus.

8. (a) Use the inequality $|\sin x| \leqslant |x|$ for all $x \in \mathbb{R}$ to prove that

$$\int_0^1 \frac{x^3}{2 - \sin^4 x} \, dx \leqslant \frac{1}{4} \log_e 2$$

(b) Use the fact that $\cos x \geqslant 0$ on $[0, \frac{1}{2}\pi]$ to prove that

$$\left| \int_0^{\pi/2} \frac{x - \frac{1}{2}\pi}{2 + \cos x} \, dx \right| \leqslant \frac{1}{16}\pi^2$$

(c) Use the inequalities $1 - x^2 \leqslant e^{-x^2} \leqslant 1 - x^2 + \frac{1}{2}x^4$ for all $x \in \mathbb{R}$ to prove that

$$\frac{2}{3} \leqslant \int_0^1 e^{-x^2} \, dx \leqslant \frac{21}{30}$$

9.   Decide which of the following improper integrals converge, and evaluate those which do.

(a) $\int_0^\infty \dfrac{x^2}{1 + x^2}\, dx$ (b) $\int_0^\infty e^{-x} \sin x\, dx$

(c) $\int_{-\infty}^\infty \dfrac{1}{1 + 3\cosh x}\, dx$ (d) $\int_0^1 x^{-2/3}\, dx$

(e) $\int_0^{\pi/2} \tan x\, dx$ (f) $\int_0^1 \dfrac{1}{\sqrt{1 - x}}\, dx$

10.   Use the comparison tests (7.3.4) and Question 5 of Exercises 7.3 to determine which of the following integrals converge:

(a) $\int_0^\infty \dfrac{x^3}{(1 + x^2)^2}\, dx$ (b) $\int_0^\infty \sin x^2\, dx$

(c) $\int_0^\infty e^{-x^2}\, dx$ (d) $\int_0^1 \dfrac{\sin^2 x}{1 + x^2}\, dx$

(e) $\int_0^\infty \dfrac{1}{\sqrt{x + x^3}}\, dx$ (f) $\int_0^1 \dfrac{1}{\log_e (1 + x)}\, dx$

(*Hint*: For (e) consider the intervals $[0, 1]$ and $[1, \infty)$ separately.)

The Elementary Functions

The end of Section 4.3 gives the power series definitions of the functions sine, cosine and exponential. The arithmetic of power series (4.3.2) enables the derivation of many of the arithmetical properties of these functions. The purpose of this appendix is to establish that sine, cosine, exponential and other functions defined in terms of them are (at least) differentiable. By 6.1.2, these elementary functions are thus continuous, and hence, by 7.1.9, they are Riemann integrable.

Consider the power series $\sum_{n=0}^{\infty} a_n x^n$ and suppose that its radius of convergence is $R > 0$. Now consider the power series

$$\sum_{n=1}^{\infty} n a_n x^{n-1} \quad \text{and} \quad \sum_{n=0}^{\infty} \frac{a_n}{n+1} x^{n+1}$$

obtained respectively from $\sum_{n=0}^{\infty} a_n x^n$ by differentiating and integrating term by term. The following theorem will be proved.

A.1 Theorem

$$\sum_{n=1}^{\infty} n a_n x^{n-1} \quad \text{and} \quad \sum_{n=0}^{\infty} \frac{a_n}{n+1} x^{n+1}$$

both have radius of convergence R.

Proof

Suppose that $\sum_{n=1}^{\infty} n a_n x^{n-1}$ has radius of convergence R_1. Let $|x| < R$ and choose a real number c such that $|x| < c < R$. Then $\sum_{n=0}^{\infty} a_n c^n$ is convergent, and so, by the vanishing condition

(4.1.2), $a_n c^n \to 0$ as $n \to \infty$. Hence there is a number M such that $|a_n c^n| \leq M$ for $n = 0, 1, 2, \ldots$. Now

$$|na_n x^{n-1}| = |a_n c^n| \frac{n}{c} \left| \frac{x}{c} \right|^{n-1} \leq \frac{M}{c} n \left| \frac{x}{c} \right|^{n-1} \leq \frac{M}{c} nr^{n-1}$$

where $0 \leq r < 1$. By the ratio test (4.2.3), $\sum_{n=0}^{\infty} nr^{n-1}$ converges, and hence $\sum_{n=1}^{\infty} na_n x^{n-1}$ is (absolutely) convergent for $|x| < R$ by the first comparison test (4.2.1). Hence $R_1 \geq R$.

Suppose that $R_1 > R$ and let y be any real number satisfying $R < y < R_1$. Then $\sum_{n=1}^{\infty} na_n y^{n-1}$ is absolutely convergent. Now

$$|a_n y^n| = |na_n y^{n-1}| \frac{y}{n} \leq y |na_n y^{n-1}|$$

and so, by the first comparison test (4.2.1),

$$\frac{1}{y} \sum_{n=1}^{\infty} a_n y^n$$

is convergent. By 4.1.4, $\sum_{n=0}^{\infty} a_n y^n$ converges, and this contradicts the definition of R. Hence $R_1 > R$ is false, and so $R_1 = R$ as claimed.

Now consider

$$\sum_{n=0}^{\infty} \frac{a_n}{n+1} x^{n+1}$$

Suppose that this series has radius of convergence R_2. Term-by-term differentiation gives the series $\sum_{n=0}^{\infty} a_n x^n$, which has radius of convergence R. By the first part of the proof, $R = R_2$. □

The key theoretical result of this appendix can now be established.

A.2 The calculus of power series

If $\sum_{n=0}^{\infty} a_n x^n$ has radius of convergence R then, for $|x| < R$,

(1) $f(x) = \sum_{n=0}^{\infty} a_n x^n$ is differentiable and

$$f'(x) = \sum_{n=1}^{\infty} na_n x^{n-1}$$

(2) $f(x) = \sum_{n=0}^{\infty} a_n x^n$ possesses the primitive

$$\int f(x)\, dx = \sum_{n=0}^{\infty} \frac{a_n}{n+1} x^{n+1}$$

Proof

For (1) let $c \in (-R, R)$ and choose $r \in (0, R)$ such that $c \in (-r, r)$. Now consider all non-zero h satisfying $|h| < r - |c|$. In other words, $c + h \in (-r, r)$. Now

$$\frac{f(c + h) - f(c)}{h} - \sum_{n=1}^{\infty} na_n c^{n-1} = \frac{1}{h} \sum_{n=1}^{\infty} a_n [(c + h)^n - c^n - hnc^{n-1}]$$

Taylor's theorem (6.3.5) applied to the function $x \mapsto x^n$ on the interval with endpoints c and $c + h$ gives

$$(c + h)^n = c^n + nhc^{n-1} + \tfrac{1}{2}n(n - 1)h^2(c + \theta h)^{n-2}$$

for some θ, $0 < \theta < 1$. Hence

$$\left| \frac{f(c + h) - f(c)}{h} - \sum_{n=1}^{\infty} na_n c^{n-1} \right|$$

$$\leqslant \tfrac{1}{2}|h| \sum_{n=1}^{\infty} |a_n| n(n - 1)|c + \theta h|^{n-2}$$

$$\leqslant \tfrac{1}{2}|h| \sum_{n=1}^{\infty} |a_n| n(n - 1) r^{n-2}$$

By A.1, $\sum_{n=1}^{\infty} na_n r^{n-1}$ has radius of convergence R, and hence so does $\sum_{n=2}^{\infty} n(n - 1)a_n r^{n-2}$. Hence $\sum_{n=2}^{\infty} n(n - 1)a_n r^{n-2}$ is absolutely convergent, and so is a fixed finite number. Therefore

$$\lim_{h \to 0} \left| \frac{f(c + h) - f(c)}{h} - \sum_{n=1}^{\infty} na_n c^{n-1} \right| = 0$$

This proves (1).

To establish (2), let

$$F(x) = \sum_{n=0}^{\infty} \frac{a_n}{n + 1} x^{n+1}$$

and apply (1) and A.1 to deduce that $F'(x) = f(x)$.   □

It follows immediately from A.1 and A.2 that if $\sum_{n=0}^{\infty} a_n x^n$ is a power series with radius of convergence R then $f(x) = \sum_{n=0}^{\infty} a_n x^n$ is infinitely differentiable on $(-R, R)$.

Recall now that the exponential, sine and cosine functions are defined as follows.

A.3 Definition

> The **exponential** function exp: $\mathbb{R} \to \mathbb{R}$ is given by
>
> $$\exp x = \sum_{n=0}^{\infty} \frac{x^n}{n!}$$

A.4 Definition

> The **sine** function sin: $\mathbb{R} \to \mathbb{R}$ is given by
>
> $$\sin x = \sum_{n=0}^{\infty} \frac{(-1)^n x^{2n+1}}{(2n+1)!}$$

A.5 Definition

> The **cosine** function cos: $\mathbb{R} \to \mathbb{R}$ is given by
>
> $$\cos x = \sum_{n=0}^{\infty} \frac{(-1)^n x^{2n}}{(2n)!}$$

Since the above power series converge for all $x \in \mathbb{R}$, the functions exp, sin and cos are infinitely differentiable. In particular, they are continuous and Riemann integrable. Term-by-term differentiation of the power series involved confirms the entries in Table 6.1. Theoretical results concerning differentiable functions can now be used to establish the familiar properties of the elementary functions. For example, a far easier proof of the identity $\cos^2 x + \sin^2 x = 1$ than that given in Example 3 of Section 4.3 may now be furnished. Let $f(x) = \cos^2 x + \sin^2 x$, a function differentiable for all x by virtue of the sum and product rules for differentiation. Differentiating gives

$$f'(x) = -2\cos x \sin x + 2 \sin x \cos x = 0$$

Hence, by the increasing–decreasing theorem (6.2.3), $f(x) = k$, a constant. Since $f(0) = 1$, this constant k is unity. In other words, $\cos^2 x + \sin^2 x = 1$ for all $x \in \mathbb{R}$. Additionally, sin and cos have the properties suggested in the graphs in Figure A.1.

To establish these, note that

$$\sin x = \left(x - \frac{x^3}{3!} \right) + \left(\frac{x^5}{5!} - \frac{x^7}{7!} \right) + \dots$$

$$= \frac{x}{3!}(6 - x^2) + \frac{x^5}{7!}(42 - x^2) + \dots$$

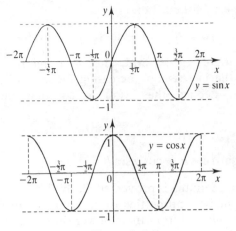

Figure A.1

Hence $\sin x > 0$ certainly for $0 < x < 2$. Since the derivative of $\cos x$ is $-\sin x$, it can be deduced from the increasing–decreasing theorem (6.2.3) that cos is strictly decreasing on $[0, 2]$. Now $\cos 0 = 1$, and inspection of the power series defining $\cos x$ soon gives that $\cos 2 < 0$. Hence, by the intermediate value property (5.3.2), $\cos x$ vanishes somewhere in $(0, 2)$. Since cos is strictly decreasing on $[0, 2]$, there is a unique value of x in $(0, 2)$ satisfying $\cos x = 0$. Hence, $\frac{1}{2}\pi$ is *defined* to be this unique value of x. In other words, $\cos \frac{1}{2}\pi = 0$. Using $\cos^2 x + \sin^2 x = 1$ and substituting $x = \frac{1}{2}\pi$ quickly gives $\sin \frac{1}{2}\pi = 1$. Now $\cos x > 0$, $0 \leqslant x < \frac{1}{2}\pi$, and so, by the increasing–decreasing theorem, sin is strictly increasing on $[0, \frac{1}{2}\pi]$. Since $\sin(-x) = -\sin x$ and $\cos(-x) = \cos x$ (Question 4(a) of Exercises 4.3), the graphs of $y = \sin x$ and $y = \cos x$ on the interval $[-\frac{1}{2}\pi, \frac{1}{2}\pi]$ are as described in Figure A.1.

To complete the graphs, it must be established that

$$\sin(x + \tfrac{1}{2}\pi) = \cos x$$

$$\cos(x + \tfrac{1}{2}\pi) = -\sin x$$

$$\sin(x + 2\pi) = \sin x$$

and

$$\cos(x + 2\pi) = \cos x$$

for all $x \in \mathbb{R}$. These all follow from the addition formulae

$$\sin(a + b) = \sin a \cos b + \cos a \sin b$$

and

$$\cos(a + b) = \cos a \cos b - \sin a \sin b$$

which will now be established. Let

$$f(x) = \sin(a + x)\cos(b - x) + \cos(a + x)\sin(b - x) \quad \text{for } x \in \mathbb{R}$$

Now f is differentiable on $\mathbb{R}$, and, after a straightforward calculation, $f'(x) = 0$. Apply the increasing–decreasing theorem (6.2.3) to f on the interval $[0, b]$ to deduce that $f(x)$ is constant on $[0, b]$. Hence $f(b) = f(0)$. Thus

$$\sin(a + b) = f(b) = f(0)$$
$$= \sin a \cos b + \cos a \sin b$$

The second addition formula is proved similarly.

Thus the key analytic properties of elementary functions such as sin and cos can be established from their power series definitions. Note also in the above that the number π is defined analytically via $\cos \frac{1}{2}\pi = 0$.

This definition of π coincides with the more familiar geometric definition that π is the ratio of the circumference of any circle to its diameter, and that, as a consequence, a circle of radius r has area πr^2. To see this, first note that the equation $x^2 + y^2 = 1$ represents a circle of unit radius, centre $(0,0)$. The portion of this circle in the positive quadrant ($x \geq 0$ and $y \geq 0$) has equation $y = \sqrt{1 - x^2}$. Now the area of this quarter-circle is given by the integral

$$\int_0^1 \sqrt{1 - x^2}\, dx$$

The substitution $x = \sin \theta$ gives

$$\int_0^1 \sqrt{1 - x^2}\, dx = \int_{\pi/2}^0 -\sin \theta \sqrt{1 - \cos^2 \theta}\, d\theta$$

since $\cos \theta = 1$ and, using the *analytic* definition of π given above, $\cos \frac{1}{2}\pi = 0$. Hence

$$\int_0^1 \sqrt{1 - x^2}\, dx = \int_{\pi/2}^0 -\sin \theta \sqrt{1 - \cos^2 \theta}\, d\theta$$

$$= \int_{\pi/2}^0 (-\sin^2 \theta)\, d\theta \qquad \text{using the formula}$$

$$\cos^2 \theta + \sin^2 \theta = 1$$

$$= \int_{\pi/2}^0 \tfrac{1}{2}(\cos 2\theta - 1)\, d\theta \qquad \text{using the addition}$$

formula for cosine

$$= [\tfrac{1}{4}\sin 2\theta - \tfrac{1}{2}\theta]_{\pi/2}^0$$

$$= \tfrac{1}{4}\pi$$

where π is defined analytically. But the area of the circle $x^2 + y^2 = 1$ is given by $\pi(1)^2 = \pi$, using the *geometric* definition of π. Hence the area of the quarter-circle is also $\frac{1}{4}\pi$, where π is the geometric ratio of the circumference of a circle to its diameter. Therefore the two definitions of π, the geometric and the analytic, do indeed coincide.

Now the following rigorous definitions can be made.

A.6   Definition

$$\tan x = \frac{\sin x}{\cos x} \quad (x \neq \tfrac{1}{2}(2n + 1)\pi)$$

A.7   Definition

$$\cot x = \frac{\cos x}{\sin x} \quad (x \neq n\pi)$$

A.8   Definition

$$\sec x = \frac{1}{\cos x} \quad (x \neq \tfrac{1}{2}(2n + 1)\pi)$$

A.9   Definition

$$\operatorname{cosec} x = \frac{1}{\sin x} \quad (x \neq n\pi)$$

By the quotient rule (6.1.4), these functions are differentiable on their domains and possess the derivatives quoted in Table 6.2. It is easy now to see that successive derivatives exist and hence that A.6–A.9 define infinitely differentiable functions.

As shown in Example 2 of Section 6.2, $\exp\colon \mathbb{R} \to \mathbb{R}$ is strictly increasing and $\exp x > 0$ for all real x. Hence the function $g\colon \mathbb{R} \to \mathbb{R}^+$ given by $g(x) = \exp x$ is a bijection.

A.10   Definition

The **logarithm** function $\log_e\colon \mathbb{R}^+ \to \mathbb{R}$ is defined as the inverse of exp.

By 5.2.7, $\log_e$ is continuous and, by Question 5 of Exercises 5.2, it is strictly increasing. Since

$$g'(x) = e^x \neq 0$$

6.1.7 shows that $\log_e$ is differentiable. As shown in Example 9 of Section 6.1,

$$\frac{d}{dx}(\log_e x) = \frac{1}{x} \quad \text{for } x > 0$$

Since $x \mapsto x^n$, $n \in \mathbb{Z}$, $n < 0$, is infinitely differentiable for $x > 0$, $\log_e$ is infinitely differentiable. Its nth derivative is given in Example 7(c) of Section 6.2.

The logarithm function can now be used to define powers of real numbers, including irrational powers.

A.11 Definition

If $a > 0$ and $x \in \mathbb{R}$ then define $a^x = \exp(x \log_e a)$.

By 6.1.6, $x \mapsto a^x$ is differentiable, and its derivative is given by

$$\frac{d}{dx}(a^x) = \exp(x \log_e x)\log_e a = a^x \log_e a$$

Thus the key analytic properties of the elementary functions used throughout the book can successfully be established.

Solutions to Exercises

Chapter 1

Exercises 1.1

1. The necessary truth tables are as follows:

(a)

P	(not P)	P or (not P)	P and (not P)	not (P and (not P))
T	F	T	F	T
F	T	T	F	T

(b)

P	Q	(not Q)	P and (not Q)	not (P and (not Q))	$P \Rightarrow Q$
T	T	F	F	T	T
T	F	T	T	F	F
F	T	F	F	T	T
F	F	T	F	T	T

(c)

P	Q	R	Q and R	$P \Rightarrow Q$	$P \Rightarrow (Q$ and $R)$	$(P \Rightarrow Q)$ and R
T	T	T	T	T	T	T
T	T	F	F	T	F	F
T	F	T	F	F	F	F
T	F	F	F	F	F	F
F	T	T	T	T	T	T
F	T	F	F	T	T	T
F	F	T	F	T	T	T
F	F	F	F	T	T	F

In (a) and (b) the statements are logically equivalent since, in both cases, the last two columns are identical. In (c) the statements are not equivalent.

2.   (a) The statement P is true.

       (b) The statement Q is true.

       (c) No conclusion can be drawn; P may be either true or false.

       (d) The statement P is false.

3.   (a) A tautology. See the truth table (a) in the answer to Question 1.

       (b) Not a tautology. See the truth table (a) in the answer to Question 1.

       (c) Not a tautology, since $(P \Rightarrow (\text{not } P))$ is false when P is true.

       (d) Not a tautology, since $((P \Rightarrow Q) \text{ or } (Q \Rightarrow P))$ and $(\text{not } Q)$ is false when Q is true.

4.   (a) and (b) No; consider $n = 9$, for example.

       (c) Yes; since $n = 12m = 6(2m)$, m a positive whole number.

       (d) Yes; since both 2 and 3 must be divisors of n.

       (e) and (f) Yes.

       (g) Yes; since $m^3 - m = (m - 1)m(m + 1)$, a product of three consecutive positive whole numbers, one of which must be divisible by 3, and at least one of which is divisible by 2.

       Now consider which of (c)–(g) are a consequence of the condition 'n is divisible by 6'.

       (c) No; consider $n = 18$, for example.

       (d) Yes; since n^2 must be divisible by 36.

       (e) No; consider $n = 6$, for example.

       (f) Yes.

       (g) No; since $m^3 - m = 6$ for $m = 2$, and $m^3 - m \geqslant 24$ for $m \geqslant 3$, and hence $m^3 - m = 12$ is impossible.

Therefore only (d) and (f) are logically equivalent to 'n is divisible by 6'.

5.   *Direct method of proof*

Write $n = p_1 p_2 \cdots p_r$, where p_1, p_2, ..., p_r are the prime factors of n. Then $n^2 = p_1^2 p_2^2 \cdots p_r^2$. Assume that n^2 is even. Then one of the p_i equals 2. Hence 2 is a prime factor of n, and therefore n is also even.

Indirect method of proof

Assume that n is odd and write $n = 2m + 1$, where m is another positive whole number. Then $n^2 = (2m + 1)^2 = 4m^2 + 4m + 1 = 2(2m^2 + 2m) + 1$; another odd number. Therefore n^2 is odd, as required.

Proof by contradiction

Assume that n^2 is even and that n is odd. Hence $n^2 + n$ is odd. But $n^2 + n = n(n + 1)$ is a product of consecutive positive whole numbers, one of which must be even. Hence $n^2 + n$ is also an even number. This is the desired contradiction.

6. Assume that mn^2 is even and, by way of contradiction, that both m and n are odd. Write $m = 2p + 1$ and $n = 2q + 1$, where p and q are positive whole numbers. Then $mn^2 = (2p + 1)(2q + 1)^2 = 8pq^2 + 8pq + 4q^2 + 2p + 4q + 1$, an odd number. This is the desired contradiction.

Exercises 1.2

1. See Figure S.1.

2. (a) $\mathscr{C}(\mathscr{C}A) = \{x : x \in \mathscr{C}(\mathscr{C}A)\}$

    $= \{x : \text{not}\,(x \in \mathscr{C}A)\}$

    $= \{x : \text{not}\,(\text{not}\,(x \in A))\}$

    $= \{x : x \in A\}$

    since not (not P) is logically equivalent to P. Hence $\mathscr{C}(\mathscr{C}A) = A$.

    (b) $A \cap (B \cup C) = \{x : x \in A \text{ and } x \in B \cup C\}$

    $= \{x : x \in A \text{ and } (x \in B \text{ or } x \in C)\}$

    $= \{x : (x \in A \text{ and } x \in B) \text{ or } ((x \in A \text{ and } x \in C)\}$

    since $(P \text{ and } (Q \text{ or } R))$ is logically equivalent to $((P \text{ and } Q) \text{ or } (P \text{ and } R))$. Hence $A \cap (B \cup C) = (A \cap B) \cup (A \cap C)$.

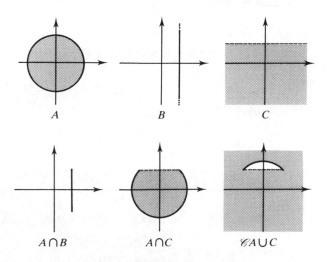

Figure S.1

(c) $\mathscr{C}(A \cap B) = \{x : \text{not}\,(x \in A \cap B)\}$

$$= \{x : \text{not}\,(x \in A \text{ and } x \in B)\}$$

$$= \{x : (\text{not}\,(x \in A)) \text{ or } (\text{not}\,(x \in B))\}$$

since $(\text{not}\,(P \text{ and } Q))$ is logically equivalent to $((\text{not } P) \text{ or } (\text{not } Q))$. Hence $\mathscr{C}(A \cap B) = \mathscr{C}A \cup \mathscr{C}B$.

The logical equivalences in (a)–(c) can easily be established using truth tables.

3.　　(a)　$A \oplus B = (A \cup B) \cap \mathscr{C}(A \cap B)$

$\qquad\qquad = (B \cup A) \cap \mathscr{C}(B \cap A)$　　　using the commutative laws

$\qquad\qquad = B \oplus A$

(b)　$A \oplus A = (A \cup A) \cap \mathscr{C}(A \cap A)$

$\qquad\qquad = (A) \cap \mathscr{C}A$　　　using the idempotent laws

$\qquad\qquad = \varnothing$　　　using a complement law

(c)　First note that

$$X \oplus Y = (X \cup Y) \cap \mathscr{C}(X \cap Y)$$

$$= (X \cup Y) \cap (\mathscr{C}X \cup \mathscr{C}Y)$$

$$= ((X \cup Y) \cap \mathscr{C}X) \cup ((X \cup Y) \cap \mathscr{C}Y)$$

$$= ((X \cup \mathscr{C}X) \cup (Y \cap \mathscr{C}X))$$

$$\qquad \cup ((X \cap \mathscr{C}Y) \cup (Y \cap \mathscr{C}Y))$$

$$= (\varnothing \cup (Y \cap \mathscr{C}X)) \cup ((X \cap \mathscr{C}Y) \cup \varnothing)$$

$$= (\mathscr{C}X \cap Y) \cup (X \cap \mathscr{C}Y)$$

freely using the laws (1.2.1). Hence

$$A \cap (B \oplus C) = A \cap ((\mathscr{C}B \cap C) \cup (B \cap \mathscr{C}C))$$

$$= (A \cap \mathscr{C}B \cap C) \cup (A \cap B \cap \mathscr{C}C)$$

and

$$(A \cap B) \oplus (A \cap C) = (\mathscr{C}(A \cap B) \cap (A \cap C))$$

$$\cup ((A \cap B) \cap \mathscr{C}(A \cap C))$$

$$= ((\mathscr{C}A \cup \mathscr{C}B) \cap (A \cap C))$$

$$\cup ((A \cap B) \cap (\mathscr{C}A \cup \mathscr{C}C))$$

$$= (\mathscr{C}A \cap (A \cap C))$$

$$\cup (\mathscr{C}B \cap (A \cap C))$$

$$\cup ((A \cap B) \cap \mathscr{C}A)$$

$$\cup ((A \cap B) \cap \mathscr{C}C))$$

$$= \varnothing \cup (A \cap \mathscr{C}B \cap C) \cup \varnothing$$
$$\cup (A \cap B \cap \mathscr{C}C)$$
$$= (A \cap \mathscr{C}B \cap C) \cup (A \cap B \cap \mathscr{C}C)$$

Therefore $A \cap (B \oplus C) = (A \cap B) \oplus (A \cap C)$.

4.   (a) False; a counterexample being $n = 2$.

(b) True; for example $n = 5$.

(c) True; since $n^2 - n = (n-1)n$, a product of consecutive whole numbers, one of which must be even.

(d) False; since $n^3 - n = (n-1)n(n+1)$, a product of three consecutive whole numbers, at least one of which must be even.

Exercises 1.3

---

1.   In this question, the required information may be gleaned from sketch-graphs of the functions involved.

(a) A suitable domain is $A = \{0\}$. The image is $f(A) = \{0\}$.

(b) A suitable domain is $A = \{x : x \in \mathbb{R} \text{ and } x \geqslant 3\} = [3, \infty)$ using a natural extension of the interval notation introduced after Example 1.3.1. The image is $f(A) = [0, \infty)$.

(c) A suitable domain is $A = \mathbb{R} - \{-1, 1\}$. The image is $f(A) = \mathbb{R} - (-1, 0]$.

(d) A suitable domain is $A = \{x : x \in \mathbb{R} \text{ and } 2n\pi < x < (2n+1)\pi, \text{ for some } n \in \mathbb{Z}\}$. The image is $f(A) = (-\infty, 0]$.

2.   All the composites involved define functions with domain and codomain $\mathbb{R}$. Since $f(x) = x^2$ for all $x \in \mathbb{R}$, the composites $f \circ f$ and $g \circ f$ are given by

$$(f \circ f)(x) = f(x^2) = x^4$$

and

$$(g \circ f)(x) = g(x^2) = -x^2$$

The other composites are calculated as follows:

$$(f \circ g)(x) = \begin{cases} f(x-1) & \text{if } x \geqslant 0 \\ f(-x) & \text{if } x < 0 \end{cases}$$

$$= \begin{cases} (x-1)^2 & \text{if } x \geqslant 0 \\ x^2 & \text{if } x < 0 \end{cases}$$

and

$$(g \circ g)(x) = \begin{cases} g(x-1) & \text{if } x \geq 0 \\ g(-x) & \text{if } x < 0 \end{cases}$$

$$= \begin{cases} (x-1)-1 & \text{if } x \geq 1 \\ -(x-1) & \text{if } 0 \leq x < 1 \\ (-x)-1 & \text{if } x < 0 \end{cases}$$

$$= \begin{cases} x-2 & \text{if } x \geq 1 \\ 1-x & \text{if } 0 \leq x < 1 \\ -(x+1) & \text{if } x < 0 \end{cases}$$

3.   As with Question 1, sketch-graphs of the functions involved are helpful.

    (a) Not bijective. For example, $f(-1) = f(1) = 1$ shows that f is not injective.

    (b) Not bijective. Since $g(x) \geq 1$ for $x \geq 0$ and $g(x) < -1$ for $x < 0$, the image of g does not contain any elements in the interval $[-1, 0)$; hence $g(\mathbb{R}) \neq \mathbb{R}$.

    (c) Bijective. Since the sign of $h(x)$ is always the same as the sign of x, if $h(x_1) = h(x_2)$ then either $x_1^2 = x_2^2$ or else $2x_1 = 2x_2$. Hence $x_1 = x_2$, and so h is injective. Also, $h(\mathbb{R}) = \mathbb{R}$.

    (d) Not bijective. There is no x satisfying $k(x) = 1$.

4.   (a) The equation $y = 1/(2x - 1)$ admits the unique solution $x = (y + 1)/2y$, for $x \neq 1/2$ and $y \neq 0$. Hence $f^{-1}: \mathbb{R} - \{0\} \to \mathbb{R} - \{1/2\}$ is given by the rule $x \mapsto (x + 1)/2x$.

    (b) The equation $y = 1/(x^2 + 1)$ admits the unique solution $x = \sqrt{(1 - y)/y}$ for $x \geq 0$ and $0 < y \leq 1$. Hence $g^{-1}: (0, 1] \to [0, \infty)$ is given by the rule $x \mapsto \sqrt{(1 - x)/x}$.

    (c) The equation $y = (3x + 2)/(x - 1)$ admits the unique solution $x = (y + 2)/(y - 3)$, for $x \neq 1$ and $y \neq 3$. Since $h(1) = 3$, the inverse function $h^{-1}: \mathbb{R} \to \mathbb{R}$ is given by the rule $x \mapsto (x + 2)/(x - 3)$, for $x \neq 3$, and $h^{-1}(3) = 1$.

5.   Since f is bijective, the inverse function f^{-1} exists. Let $a \in A$. Then $f(a) = b \in A$ and $f^{-1}(b) = a$. Hence $(f^{-1} \circ f)(a) = f^{-1}(f(a)) = f^{-1}(b) = a$, and so $f^{-1} \circ f = \text{id}_A$. Now f^{-1} is a bijective function with inverse $(f^{-1})^{-1} = f$. Hence an analogous argument gives $f \circ f^{-1} = \text{id}_A$.

Chapter 2

Exercises 2.1

1.   Suppose that $\sqrt{5}$ is rational, and write $\sqrt{5} = p/q$, p, q integers with no common factors greater than one and $q \neq 0$. Hence $q^2 = 5p^2$, and so q^2

and hence q are divisible by 5. Write $q = 5q_1$. Then $p^2 = 5q_1^2$, and so p is also divisible by 5. This is a contradiction, and so $\sqrt{5}$ is in fact irrational.

Suppose now that $a + b\sqrt{5}$ is rational, where a, b are irrational. Then $a + b\sqrt{5} = c \in \mathbb{Q}$. Hence $\sqrt{5} = (c - a)/b$ is rational, which contradicts the first part of the solution. Hence $a + b\sqrt{5}$ is irrational.

If $r = 1 + 1/r$, $r > 0$, then r is the positive root of $r^2 - r - 1 = 0$. Solving gives $r = \frac{1}{2}(1 + \sqrt{5})$. Thus r is irrational.

2.   (a) True. Let $x = m/n$ where m, n are integers, $n \neq 0$. Let y be an irrational and suppose, by way of contradiction, that $x + y$ is rational. Then there exist integers p, q, $q \neq 0$, such that $x + y = p/q$. Hence

$$y = \frac{p}{q} - \frac{m}{n} = \frac{pn - mq}{qn}$$

which is rational. So y is both irrational and rational, and thus the assumption that $x + y$ was rational is false. Hence $x + y$ is irrational.

(b) True. Let $x = m/n$ and $y = p/q$, where m, n, p, q are integers, $n \neq 0$, $q \neq 0$. Thus

$$x + y = \frac{m}{n} + \frac{p}{q} = \frac{mq + np}{nq}$$

is a rational.

(c) False. For example, $\sqrt{2} + (-\sqrt{2}) = 0$ is rational.

3.   Let a, b be real numbers with $a < b$.

(1) There is an irrational between a and b. This is proved in the text for a and b both rational. If a is rational and b is irrational, or vice versa, then $x = \frac{1}{2}(a + b)$ is an irrational between a and b. If a and b are both irrational then, as shown in the text, there is a rational c between a and b. Since there is an irrational between a and c, the result follows.

(2) There is a rational between a and b. This is proved in the text for a and b both irrational. If a is rational and b is irrational, or vice versa, then there is an irrational c between a and b by (1) above. Since c and b are both irrational, there is a rational between c and b by the first part of (2) above. Hence there is a rational between a and b. Finally, if a and b are both rational then $x = \frac{1}{2}(a + b)$ is a rational between a and b.

Between any two real numbers a and b there is a rational a_1 and an irrational b_1. Between a_1 and b_1 there is a rational a_2 and an irrational b_2 ..., so there are infinitely many rationals and irrationals between any two distinct real numbers.

4.   Suppose that there is a rational number m/n with $10^{m/n} = 2$. Hence $10^m = 2^n$, giving $2^m \cdot 5^m = 2^n$. Thus $m = 0$ and $m = n$, which is nonsense. Hence $10^x = 2$ implies that $x = \log_{10} 2$ is irrational.

5.
$$x^2 = (\sqrt{3 + 2\sqrt{2}})^2 - 2(\sqrt{3 + 2\sqrt{2}})(\sqrt{3 - 2\sqrt{2}})$$
$$+ (\sqrt{3 - 2\sqrt{2}})^2$$
$$= (3 + 2\sqrt{2}) - 2\sqrt{(3 + 2\sqrt{2})(3 - 2\sqrt{2})} + (3 - 2\sqrt{2})$$
$$= (3 + 2\sqrt{2}) - 2\sqrt{(9 - (2\sqrt{2})^2)} + (3 - 2\sqrt{2})$$
$$= 6 - 2\sqrt{9 - 8} = 4$$

Since x is clearly positive, $x = 2$, a rational.

Exercises 2.2

1.   (a) $(-x)\cdot(-y) = -((-x)\cdot y)$ by Example 1(b) of Section 2.2
$$= -(y\cdot(-x)) \quad \text{by A6}$$
$$= -(-(y\cdot x)) \quad \text{by Example 1(b) of Section 2.2}$$
$$= -(-(x\cdot y)) \quad \text{by A6}$$

To complete the proof, it suffices to show that $-(-a) = a$ for all $a \in \mathbb{R}$. Now $(-a) + (-(-a)) = 0$ by A4, and $(-a) + a = 0$ by A4 and A2. Hence the uniqueness condition in A4 implies that $-(-a) = a$.

(b) By A8 and part (a),

$$(-x)\cdot(-(x)^{-1}) = x\cdot x^{-1} = 1$$

Hence, by the uniqueness condition in A8, $-(x)^{-1} = (-x)^{-1}$.

(c) By A8 and A6,

$$x^{-1}\cdot x = x\cdot x^{-1} = 1$$

Hence, by the uniqueness condition in A8, $(x^{-1})^{-1} = x$.

2.   (a) $0 \leqslant x$ and $0 \leqslant y \Rightarrow 0\cdot y \leqslant x\cdot y$ by A14
$$\Rightarrow 0 \leqslant x\cdot y \quad \text{by A6 and Example 1(a)}$$
$$\text{of Section 2.2}$$

(b) $0 \leqslant x$ and $y \leqslant 0 \Rightarrow x\cdot y \leqslant 0\cdot y$ by Example 4 of Section 2.2
$$\Rightarrow x\cdot y \leqslant 0 \quad \text{by A6 and Example 1(a)}$$
$$\text{of Section 2.2}$$

(c) $x \leqslant 0$ and $y \leqslant 0 \Rightarrow 0\cdot y \leqslant x\cdot y$ by Example 4 of Section 2.2
$$\Rightarrow 0 \leqslant x\cdot y \quad \text{by A6 and Example 1(a)}$$
$$\text{of Section 2.2}$$

3.   (a) $\dfrac{1}{4} < \dfrac{1}{x+3} \Leftrightarrow \dfrac{1}{x+3} - \dfrac{1}{4} > 0$

$\Leftrightarrow \dfrac{1-x}{x+3} > 0$

$\Leftrightarrow 1-x$ and $x+3$ have the same sign

$\Leftrightarrow -3 < x < 1$

(b) $\dfrac{x-1}{x+1} \leqslant \dfrac{x+1}{x-1} \Leftrightarrow \dfrac{x+1}{x-1} - \dfrac{x-1}{x+1} \geqslant 0$

$\Leftrightarrow \dfrac{(x+1)^2 - (x-1)^2}{x^2-1} \geqslant 0$

$\Leftrightarrow \dfrac{4x}{x^2-1} \geqslant 0$

$\Leftrightarrow x = 0$ or (x and $x^2 - 1$ have the same sign)

$\Leftrightarrow x = 0$ or $1 < x$ or $-1 < x < 0$

$\Leftrightarrow -1 < x \leqslant 0$ or $1 < x$

4.   By the triangle inequality

$$|x| = |(x-y) + y| \leqslant |x-y| + |y|$$

Hence

$$|x| - |y| \leqslant |x-y| \quad \text{for all} \ x, y \in \mathbb{R}$$

Reversing the roles of x and y gives that

$$|y| - |x| \leqslant |y-x| = |x-y|$$

Hence $\big||x| - |y|\big| \leqslant |x-y|$.

5.   Since S is bounded above, there exists a real number M such that $x \leqslant M$ for all $x \in S$. But then $-x \geqslant -M$ for all $-x \in T$. In other words, T is bounded below.

Suppose now that M is the supremum of S. Then $-M$ is certainly a lower bound for T. Let m be the infimum of T. Then $-x \geqslant m \geqslant -M$ for all $-x \in T$. Hence $x \leqslant -m \leqslant M$ for all $x \in S$, and so $-m = M$ by the definition of M. In other words, $\inf T = -\sup S$.

Exercises 2.3

1.   Axioms A1, A2, A5, A6, A9 and A10–A14 hold for *any* non-empty subset S of $\mathbb{R}$ for which $x + y \in S$ and $xy \in S$ whenever $x, y \in S$. Let $x = a + b\sqrt{2}$ and $y = c + d\sqrt{2}$, where a, b, c and d are rationals. Now

$$x + y = (a+c) + (b+d)\sqrt{2} \in S$$

and

$$xy = (a + b\sqrt{2})(c + d\sqrt{2})$$
$$= (ac + 2bd) + (ad + bc)\sqrt{2} \in S$$

since $a + c$, $b + d$, $ac + 2bd$ and $ad + bc$ are all rational.
A3 holds since

$$0 = 0 + 0\sqrt{2} \in S$$

A4 holds since

$$-(a + b\sqrt{2}) = (-a) + (-b)\sqrt{2} \in S$$

A7 holds since

$$1 = 1 + 0\sqrt{2} \in S$$

A8 holds since

$$(a + b\sqrt{2})^{-1} = \frac{1}{a + b\sqrt{2}} = \frac{a - b\sqrt{2}}{(a + b\sqrt{2})(a - b\sqrt{2})}$$
$$= \frac{a}{a^2 + 2b^2} + \frac{(-b)}{a^2 + 2b^2}\sqrt{2} \in S$$

Thus S satisfies A1–A14, and so forms an ordered field. Suppose that $\sqrt{3} \in S$. Then $\sqrt{3} = a + b\sqrt{2}$ for some rationals a and b. Hence

$$3 = (a + b\sqrt{2})^2 = a^2 + 2b^2 + 2ab\sqrt{2}$$

and so

$$\sqrt{2} = \frac{3 - a^2 - 2b^2}{2ab}, \quad \text{a rational}$$

But $\sqrt{2}$ is irrational, and so we have a contradiction. In other words, $\sqrt{3} \notin S$. Thus S is *not* complete.

2.   (a) Now

$$|2x - 1| < 11 \Leftrightarrow -11 < 2x - 1 < 11$$
$$\Leftrightarrow -5 < x < 6$$

Hence sup $S = 6$ and inf $S = -5$.
(b) For $x \geqslant 1$

$$x + |x - 1| = 2x - 1$$

which can take arbitrarily large values. If $x < 1$ then

$$x + |x - 1| = 1$$

Hence sup S does not exist and inf $S = 1$.

(c) $S = \{\ldots,\ 1 + \frac{1}{3},\ 1 + \frac{1}{2},\ 1 + 1,\ 1 - 1,\ 1 - \frac{1}{2},\ 1 - \frac{1}{3},\ \ldots\}$ and so sup $S = 2$ and inf $S = 0$.

(d) $2^{-m} + 3^{-n} + 5^{-p} = 1/2^m + 1/3^n + 1/5^p$ decreases in value as m, n and p take larger values. Hence

$$\sup S = \frac{1}{2} + \frac{1}{3} + \frac{1}{5} = \frac{31}{30}$$

and inf $S = 0$.

(e) For even n

$$\frac{(-1)^n n}{2n + 1} = \frac{n}{2n + 1} = \frac{1}{2}\left(1 - \frac{1}{2n + 1}\right)$$

which increases as n does. For odd n

$$\frac{(-1)^n n}{2n + 1} = \frac{-1}{2n + 1} = -\frac{1}{2}\left(1 - \frac{1}{2n + 1}\right)$$

which decreases as n increases. Hence sup $S = \frac{1}{2}$ and inf $S = -\frac{1}{2}$.

3. From the formulae in (e) above, S is bounded above by $\frac{1}{2}$ and below by $-\frac{1}{2}$. Suppose that $M < \frac{1}{2}$ is an upper bound for S. Now

$$\frac{(-1)^n n}{2n + 1} \leqslant M \quad \text{for all } n \Rightarrow \frac{2m}{4m + 1} \leqslant M \quad \text{for all } m(n = 2m)$$

$$\Rightarrow 2m \leqslant M(4m + 1) \quad \text{for all } m$$

$$\Rightarrow (2 - 4M)m \leqslant M \quad \text{for all } m$$

$$\Rightarrow m \leqslant \frac{M}{2 - 4M} \quad \text{for all } m,$$

$$\text{since } 2 - 4M > 0$$

This contradicts the Archimedean postulate, and so no such M exists. In other other words, sup $S = \frac{1}{2}$.

Now suppose that $M' > -\frac{1}{2}$ is a lower bound for S. Now

$$\frac{(-1)^n n}{2n + 1} \geqslant M' \quad \text{for all } n \Rightarrow \frac{-(2m + 1)}{4m + 3} \geqslant M'$$

$$\text{for all } m(n = 2m + 1)$$

$$\Rightarrow -(2m + 1) \geqslant M'(4m + 3) \quad \text{for all } m$$

$$\Rightarrow -(3M' + 1) \geqslant m(4M' + 2) \quad \text{for all } m$$

$$\Rightarrow m \leqslant \frac{-3M' - 1}{4M' + 2} \quad \text{for all } m,$$

$$\text{since } 4M' + 2 > 0$$

This contradicts the Archimedean postulate, and so no such M' exists. In other words, $\inf S = -\frac{1}{2}$.

4.　From the given information, $\sup A$ and $\sup B$ exist. Moreover, $\sup A \cdot \sup B > 0$. Now $x \cdot y \leqslant \sup A \cdot \sup B$ for all $x \in A$ and all $y \in B$. Hence C is bounded above. Let $K = \sup C$. Then $K \leqslant \sup A \cdot \sup B$. Suppose that $K < \sup A \cdot \sup B$. Then $K/\sup A < \sup B$. Hence there exists a $y' \in B$ with $K/\sup A < y'$. Now $K/y' < \sup A$, and so there exists an $x' \in A$ with $K/y' < x'$. Hence $x'y' > K$, and this contradicts the definition of K. Therefore $K = \sup C \geqslant \sup A \cdot \sup B$, and so $\sup C = \sup A \cdot \sup B$. A similar argument with $\inf A$ and $\inf B$ shows that $\inf C$ exists and that $\inf C = \inf A \cdot \inf B$, although care is needed with the case $\inf A \cdot \inf B = 0$.

5.　(a)　$\sum_{r=1}^{1} r^2 = 1$ and $\frac{1}{6}(1+1)(2+1) = 1$, so the statement is true when $n = 1$. If

$$\sum_{r=1}^{k} r^2 = \frac{1}{6}k(k+1)(2k+1) \quad \text{for some } k \geqslant 1$$

then

$$\sum_{r=1}^{k+1} r^2 = \sum_{r=1}^{k} r^2 + (k+1)^2 = \frac{1}{6}k(k+1)(2k+1)$$

$$+ (k+1)^2$$

$$= \frac{1}{6}(k+1)[(2k^2 + k)$$

$$+ 6(k+1)]$$

$$= \frac{1}{6}(k+1)(2k^2 + 7k + 6)$$

$$= \frac{1}{6}(k+1)(k+2)(2k+3)$$

and so the formula holds for $n = k + 1$. By induction,

$$\sum_{r=1}^{n} r^2 = \frac{1}{6}n(n+1)(2n+1) \quad \text{for all } n \in \mathbb{N}$$

(b)　Since $x^2 > 0$ for $x \neq 0$,

$$1 + 2x < 1 + 2x + x^2 = (1+x)^2$$

Hence the statement holds for $n = 2$. If $1 + kx < (1+x)^k$ for some $k \geqslant 2$ then

$$(1+x)^{k+1} = (1+x)^k(1+x) > (1+kx)(1+x)$$

$$\text{since } x > -1$$

$$= 1 + (k+1)x + kx^2 > 1 + (k+1)x$$

$$\text{since } kx^2 > 0$$

Thus the formula holds for $n = k + 1$ also, and the result follows by induction.

(c) $2^4 = 16$ and $4! = 24$ verifies the given inequality when $n = 4$. Suppose that $2^k < k!$ for some $k \geqslant 4$. Then

$$2^{k+1} = 2 \cdot 2^k < 2 \cdot k! < (k + 1)k! \quad \text{since } k + 1 > 2$$

Hence $2^{k+1} < (k + 1)!$ By induction, $2^n < n!$ for all $n \geqslant 4$.

6.   The result is clear for a set containing one element. Suppose the result holds for sets containing k elements for some $k \geqslant 1$. Let

$$S = \{a_1, a_2, \ldots, a_{k+1}\}$$

a set containing $k + 1$ elements. By the inductive hypothesis, the set $\{a_2, \ldots, a_k\}$ contains a minimum element. Call this element a. By Axiom A10, $a \leqslant a_1$ or $a_1 \leqslant a$, and so, by A12, either a or a_1 is the minimum element of S. Hence every set containing $k + 1$ elements has a minimum element. The induction proof is complete.

7.   By the triangle inequality, the result holds for $n = 2$. Now, by the triangle inequality,

$$|a_1 + a_2 + \ldots + a_{k+1}| \leqslant |a_1 + a_2 + \ldots + a_k| + |a_{k+1}|$$

Hence if

$$|a_1 + a_2 + \ldots + a_k| \leqslant |a_1| + |a_2| + \ldots + |a_k|$$

for some $k \geqslant 2$ then

$$|a_1 + a_2 + \ldots + a_{k+1}| \leqslant |a_1| + |a_2| + \ldots + |a_k| + |a_{k+1}|$$

as required. By induction, the result follows.

Chapter 3

Exercises 3.1

1.   (a) Now

$$\left| \frac{n - 1}{2n} - \frac{1}{2} \right| < \varepsilon$$

provided that $|-1/2n| < \varepsilon$, which requires $n > 1/2\varepsilon$. So, given any

$\varepsilon > 0$, choose N to be an integer exceeding $1/2\varepsilon$. Thus

$$n > N \Rightarrow \left| \frac{n-1}{2} - \frac{1}{2} \right| < \varepsilon$$

and so

$$\frac{n-1}{2n} \to \frac{1}{2} \quad \text{as } n \to \infty$$

(b) Now

$$\left| \frac{(-1)^n}{n^2} - 0 \right| < \varepsilon$$

provided that $1/n^2 < \varepsilon$, which requires $n > 1/\sqrt{\varepsilon}$. So, given $\varepsilon > 0$, choose N to be an integer exceeding $1/\sqrt{\varepsilon}$. Thus

$$n > N \Rightarrow \left| \frac{(-1)^n}{n^2} - 0 \right| < \varepsilon$$

and so

$$\frac{(-1)^n}{n^2} \to 0 \quad \text{as } n \to \infty$$

2. If $a_n \to A$ as $n \to \infty$ then, by the scalar product rule, $\alpha a_n \to \alpha A$ as $n \to \infty$. Similarly, $\beta b_n \to \beta B$ as $n \to \infty$. Hence, by the sum rule,

$$\alpha a_n + \beta b_n \to \alpha A + \beta B \quad \text{as } n \to \infty$$

3. (a) $\dfrac{4n^3 + 6n - 7}{n^3 - 2n^2 + 1} = \dfrac{4 + 6/n^2 - 7/n^3}{1 - 2/n + 1/n^3}$

$$\to \frac{4 + 0 - 0}{1 - 0 + 0} = 4 \quad \text{as } n \to \infty$$

using the sum, product, scalar product and quotient rules freely, together with the basic fact that $1/n \to 0$ as $n \to \infty$.

(b) $\dfrac{6 - n^2}{n^2 + 5n} = \dfrac{6/n^2 - 1}{1 + 5/n} \to \dfrac{0 - 1}{1 + 0} = -1 \quad \text{as } n \to \infty$

(c) $\log_e (n + 1) - \log_e n = \log_e \left(\dfrac{n+1}{n} \right) = \log_e \left(1 + \dfrac{1}{n} \right)$

Now $1 + 1/n \to 1$ as $n \to \infty$, by the sum rule, and so, by the composite rule,

$$\log_e \left(1 + \frac{1}{n} \right) \to \log_e 1 = 0 \quad \text{as } n \to \infty$$

4.   (a) $\left|\dfrac{(-1)^n n}{\sqrt{n^3+1}}\right| = \dfrac{n}{\sqrt{n^3+1}} \leqslant \dfrac{n}{\sqrt{n^3}} = \left(\dfrac{1}{n}\right)^{1/2}$

Since $1/n \to 0$ as $n \to \infty$, the composite rule gives that $(1/n)^{1/2} \to 0$ as $n \to \infty$, and thus $-(1/n)^{1/2} \to 0$ as $n \to \infty$, by the scalar product rule. Now

$$-\left(\frac{1}{n}\right)^{1/2} \leqslant \frac{(-1)^n n}{\sqrt{n^3+1}} \leqslant \left(\frac{1}{n}\right)^{1/2}$$

and so, by the sandwich rule,

$$\frac{(-1)^n n}{\sqrt{n^3+1}} \to 0 \quad \text{as } n \to \infty$$

(b) Now

$$\left|\frac{\cos n}{n}\right| \leqslant \frac{1}{n}$$

since $-1 \leqslant \cos x \leqslant 1$ for all $x \in \mathbb{R}$. Hence

$$-\frac{1}{n} \leqslant \frac{\cos n}{n} \leqslant \frac{1}{n} \quad \text{for all } n \in \mathbb{N}$$

and so, by the sandwich rule, $(\cos n)/n \to 0$ as $n \to \infty$.

Exercises 3.2

1.   Suppose that the sequence (a_n) is null. Then for every $\varepsilon > 0$ there exists a natural number N such that $n > N \Rightarrow |a_n| < \varepsilon$. For this value of N, $n > N \Rightarrow \|a_n\| = |a_n| < \varepsilon$, and so $(|a_n|)$ is a null sequence. The converse is proved in a similar fashion. Hence (a_n) is null if and only if $(|a_n|)$ is null.

If the word 'null' is replaced by the word 'convergent', the result is false. For example, the sequence (a_n) given by $a_n = (-1)^n$ is divergent, but the sequence $(|a_n|)$ is the convergent sequence $1, 1, 1, \ldots$. What can be said is that if $a_n \to L$ as $n \to \infty$ then $|a_n| \to |L|$ as $n \to \infty$. To see this, note that $\|a_n| - |L\| < |a_n - L|$ by Question 4 of Exercises 2.2, and appeal directly to Definition 3.1.1.

2.   (a) The dominant term is $n!$. Hence

$$\frac{n^2+2^n}{n!+3n^3} = \frac{n^2/n! + 2^n/n!}{1 + 3(n^3/n!)}$$

Now using the basic null sequences (3.2.2) and the rules (3.1.2),

$$\frac{n^2+2^n}{n!+3n^3} \to \frac{0+0}{1+0} = 0 \quad \text{as } n \to \infty$$

(b) The dominant term is $(n + 1)!$. Hence

$$\frac{2n!}{(n+1)! + (n-1)!} = \frac{2/(n+1)}{1 + 1/n(n+1)}$$

Now using the basic null sequences (3.2.2) and the rules (3.1.2),

$$\frac{2n!}{(n+1)! + (n-1)!} \to \frac{0}{1+0} = 0 \quad \text{as } n \to \infty$$

(c) Write

$$\frac{n^4 4^n}{n!} = \left(\frac{n^4}{4^n}\right)\left(\frac{16^n}{n!}\right) = \left(n^4\left(\frac{1}{4}\right)^n\right)\left(\frac{16^n}{n!}\right)$$

Hence $(n^4 4^n/n!)$ is a null sequence, since it is the product of two basic null sequences.

3. Since (b_n) is bounded, there exists a real number $M > 0$ with $|b_n| \leq M$ for all $n \in \mathbb{N}$. For any $\varepsilon > 0$, set $\varepsilon' = \varepsilon/M$. Since (a_n) is null and $\varepsilon' > 0$, there exists a natural number N such that $n > N \Rightarrow |a_n| < \varepsilon'$. But then $n > N \Rightarrow |a_n b_n| \leq |a_n| M < \varepsilon' M = \varepsilon$. In other words, $(a_n b_n)$ is null.

4. (a) The dominant term is 2^n. Hence

$$\frac{n^2 - 2^n}{2^n + n} = \frac{n^2/2^n - 1}{1 + n/2^n} \to \frac{0 - 1}{1 + 0} = -1 \quad \text{as } n \to \infty$$

(b) The dominant term is $n!$. Hence

$$\frac{3n! + 3^n}{n! + n^3} = \frac{3 + 3^n/n!}{1 + n^3/n!} \to \frac{3 + 0}{1 + 0} = 3 \quad \text{as } n \to \infty$$

(c) Since $3^n \leq 2^n + 3^n \leq 2(3^n)$, the following inequalities hold:

$$3 \leq (2^n + 3^n)^{1/n} \leq 3(2^{1/n})$$

Now $2^{1/n} \to 1$ as $n \to \infty$, by Example 3(b) of Section 3.2. Hence, by the sandwich rule (3.1.3), $(2^n + 3^n)^{1/n} \to 3$ as $n \to \infty$.

5. Let $x = \sqrt{2/(n-1)}$, $n \geq 2$, in the inequality $(1 + x)^n \geq \frac{1}{2}n(n-1)x^2$. Hence $[1 + \sqrt{2/(n-1)}]^n \geq n$ for $n \geq 2$. This gives

$$n^{1/n} \leq 1 + \sqrt{\frac{2}{n-1}} \quad \text{for } n \geq 2$$

Since $n \geq 2$, $n^{1/n} \geq 2^{1/n}$, giving the inequality

$$2^{1/n} \leq n^{1/n} \leq 1 + \sqrt{\frac{2}{n-1}} \quad \text{for } n \geq 2$$

Now apply the sandwich rule (3.1.3), to deduce that $\lim_{n \to \infty} n^{1/n} = 1$.

Exercises 3.3

1. (a) The sequence $(n^3/2^n) = (n^3(\frac{1}{2})^n)$ is a basic null sequence. By the reciprocal rule (3.3.4), $(2^n/n^3)$ tends to infinity, and hence is divergent.

    (b) Since $n! > n^3$ for $n \geqslant 6$ (proved by induction on n), the sequence $(n! - n^3)$ is eventually positive. Hence the sequence $((n! - n^3)/3^n)$ is eventually positive. Now

$$\frac{3^n}{n! - n^3} = \frac{3^n/n!}{1 - n^3/n!} \to \frac{0}{1 - 0} = 0 \quad \text{as } n \to \infty$$

    By the reciprocal rule (3.3.4), $((n! - n^3)/3^n)$ is divergent.

    (c) By Question 5 of Exercises 3.2, $n^{1/n} \to 1$ as $n \to \infty$. Hence

$$\frac{1}{n^{1+1/n}} = \left(\frac{1}{n}\right)\left(\frac{1}{n^{1/n}}\right)$$

    and so $1/n^{1+1/n} \to 0.1 = 0$ as $n \to \infty$. By the reciprocal rule (3.3.4), $(n^{1+1/n})$ is divergent.

2. Let $\varepsilon > 0$ and $K = 1/\varepsilon$. Since $a_n \to \infty$ as $n \to \infty$, there exists an integer N such that $a_n > K > 0$ for all $n > N$. Hence

$$\left|\frac{1}{a_n}\right| = \frac{1}{a_n} < \frac{1}{K} = \varepsilon \quad \text{for } n > N$$

    In other words, $1/a_n \to 0$ as $n \to \infty$.

3. (a) If $a_n = 2n + (-1)^n$, the subsequences given by setting $n = 2k$ and $n = 2k - 1$ are $(a_{2k}) = (4k + 1)$ and $(a_{2k-1}) = (4k - 3)$ respectively. A straightforward application of the reciprocal rule shows that both these subsequences tend to infinity. Hence, by Strategy 2, $(2n + (-1)^n)$ is divergent.

    (b) If $a_n = (-1)^n n/(2n + 1)$, the subsequences given by setting $n = 2k$ and $n = 2k - 1$ are $(a_{2k}) = (2k/(4k + 1))$ and $(a_{2k-1}) = ((-2k + 1)/(4k - 1))$ respectively. Both of these subsequences converge, the first to $\frac{1}{2}$ and the second to $-\frac{1}{2}$. Hence, by Strategy 1, $((-1)^n n/(2n + 1))$ is divergent.

    (c) If $a_n = \sin\frac{1}{3}n\pi$, the subsequence (a_{3k}) is the null sequence $0, 0, 0,$ .... The subsequence (a_{6k+1}) is the convergent sequence $\sin\frac{1}{3}\pi$, $\sin\frac{1}{3}\pi$, $\sin\frac{1}{3}\pi$, ... whose limit, $\sin\frac{1}{3}\pi \neq 0$. By strategy 1, $(\sin\frac{1}{3}n\pi)$ is divergent.

4. Since (a_{2k}) and (a_{2k-1}) both converge to L, for every $\varepsilon > 0$ there exist natural numbers K_1 and K_2 such that

$$k > K_1 \Rightarrow |a_{2k} - L| < \varepsilon$$

and

$$k > K_2 \Rightarrow |a_{2k-1} - L| < \varepsilon.$$

Now let N be the maximum of $2K_1$ and $2K_2 - 1$. Since each $n > N$ is either of the form $2k$, with $k > K_1$, or of the form $2k - 1$, with $k > K_2$, it follows that

$$n > N \Rightarrow |a_n - L| < \varepsilon$$

In other words, $\lim_{n \to \infty} a_n = L$.

Exercises 3.4

1.   (a) Consider

$$a_{n+1} - a_n = \frac{n + 2}{n + 3} - \frac{n + 1}{n + 2}$$

$$= \frac{1}{(n + 3)(n + 2)} > 0 \quad \text{for } n \geq 1$$

Hence $((n + 1)/(n + 2))$ is monotone increasing.

    (b) Consider

$$a_{n+1} - a_n = n + 1 + \frac{8}{n + 1} - n - \frac{8}{n} = 1 - \frac{8}{n(n + 1)}$$

Now

$$\frac{8}{n(n + 1)} \leq 1 \Leftrightarrow n(n + 1) \geq 8$$

and this last inequality holds for $n \geq 3$. Hence $(n + 8/n) = (9, 6, 5\frac{2}{3}, 6, \ldots)$ is *eventually* monotone increasing.

    (c) Consider $a_{n+1} - a_n = n + 1 + (-1)^{n+1} - n - (-1)^n = 1 - 2(-1)^n$. Hence $a_{n+1} < a_n$ for even and $a_{n+1} > a_n$ for n odd. Hence $(n + (-1)^n)$ is not monotone.

    (d) Consider $a_{n+1} - a_n = 2(n + 1) + (-1)^{n+1} - 2n - (-1)^n = 2(1 - (-1)^n) \geq 0$ for all n. Hence $(2n + (-1)^n)$ is monotone increasing.

2.   (a) Since $a_{n+1} = a_n + 1/(n + 1)^2 > a_n$, (a_n) is an increasing sequence.

    (b) $a_1 = 1 \leq 2 - 1/1$, and so the statement holds for $n = 1$. If $a_k \leq 2 - 1/k$ for some $k \geq 1$ then

$$a_{k+1} = a_k + \frac{1}{(k+1)^2} \le 2 - \frac{1}{k} + \frac{1}{(k+1)^2}$$

$$= 2 - \frac{1}{k+1} + \frac{1}{k+1} - \frac{1}{k} + \frac{1}{(k+1)^2}$$

$$= 2 - \frac{1}{k+1} - \frac{1}{k(k+1)^2}$$

$$\le 2 - \frac{1}{k+1}$$

and so the statement holds for $n = k+1$. By induction, $a_n \le 2 - 1/n$ for all $n \in \mathbb{N}$.

(c) By (b), $a_n \le 2$ for all n, and so (a_n) is an increasing sequence that is bounded above. By the principle of monotone sequences, (a_n) converges.

3. The first n terms of the sequence $((1 + 1/n)^n)$ are $(2/1)^1$, $(3/2)^2$, $(4/3)^3$, $\ldots$, $[(n+1)/n]^n$, and each is less than e since $((1 + 1/n)^n)$ is monotone increasing, with limit e. Taking the product of these terms gives

$$\frac{2^1 3^2 4^3 \cdots (n+1)^n}{2^2 3^3 \cdots n^n} < e^n$$

By cancellation,

$$\frac{(n+1)^n}{n!} < e^n$$

and so

$$n! > \left(\frac{n+1}{e}\right)^n$$

Chapter 4

Exercises 4.1

1. Since

$$\frac{1}{r!} - \frac{1}{(r+1)!} = \frac{(r+1) - 1}{(r+1)!} = \frac{r}{(r+1)!}$$

the nth partial sum s_n of $\sum_{r=1}^{\infty} r/(r+1)!$ may be written as

$$s_n = \sum_{r=1}^{n} \left[\frac{1}{r!} - \frac{1}{(r+1)!}\right]$$

Hence

$$s_n = \left(\frac{1}{1!} - \frac{1}{2!}\right) + \left(\frac{1}{2!} - \frac{1}{3!}\right) + \ldots + \left(\frac{1}{n!} - \frac{1}{(n+1)!}\right)$$

$$= 1 - \frac{1}{(n+1)!}$$

Now $s_n \to 1$ as $n \to \infty$, and so

$$\sum_{r=1}^{\infty} \frac{r}{(r+1)!}$$

converges and

$$\sum_{r=1}^{\infty} \frac{r}{(r+1)!} = 1$$

2.   (a) $\dfrac{n}{n+1} = \dfrac{1}{(1+1/n)} \to 1$ as $n \to \infty$

By the vanishing condition,

$$\sum_{r=1}^{\infty} \frac{r}{(r+1)}$$

is divergent.

(b) $n - \sqrt{n(n-1)} = \dfrac{[n - \sqrt{n(n-1)}][n + \sqrt{n(n-1)}]}{n + \sqrt{n(n-1)}}$

$$= \frac{n^2 - n(n-1)}{n + \sqrt{n(n-1)}}$$

$$= \frac{1}{1 + \sqrt{(1 - 1/n)}} \to \frac{1}{2} \quad \text{as } n \to \infty$$

By the vanishing condition, $\sum_{r=1}^{\infty}[r - \sqrt{r(r-1)}]$ is divergent.

3.   Let $s_n = \sum_{r=1}^{n} a_r$ and $t_n = \sum_{r=1}^{n} b_r$. Then $s_n \to s$ and $t_n \to t$ as $n \to \infty$. By the sum rule for sequences (3.1.2), $s_n + t_n \to s + t$ as $n \to \infty$. But $s_n + t_n$ is the nth partial sum of $\sum_{r=1}^{\infty}(a_r + b_r)$. Hence $\sum_{r=1}^{\infty}(a_r + b_r)$ converges to $\sum_{r=1}^{\infty} a_r + \sum_{r=1}^{\infty} b_r$.

By the scalar product rule for sequences (3.1.2), $ks_n \to ks$ as $n \to \infty$, and hence $\sum_{r=1}^{\infty} ka_r$ converges to $k\sum_{r=1}^{\infty} a_r$.

4.   Suppose that $\sum_{r=1}^{\infty}(a_r + b_r)$ is convergent. Since $\sum_{r=1}^{\infty} a_r$ is convergent, so is $\sum_{r=1}^{\infty}(-a_r)$, by 4.1.4. Hence, by 4.1.3,

$$\sum_{r=1}^{\infty}(-a_r) + \sum_{r=1}^{\infty}(a_r + b_r) \quad \text{converges to} \quad \sum_{r=1}^{\infty} b_r$$

This contradicts the hypothesis on $\sum_{r=1}^{\infty} b_r$. Hence $\sum_{r=1}^{\infty} a_r$ convergent and $\sum_{r=1}^{\infty} b_r$ divergent implies that $\sum_{r=1}^{\infty}(a_r + b_r)$ is divergent.

Exercises 4.2

1. (a) If

$$a_n = \frac{1}{2^n n!} \quad \text{and} \quad b_n = \frac{1}{2^n}$$

then $0 \le a_n \le b_n$ for all $n \in \mathbb{N}$. Now $\sum_{n=1}^{\infty} b_n$ converges, since it is a geometric series with $x = \frac{1}{2}$. By the first comparison test, $\sum_{n=1}^{\infty} a_n$ also converges.

(b) Suppose, by way of contradiction, that

$$\sum_{n=2}^{\infty} \frac{1}{\log_e n}$$

is convergent. Since

$$0 \le \frac{1}{n} \le \frac{1}{\log_e n} \quad \text{for } n \ge 2$$

the first comparison test implies that $\sum_{n=2}^{\infty} 1/n$ is convergent. But $\sum_{n=1}^{\infty} 1/n$ is a divergent p-series $(p = 1)$. This contradiction means that

$$\sum_{n=2}^{\infty} \frac{1}{\log_e n}$$

is divergent.

2. Let

$$a_n = \frac{n + 3}{n^2 + n} \quad \text{and} \quad b_n = \frac{1}{n}$$

then

$$\frac{a_n}{b_n} = \frac{n(n + 3)}{n^2 + n} = \frac{1 + 3/n}{1 + 1/n} \to 1 \quad \text{as } n \to \infty$$

Since $\sum_{n=1}^{\infty} 1/n$ diverges, the second comparison test gives that $\sum_{n=1}^{\infty} a_n$ also diverges.

3. (a) If $a_n = n/3^n$ then

$$\left| \frac{a_{n+1}}{a_n} \right| = \frac{n + 1}{3^{n+1}} \frac{3^n}{n}$$

$$= \frac{1}{3} \left(1 + \frac{1}{n} \right) \to \frac{1}{3} \quad \text{as } n \to \infty$$

Hence, by the ratio test, $\sum_{n=1}^{\infty} a_n$ converges.

(b) If $a_n = n^3/n!$ then

$$\left|\frac{a_{n+1}}{a_n}\right| = \frac{(n+1)^3}{(n+1)!} \frac{n!}{n^3}$$

$$= \frac{1}{n+1}\left(1 + \frac{1}{n}\right)^3 \rightarrow 0 \quad \text{as } n \rightarrow \infty$$

Hence, by the ratio test, $\sum_{n=1}^{\infty} a_n$ converges.

(c) If $a_n = 3^n/(2^n + 1)$ then

$$\left|\frac{a_{n+1}}{a_n}\right| = \frac{3^{n+1}}{2^{n+1} + 1} \frac{2^n + 1}{3^n}$$

$$= 3\left(\frac{1 + 1/2^n}{2 + 1/2^n}\right) \rightarrow \frac{3}{2} \quad \text{as } n \rightarrow \infty$$

Hence, by the ratio test, $\sum_{n=1}^{\infty} a_n$ diverges.

4. (a) $1 - \frac{1}{3} + \frac{1}{5} - \frac{1}{7} + \ldots = \sum_{n=1}^{\infty}(-1)^{n-1}/(2n-1)$ is an alternating series of the form $\sum_{n=1}^{\infty}(-1)^{n-1}b_n$, where

$$b_n = \frac{1}{2n - 1}$$

Since (b_n) is a decreasing sequence with limit zero, the series $\sum_{n=1}^{\infty}(-1)^{n-1}b_n$ converges, by the alternating series test. The associated series of absolute values is

$$\sum_{n=1}^{\infty} \frac{1}{2n - 1}$$

which diverges by comparison with

$$\sum_{n=1}^{\infty} \frac{1}{n}$$

using the second comparison test. Hence

$$\sum_{n=1}^{\infty} \frac{(-1)^{n-1}}{2n - 1}$$

is conditionally convergent.

(b) $\sum_{r=1}^{\infty}(\cos r\pi/r \sqrt{r}) = \sum_{r=1}^{\infty}(-1)^r/r \sqrt{r}$, which is an alternating series. The sequence

$$(b_n) = \left(\frac{1}{n\sqrt{n}}\right) = \left(\frac{1}{n}\right)^{3/2}$$

is decreasing with limit zero. Hence

$$\sum_{r=1}^{\infty} \frac{\cos r\pi}{r\sqrt{r}}$$

is convergent, by the alternating series test. Since

$$\sum_{r=1}^{\infty} \left| \frac{\cos r\pi}{r\sqrt{r}} \right| = \sum_{r=1}^{\infty} \frac{1}{r\sqrt{r}}$$

is a convergent p-series (with $p = \frac{3}{2}$),

$$\sum_{r=1}^{\infty} \frac{\cos r\pi}{r\sqrt{r}}$$

is absolutely convergent.

5. $\dfrac{d}{dx}(\log_e(\log_e x)) = \dfrac{1}{\log_e x}\dfrac{1}{x}$

Now $f: \mathbb{R}^+ \to \mathbb{R}^+$, given by $f(x) = 1/(x\log_e x)$, satisfies the hypotheses of the integral test. Let

$$j_n = \int_2^n \frac{1}{x\log_e x}\,dx$$

then

$$\begin{aligned}
j_n &= [\log_e(\log_e x)]_2^n \\
&= \log_e(\log_e n) - \log_e(\log_e 2)
\end{aligned}$$

Now (j_n) is a divergent sequence, since $\log_e$ is an increasing function. By the integral test,

$$\sum_{r=2}^{\infty} \frac{1}{r\log_e r}$$

is a divergent series.

6. Since $\sum_{r=1}^{\infty}a_r^+$ and $\sum_{r=1}^{\infty}a_r^-$ are (absolutely) convergent,

$$\sum_{r=1}^{\infty}a_r^+ + \sum_{r=1}^{\infty}a_r^- = \sum_{r=1}^{\infty}|a_r|$$

converges, by 4.1.3. Hence $\sum_{r=1}^{\infty}a_r$ is absolutely convergent.

Exercises 4.3

1.   (a) $\left| \dfrac{[(n + 1)!]^2 x^{n+1}}{[2(n + 1)]!} \dfrac{(2n)!}{(n!)^2 x^n} \right| = \dfrac{(n + 1)^2}{(2n + 2)(2n + 1)} |x|$ $(x \neq 0)$

$$= \frac{1 + 1/n}{2(2 + 1/n)} |x| \to \tfrac{1}{4}|x| \qquad \text{as } n \to \infty$$

Hence

$$\sum_{n=0}^{\infty} \frac{(n!)^2 x^n}{(2n)!}$$

has radius of convergence $R = 4$.

(b) $\left| \dfrac{(n + 1)^3 x^{n+1}}{(n + 1)!} \dfrac{n!}{n^3 x^n} \right| = \dfrac{1}{n + 1}\left(1 + \dfrac{1}{n}\right)^3 |x| \to 0$

$$\text{as } n \to \infty \quad (x \neq 0)$$

Hence

$$\sum_{n=0}^{\infty} \frac{n^3 x^n}{n!}$$

converges for all $x \in \mathbb{R}$.

2.   The geometric series $\sum_{n=0}^{\infty} x^n$ is absolutely convergent for $|x| < 1$. Hence the Cauchy product result (4.2.9) gives

$$\left(\sum_{n=0}^{\infty} x^n\right)\left(\sum_{n=0}^{\infty} x^n\right) = \sum_{n=0}^{\infty} c_n x^n$$

where

$$c_n = \sum_{r=0}^{n} a_r b_{n-r} = \sum_{r=0}^{n} 1 = n + 1$$

Hence

$$\left(\sum_{n=0}^{\infty} x^n\right)^2 = \sum_{n=0}^{\infty} (n + 1)x^n$$

Now

$$\sum_{n=0}^{\infty} x^n = \frac{1}{1 - x} \qquad \text{for } |x| < 1$$

by 4.1.1. Hence

$$\sum_{n=0}^{\infty}(n+1)x^n = \frac{1}{(1-x)^2} \quad \text{for } |x| < 1$$

3.   Now, for $x \neq 0$,

$$\left|\frac{x^{n+1}}{(n+1)!}\frac{n!}{x^n}\right| = \frac{|x|}{n+1} \to 0 \quad \text{as } n \to \infty$$

for any $x \in \mathbb{R}$, and

$$\left|\frac{(-1)^{n+1}x^{2n+3}}{(2n+3)!}\frac{(2n+1)!}{(-1)^n x^{2n+1}}\right| = \frac{x^2}{(2n+3)(2n+2)} \to 0 \quad \text{as } n \to \infty$$

for any $x \in \mathbb{R}$, and

$$\left|\frac{(-1)^{n+1}x^{2n+2}}{(2n+2)!}\frac{(2n)!}{(-1)^n x^{2n}}\right| = \frac{x^2}{(2n+2)(2n+1)} \to 0 \quad \text{as } n \to \infty$$

for any $x \in \mathbb{R}$. Hence the series defining the sine, cosine and the exponential functions are convergent for all $x \in \mathbb{R}$.

4.   (a) $\cos(-x) = \displaystyle\sum_{n=0}^{\infty}\frac{(-1)^n(-x)^{2n}}{(2n)!} = \sum_{n=0}^{\infty}\frac{(-1)^n x^{2n}}{(2n)!} = \cos x$

$\sin(-x) = \displaystyle\sum_{n=0}^{\infty}\frac{(-1)^n(-x)^{2n+1}}{(2n+1)!} = -\sum_{n=0}^{\infty}\frac{(-1)^n x^{2n+1}}{(2n+1)!}$

$\qquad = -\sin x$

(b) Since $\cos^2 x + \sin^2 x = 1$ for all $x \in \mathbb{R}$, $0 \leqslant \cos^2 x \leqslant 1$ and $0 \leqslant \sin^2 x \leqslant 1$ for all $x \in \mathbb{R}$. Hence $|\cos x| \leqslant 1$ and $|\sin x| \leqslant 1$.

5.   Let $\sum_{n=0}^{\infty}a_n x^n$ have radius of convergence $R > 0$. By the ratio test,

$$\left|\frac{a_{n+1}}{a_n}\right| \to \frac{1}{R} \quad \text{as } n \to \infty$$

Now

$$\left|\frac{(n+1)a_{n+1}x^n}{na_n x^{n-1}}\right| = |x|\left(1+\frac{1}{n}\right)\left|\frac{a_{n+1}}{a_n}\right| \to \frac{|x|}{R} \quad \text{as } n \to \infty$$

Hence $\sum_{n=1}^{\infty}na_n x^{n-1}$ converges for $|x| < R$. By 4.1.1,

$$f(x) = \sum_{n=0}^{\infty}x^n = \frac{1}{1-x} \quad \text{for } |x| < 1$$

Now

$$f'(x) = \frac{1}{(1-x)^2} = \sum_{n=0}^{\infty}(n+1)x^n = \sum_{n=1}^{\infty} nx^{n-1}$$

for $|x| < 1$ by Question 2.

This is a special case of A.2 proved in the Appendix.

Chapter 5

Exercises 5.1

1.   (a) $\dfrac{x^3 + x + 1}{2 - 3x^3} = \dfrac{1 + 1/x^2 + 1/x^3}{2/x^3 - 3} \to \dfrac{-1}{3}$ as $x \to \pm\infty$

See Figure S.2(a).

(b) $|e^{-x}\sin x| \leq e^{-x} \to 0$ as $x \to \infty$. Hence, by the sandwich rule for limits, $e^{-x}\sin x \to 0$ as $x \to \infty$. If $t = -x$ then $e^{-x}\sin x = -e^t \sin t$, which is unbounded for large t. Hence $e^{-x}\sin x$ is unbounded as $x \to -\infty$. See Figure S.2(b).

(c) Now $[x] = n$ for $n \leq x < n+1$, and so, for such x,

$$\frac{[x]}{x} = \frac{n}{x}$$

Hence

$$n \leq x < n+1 \Rightarrow \frac{n}{n+1} \leq \frac{[x]}{x} \leq \frac{n}{n}$$

$$\Rightarrow 1 - \frac{1}{n+1} \leq \frac{[x]}{x} \leq 1$$

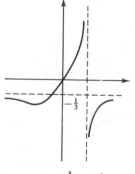

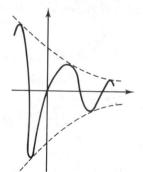

 (a) $x \mapsto \dfrac{x^3 + x + 1}{2 - 3x^3}$

 (b) $x \mapsto e^{-x}\sin x$

(c) $x \mapsto \dfrac{[x]}{x}$

Figure S.2

By the sandwich rule for limits, $[x]/x \to 1$ as $x \to \infty$. Similarly, $[x]/x \to 1$ as $x \to -\infty$. See Figure S.2(c).

2.   Since $|\sin t| \leq 1$ for all $t \in \mathbb{R}$,

$$\left| 2x \sin\left(\frac{1}{x}\right) - 0 \right| \leq 2|x| \quad \text{for all } x \neq 0$$

Now $2|x| < \varepsilon$ provided $|x| < \frac{1}{2}\varepsilon$. So, given $\varepsilon > 0$, choose $\delta = \frac{1}{2}\varepsilon$; then

$$|x - 0| < \delta \Rightarrow \left| 2x \sin\left(\frac{1}{x}\right) - 0 \right| < \varepsilon$$

Hence

$$\lim_{x \to 0} 2x \sin\left(\frac{1}{x}\right) = 0$$

3.   It suffices to show that if $g(x) \to M \neq 0$ as $x \to a$ then

$$\frac{1}{g(x)} \to \frac{1}{M} \quad \text{as } x \to a$$

The quotient rule then follows from the product rule. Now

$$\left| \frac{1}{g(x)} - \frac{1}{M} \right| = \frac{|g(x) - M|}{|g(x)||M|}$$

If $\varepsilon = \frac{1}{2}|M|$ then $\varepsilon > 0$, and so there exists a $\delta_1 > 0$ such that

$$0 < |x - a| < \delta_1 \Rightarrow |g(x) - M| < \tfrac{1}{2}|M|$$
$$\Rightarrow \tfrac{1}{2}|M| < |g(x)| < \tfrac{3}{2}|M|$$
$$\Rightarrow \frac{1}{|g(x)|} < \frac{2}{|M|}$$

Now, for any $\varepsilon > 0$,

$$\varepsilon' = \tfrac{1}{2}\varepsilon|M|^2 > 0$$

Hence there exists a $\delta_2 > 0$ such that

$$0 < |x - a| < \delta_2 \Rightarrow |g(x) - M| < \varepsilon'$$

Now let δ be the minimum of δ_1 and δ_2. Then

$$0 < |x - a| < \delta \Rightarrow \left| \frac{1}{g(x)} - \frac{1}{M} \right| < \frac{2\varepsilon'}{|M|^2} = \varepsilon$$

Hence

$$\frac{1}{g(x)} \to \frac{1}{M} \quad \text{as } x \to a$$

4.   (a) Using the sum, product and quotient rules,

$$\frac{x^3 - 1}{x^2 + 1} \to \frac{0^3 - 1}{0^2 + 1} = -1 \quad \text{as } x \to 0$$

(b) $\dfrac{x - 1}{x^3 - 1} = \dfrac{x - 1}{(x - 1)(x^2 + x + 1)}$

$$= \frac{1}{x^2 + x + 1} \quad \text{for } x \neq 1$$

Using the sum, product and quotient rules,

$$\frac{1}{x^2 + x + 1} \to \tfrac{1}{3} \quad \text{as } x \to 1$$

Hence

$$\lim_{x \to 1} \frac{x - 1}{x^3 - 1} = \tfrac{1}{3}$$

(c) By the sum, product and quotient rules,

$$\frac{\pi x}{x + 1} \to \tfrac{1}{2}\pi \quad \text{as } x \to 1$$

By the composite rule,

$$\cos\left(\frac{\pi x}{x + 1}\right) \to \cos \tfrac{1}{2}\pi$$

and so

$$\lim_{x \to 1} \cos\left(\frac{\pi x}{x + 1}\right) = 0$$

(d) Now

$$\left| x^2 \cos\left(\frac{1}{x}\right) \right| \leqslant x^2 \quad \text{for all } x \neq 0$$

Hence

$$-x^2 \leqslant x^2 \cos\left(\frac{1}{x}\right) \leqslant x^2 \quad \text{for } x \neq 0$$

Clearly $\pm x^2 \to 0$ as $x \to 0$ and so by the sandwich rule

$$\lim_{x \to 0} x^2 \cos\left(\frac{1}{x}\right) = 0$$

5.   (a) For $1 \leqslant x < 2$, $[x] = 1$, and so $[x] \to 1$ as $x \to 1+$. Hence

$$\lim_{x \to 1+} \frac{1}{[x] + 1} = \tfrac{1}{2}$$

(b) If $4 \leqslant x < 9$ then $2 \leqslant \sqrt{x} < 3$. As $x \to 4+$, $\sqrt{x} \to 2+$, and so

$$\lim_{x \to 4+} [\sqrt{x}] = 2$$

(c) If $1 \leqslant x < 4$ then $1 \leqslant \sqrt{x} < 2$. As $x \to 4-$, $\sqrt{x} \to 2-$, and so

$$\lim_{x \to 4-} [\sqrt{x}] = 1$$

6.   Suppose that $\lim_{x \to 0+} f(x) = L$. Then for any $\varepsilon > 0$ there exists a $\delta > 0$ such that

$$0 < x < \delta \Rightarrow |f(x) - L| < \varepsilon$$

Now let $T = 1/\delta$ and $t = 1/x$. Then

$$
\begin{aligned}
t > T &\Rightarrow 1/x > 1/\delta > 0 \\
&\Rightarrow 0 < x < \delta \\
&\Rightarrow |f(x) - L| < \varepsilon \\
&\Rightarrow \left|f\left(\frac{1}{t}\right) - L\right| < \varepsilon
\end{aligned}
$$

In other words,

$$\lim_{t \to \infty} f\left(\frac{1}{t}\right) = L$$

The converse is proved similarly.

Exercises 5.2

1.   (a) $f_1(x)$ is defined for $x \neq \pm 1$. The singularities are infinite ones.

(b) $f_2(x)$ is defined for $x \neq \pm 1$. At $x = -1$ the singularity is an infinite one. However, if $x = 1 + h$, $h \neq 0$, then

$$f_2(x) = \frac{h}{h^2 + 2h} = \frac{1}{h + 2}$$

Hence $f_2(x) \to \frac{1}{2}$ as $h \to 0$ (that is, $x \to 1$). Thus f_2 has a removable singularity at $x = 1$.

(c) $f_3(x) = 1$ for $|x| > 1$ and $f_3(x) = -1$ for $|x| < 1$. Hence $f_3(x)$ is undefined for $x = \pm 1$. These are finite singularities.

2. (a) If c is irrational then for each $\delta > 0$ there exists a rational d in the interval $(c - \delta, c + \delta)$ by 2.3.2. Now $|f(d) - f(c)| = 1$, and so there is no $\delta > 0$ for which

$$|x - c| < \delta \Rightarrow |f(x) - f(c)| < \varepsilon \quad \text{when } \varepsilon < 1$$

A similar argument applies when c is rational. Hence f is discontinuous everywhere.

(b) If c is rational and $c > \frac{1}{2}$ then for any $\delta > 0$ there is an irrational d_1, $\frac{1}{2} < c < d_1 < c + \delta$. Then

$$|g(d_1) - g(c)| = |1 - (d_1 + c)|$$
$$= c + d_1 - 1 > 2c - 1$$

If c is rational and $c < \frac{1}{2}$ then for any $\delta > 0$ there is an irrational d_2, $c - \delta < d_2 < c < \frac{1}{2}$. Then

$$|g(d_2) - g(c)| = 1 - (c + d_2) > 1 - 2c$$

Now let $\varepsilon = \frac{1}{2}|2c - 1|$. If c is rational and $c \neq \frac{1}{2}$ then $\varepsilon > 0$ and there is no $\delta > 0$ such that

$$|x - c| < \delta \Rightarrow |g(x) - g(c)| < \varepsilon$$

Hence g is discontinuous at rational $c \neq \frac{1}{2}$. A similar argument shows that g is discontinuous at all irrationals. However, g is continuous at $c = \frac{1}{2}$. To see this, note that $|g(x) - g(\frac{1}{2})| = |x - \frac{1}{2}|$ for any $x \in \mathbb{R}$. So if $\varepsilon > 0$ then, with $\delta = \varepsilon > 0$,

$$|x - \tfrac{1}{2}| < \delta \Rightarrow |g(x) - g(\tfrac{1}{2})| < \varepsilon$$

3. Since $|f(x)| \leq M|x|$ for all $x \in \mathbb{R}$, $f(0) = 0$. Hence

$$|f(x) - f(0)| \leq M|x| \quad \text{for all } x \in \mathbb{R}$$

If $\varepsilon > 0$ then let

$$\delta = \frac{\varepsilon}{M} > 0$$

Then

$$|x| < \delta \Rightarrow |f(x) - f(0)| \leq M|x| < \varepsilon$$

Hence f is continuous at 0.

4.   (a) By the sum rule, $x \mapsto x - 2$ is continuous for all x. So, for $x \neq 2$,

$$x \mapsto \frac{1}{x - 2}$$

is continuous by the reciprocal rule. By the product rule, $x \mapsto x^2$ is continuous for all x, and so, by the product rule again,

$$x \mapsto \frac{x^2}{x - 2}$$

is continuous for $x \neq 2$.

(b) By the product rule (twice), $x \mapsto x^3$ is continuous for all x. By the product and composite rules, $x \mapsto e^{2 \sin x}$ is continuous for all x. Finally the sum rule yields that $x \mapsto x^3 + e^{2 \sin x}$ is continuous.

(c) Call the given function f. For $x \neq 0$, $x \mapsto x^2$ is continuous by the product rule, and $x \mapsto 1/x$ is continuous by the reciprocal rule. By the composite and product rules, $x \mapsto x^2 \cos(1/x)$ is continuous for $x \neq 0$. Since

$$\left| x^2 \cos\left(\frac{1}{x}\right) \right| \leq x^2 \quad \text{for } x \neq 0$$

and because $f(0) = 0$, we have that $-x^2 \leq f(x) \leq x^2$ for all x. Now $x \mapsto \pm x^2$ are continuous at $x = 0$ and $f(0) = \pm 0^2$, and so, by the sandwich rule, f is continuous at $x = 0$. Hence f is continuous on $\mathbb{R}$.

(d) The function $f: \mathbb{R}^+ \to \mathbb{R}^+$ given by $f(x) = x^2$ is a continuous bijection. Its inverse is the function $x \mapsto \sqrt{x}$, $x > 0$, which is thus continuous by the inverse rule.

(e) The function $x \mapsto \cosh^{-1} x$, $x > 1$, is the inverse of the function $f: \mathbb{R}^+ \to (1, \infty)$ given by $f(x) = \cosh x$. The rules show that f is continuous. By the product and composite rules, $x \mapsto e^{-x}$ is continuous for $x > 1$. By the sum and product rules, $x \mapsto \frac{1}{2}(e^x + e^{-x}) = \cosh x$ is continuous for $x > 1$. Finally by the inverse rule $x \mapsto \cosh^{-1} x$ is continuous for $x > 1$.

5.   Let $b_1 < b_2$, where b_1, $b_2 \in B$. Now there exist a_1, $a_2 \in A$ such that $f(a_1) = b_1$, and $f(a_2) = b_2$. Since f is strictly increasing, $a_1 < a_2$. In other words, $f^{-1}(b_1) < f^{-1}(b_2)$. Hence f^{-1} is strictly increasing. Define $f: \mathbb{R}^+ \to \mathbb{R}^+$ by $f(x) = x^n$, $n \in \mathbb{N}$. Now f is continuous by the product rule and f is strictly increasing on $\mathbb{R}^+$. Hence f is a continuous bijection. By the inverse rule $f^{-1}: \mathbb{R}^+ \to \mathbb{R}^+$ given by $f(x) = x^{1/n}$ is continuous.

Exercises 5.3

1.   If $P(x) = x^3 - 3x + 1$ then $P(0) = 1$ and $P(1) = -1$. Since P is continuous on $[0, 1]$, and $P(0) > 0$ and $P(1) < 0$, then $P(x) = 0$ for some x in

[0, 1] by the intermediate value property. Now $P(0.5) = -0.3750$, and so, by similar reasoning, P has a root in the interval $[0, 0.5]$. If we now evaluate $P(0.25)$, P has a root in either the left-hand half or the right-hand half of $[0, 0.5]$, depending on the sign of $P(0.25)$. This *bisection process* can be continued indefinitely in order to 'trap' the root of P to any desired degree of accuracy. In fact, $P(x) = 0$ in $[0, 1]$ for $x \approx 0.3473$.

2. The interval theorem states that, for a continuous function f, $f(J)$ is an interval. To determine this interval, we have to find the maximum and minimum values of f on J.

(a) On $J = [-1, 1]$, $f(x) = x^2$ has a minimum value of 0 at $x = 0$ and a maximum value of 1 at $x = \pm 1$. Hence $f(J) = [0, 1]$. See Figure S.3(a).

(b) On $J = [0, \frac{1}{6}\pi]$, $f(x) = 3 \sin x$ has a minimum value of 0 at $x = 0$ and a maximum value of $3 \sin \frac{1}{6}\pi$ at $x = \frac{1}{6}\pi$. Hence $f(J) = [0, \frac{3}{2}]$. See Figure S.3(b).

(c) If $x < 0$ then $f(x) = x + |x| = 0$, and if $x \geqslant 0$ then $f(x) = 2x$. See Figure S.3(c). On $J = [-1, 2]$, $f(x)$ has a minimum value of 0, for $-1 \leqslant x \leqslant 0$, and a maximum value of 4 at $x = 2$. Hence $f(J) = [0, 4]$.

(d) Now $f(x) = x^2 - x^4 = x^2(1 - x^2) \geqslant 0$ on $J = [0, 1]$. Hence $f(x)$ has a minimum value of 0 at $x = 0$, 1. For the maximum value note that

$$x^2 - x^4 = \tfrac{1}{4} - (x^2 - \tfrac{1}{2})^2$$

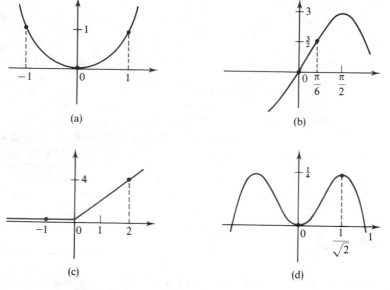

(a)

(b)

(c)

(d)

Figure S.3

by completing the square. Hence the maximum value on J is $\frac{1}{4}$ occurring when $x = 1/\sqrt{2}$. Hence $f(J) = [0, \frac{1}{4}]$. See Figure S.3(d).

3. With $r(x)$ defined as in Example 1, $P(x) = x^n[1 + r(x)]$. Hence $P(x)$ has the same sign as x^n for $|x| > 1 + M$, where M is the maximum of $|a_{n-1}|$, $\ldots$, $|a_0|$. Since n is even, there exist $\alpha, \beta \in \mathbb{R}$, $\alpha > 0$, $\beta < 0$ with $P(\alpha) > 0$ and $P(\beta) > 0$ (choose $\alpha > 1 + M$ and $\beta < -(1 + M)$). Now $P(0) = a_0 < 0$, and so, by the intermediate value property, P vanishes between β and 0 and between 0 and α. Hence P has at least two real roots.

4. If $x \in J = [0, 1]$ then $f(x) \in J$ since $f: J \to J$. Hence

$$g(x) = [f(x)]^2 \in J$$

Thus $g: J \to J$ and, by the product rule, g is continuous on J. By the fixed point theorem, $g(c) = c$ for some $c \in J$. Hence $f(c) = \sqrt{c}$ for some $c \in J$ as required.

Chapter 6

Exercises 6.1

1. (a)
$$\frac{f(x) - f(c)}{x - c} = \frac{x^4 + 3x - c^4 - 3c}{x - c}$$

$$= 3 + \frac{x^4 - c^4}{x - c}$$

$$= 3 + \frac{(x^2 - c^2)(x^2 + c^2)}{x - c}$$

$$= 3 + (x + c)(x^2 + c^2) \quad \text{for } x \neq c$$

Hence

$$\lim_{x \to c} \frac{f(x) - f(c)}{x - c} = 3 + 4c^3$$

and so f is differentiable at c and $f'(c) = 4c^3 + 3$.

(b)
$$\frac{f(x) - f(c)}{x - c} = \frac{1/x - 1/c}{x - c}$$

$$= \frac{c - x}{xc(x - c)} = \frac{-1}{xc} \quad \text{for } x \neq c$$

Hence

$$\lim_{x \to c} \frac{f(x) - f(c)}{x - c} = \frac{-1}{c^2}$$

and so f is differentiable at c and $f'(c) = -1/c^2$.

2. Let f and g be differentiable at c. Then

$$\frac{[f(x) + g(x)] - [f(c) + g(c)]}{x - c} = \frac{f(x) - f(c)}{x - c}$$

$$+ \frac{g(x) - g(c)}{x - c} \quad (x \neq c)$$

$$\to f'(c) + g'(c) \quad \text{as } x \to c$$

Hence $f + g$ is differentiable at c and $(f + g)'(c) = f'(c) + g'(c)$. Now

$$\frac{1/f(x) - 1/f(c)}{x - c} = \frac{-[f(x) - f(c)]}{x - c} \frac{1}{f(x)f(c)} \quad \text{for } x \neq c$$

Since f is differentiable at c,

$$\frac{f(x) - f(c)}{x - c} \to f'(c) \quad \text{as } x \to c$$

But f is also continuous at c (6.1.2), and so $f(x) \to f(c)$ as $x \to c$. Thus

$$\frac{1/f(x) - 1/f(c)}{x - c} \to \frac{-f'(c)}{[f(c)]^2} \quad \text{as } x \to c$$

This proves the reciprocal rule.

3. (a) $\dfrac{f(x) - f(2)}{x - 2} = \dfrac{x - [x]}{x - 2}$ for $x \neq 2$

If $2 < x < 3$ then $[x] = 2$, and so

$$\frac{(x - [x])}{x - 2} = \frac{x - 2}{x - 2} = 1 \to 1 \quad \text{as } x \to 2+$$

If $1 < x < 2$ then $[x] = 1$, and so

$$\frac{x - [x]}{x - 2} = \frac{x - 1}{x - 2} = 1 + \frac{1}{x - 2} \to -\infty \quad \text{as } x \to 2-$$

Hence $f'_+(2) = 1$ and $f'_-(2)$ does not exist.

(b) $\dfrac{f(x) - f(0)}{x - 0} = \dfrac{1}{1 + e^{1/x}}$ for $x \neq 0$

Now $e^{1/x} \to \infty$ as $x \to 0+$ and $e^{1/x} \to 0$ as $x \to 0-$. Hence $f'_+(0) = 0$ and $f'_-(0) = 1$.

4.   (a) By the product rule $x \mapsto e^x \cos x$ is differentiable for all x and so by the sum rule $x \mapsto e^x \cos x + 1$ is differentiable for all x.

(b) By the product rule (twice), $x \mapsto x^4$ is differentiable for all x, and so, by the sum rule, $x \mapsto x^4 + 1$ is differentiable. Since $x^4 + 1 \neq 0$, the reciprocal rule gives that $x \mapsto 1/(x^4 + 1)$ is differentiable for all x.

(c) By the sum and product rules, $x \mapsto x^2 + 1$ is differentiable for all x. Using (b) and the product rule, we have that

$$x \mapsto \frac{x^2 + 1}{x^4 + 1}$$

is differentiable for all x.

(d) Since $\cos x \neq 0$ for $x \neq \frac{1}{2}(2n + 1)\pi$, the reciprocal rule gives that

$$x \mapsto \frac{1}{\cos x}$$

is differentiable for such x. By the product rule,

$$x \mapsto \frac{\sin x}{\cos x} = \tan x$$

is differentiable for $x \neq \frac{1}{2}(2n + 1)\pi$. Finally, two applications of the product rule give that $x \mapsto \tan^3 x$ is differentiable for $x \neq \frac{1}{2}(2n + 1)\pi$.

(e) If $x \neq 0$ then $x \mapsto 1/x$ is differentiable, by the reciprocal rule, and hence, by the composite rule, $x \mapsto \cos(1/x)$ is differentiable. By the product rule, $x \mapsto x^2$ is differentiable, and so

$$x \mapsto x^2 \cos\left(\frac{1}{x}\right)$$

is differentiable for $x \neq 0$.
For $x = 0$ we need the sandwich rule. Since

$$\left| x^2 \cos\left(\frac{1}{x}\right) \right| \leqslant x^2$$

and $f(0) = 0$, f is sandwiched between $\pm x^2$ at $x = 0$. Hence f is also differentiable at $x = 0$.

(f) The function $g : \mathbb{R}^+ \to \mathbb{R}^+$ given by $g(x) = x^5$ is a bijection, which is differentiable for $x > 0$ by several applications of the product rule. By the inverse rule, its inverse is differentiable for $y > 0$, provided that $g'(x) \neq 0$ for $x > 0$. Since $g'(x) = 5x^4 > 0$ for $x > 0$, $g^{-1} = f$ is differentiable on $\mathbb{R}^+$.

(g) The function $g: \mathbb{R} \to \mathbb{R}$ given by $g(x) = \sinh x$ is a bijection. Since $x \mapsto e^x$ is differentiable, the reciprocal rule gives that $x \mapsto e^{-x}$ is also differentiable. By the sum and product rules,

$$x \mapsto \tfrac{1}{2}(e^x - e^{-x}) = \sinh x$$

is differentiable for all x. Hence by the inverse rule f is differentiable provided that $g'(x) \neq 0$ for all x. Since $g'(x) = \cosh x \geq 1$ for all x, $g^{-1} = f$ is differentiable on $\mathbb{R}$.

5.   (a) $f'(x) = e^x(\cos x - \sin x)$ (b) $f'(x) = \dfrac{-4x^3}{(1 + x^4)^2}$

(c) $f'(x) = \dfrac{2x(1 - 2x^2 - x^4)}{(1 + x^4)^2}$ (d) $f'(x) = 3(\sec x \tan x)^2$

(e) $f'(x) = 2x \cos(1/x) + \sin(1/x)$ for $x \neq 0$, and $f'(0) = 0$

(f) $f'(x) = \tfrac{1}{5}x^{-4/5}$ (g) $f'(x) = \dfrac{1}{\sqrt{1 + x^2}}$

Exercises 6.2

1. P has at least one real root, by Example 1 of Section 5.3. Suppose that P has two (or more) real roots. Thus there exist $x_1, x_2 \in \mathbb{R}$, $x_1 \neq x_2$ and $P(x_1) = P(x_2) = 0$. By Rolle's theorem, $P'(c) = 0$ for some c between x_1 and x_2. But

$$P'(x) = 3x^2 + a > 0 \quad \text{for all } x$$

Hence $P'(c) \neq 0$. This contradiction shows that P has precisely one real root.

2.   Suppose that c_1 and c_2 are distinct fixed points of f. Then $h(c_1) = h(c_2) = 0$. Since h is differentiable on $(0, 1)$ and continuous on $[0, 1]$, Rolle's theorem implies that $h'(c) = 0$ for some c between c_1 and c_2. Hence $f'(c) = 1$. But $f'(x) \neq 1$ on $(0, 1)$, and this is the desired contradiction. Therefore f has precisely one fixed point.

3.   By Rolle's theorem $f'(c_1) = 0$ for some c_1, $a_1 < c_1 < a_2$ and $f'(c_2) = 0$ for some c_2, $a_2 < c_2 < a_3$. Now apply Rolle's theorem to f' on the interval $[c_1, c_2]$ to deduce that $f''(c) = 0$ for some c between c_1 and c_2. Clearly, $a_1 < c < a_3$.

4.   (a) Let $f(x) = \sin x$ for all $x \in \mathbb{R}$. Let $a, b \in \mathbb{R}$ and for simplicity assume that $a < b$. Apply the mean value theorem (6.2.2) to f on the interval $[a, b]$. Hence

$$f'(c) = \frac{f(b) - f(a)}{b - a} \quad \text{for some } c, \; a < c < b$$

Thus

$$\frac{\sin b - \sin a}{b - a} = \cos c$$

Since $|\cos c| \leqslant 1$,

$$\left| \frac{\sin b - \sin a}{b - a} \right| \leqslant 1$$

Hence

$$|\sin a - \sin b| \leqslant |a - b| \quad \text{for all } a, b \in \mathbb{R}$$

(b) Consider $f(x) = \sqrt{x}$, $x > 0$. Apply the mean value theorem to f on the interval $[81, 83]$. Hence there is a c, $81 < c < 83$, such that

$$f'(c) = \frac{f(83) - f(81)}{83 - 81}$$

Thus

$$\frac{1}{2\sqrt{c}} = \frac{\sqrt{83} - 9}{2}$$

and so

$$\sqrt{83} - 9 = 1/\sqrt{c}$$

Since $81 < c < 83$, $81 < c < 100$, and so $9 < \sqrt{c} < 10$. Hence

$$\tfrac{1}{10} < 1/\sqrt{c} < \tfrac{1}{9}$$

and so

$$\tfrac{1}{10} < \sqrt{83} - 9 < \tfrac{1}{9}$$

5. (a) If $f(x) = x + 1/x$ then

$$f'(x) = 1 - \frac{1}{x^2}$$

and so local extrema occur at $x = \pm 1$. Since $f'(x) > 0$ for $|x| > 1$ and $f'(x) < 0$ for $|x| < 1$, f is increasing on $(-\infty, -1)$, decreasing on $(-1, 1)$ and increasing again on $(1, \infty)$. Hence f has a local maximum at $x = -1$ and a local minimum at $x = 1$. See Figure S.4(a).

(b) If $f(x) = e^{-x^2}$ then $f'(x) = -2xe^{-x^2}$ and so f has one local extremum at $x = 0$. Since $f'(x) > 0$ for $x < 0$ and $f'(x) < 0$ for $x > 0$, f is increasing on $(-\infty, 0)$ and decreasing on $(0, \infty)$. Hence f has a local maximum at $x = 0$. See Figure S.4(b).

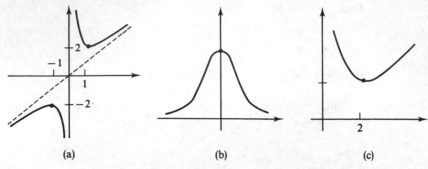

(a)   (b)   (c)

Figure S.4

(c) The function

$$f(x) = \frac{2}{x} + \log_e x$$

is only defined for $x > 0$. Since

$$f'(x) = \frac{-2}{x^2} + \frac{1}{x} = \frac{x-2}{x^2}$$

the sole local extremum occurs at $x = 2$. Since $f'(x) > 0$ for $x > 2$ and $f'(x) < 0$ for $x < 2$, f is decreasing on $(-\infty, 2)$ and increasing on $(2, \infty)$. Hence f has a local minimum at $x = 2$. See Figure S.4(c).

6.   (a) Let $f(x) = \log_e (1 + x) - x + \frac{1}{2}x^2$, $x > 0$. Now

$$f'(x) = \frac{1}{1+x} - 1 + x = \frac{x^2}{1+x} > 0 \quad \text{for } x > 0$$

Hence f is strictly increasing on $(0, \infty)$. In particular, $f(x) > f(0)$ for any $x > 0$. But $f(0) = 0$, and so $f(x) > 0$ for $x > 0$. Thus

$$\log_e (1 + x) - x + \frac{1}{2}x^2 > 0 \quad \text{for } x > 0$$

and so $x - \frac{1}{2}x^2 < \log_e (1 + x)$.

(b) Let

$$f(x) = \tan^{-1} x - \frac{x}{1 + \frac{1}{3}x^2} \quad \text{for } x > 0$$

Now

$$f'(x) = \frac{1}{1+x^2} - \frac{(1 + \frac{1}{3}x^2) - x \cdot \frac{2}{3}x}{(1 + \frac{1}{3}x^2)^2}$$

$$= \frac{\frac{4}{9}x^4}{(1+x^2)(1 + \frac{1}{3}x^2)^2} > 0 \quad \text{for } x > 0$$

Hence f is strictly increasing on $(0, \infty)$, and so $f(x) > f(0) = 0$ for $x > 0$. Thus

$$\tan^{-1} x > \frac{x}{1 + \frac{1}{3}x^2} \quad \text{for } x > 0$$

(c) Let $f(x) = x - \sin x$ and $g(x) = \tan x - x$ for $0 < x < \frac{1}{2}\pi$. Now

$$f'(x) = 1 - \cos x$$

and

$$g'(x) = \sec^2 x - 1 \quad \text{for } 0 < x < \frac{1}{2}\pi$$

Since $0 < \cos x < 1$ for $0 < x < \frac{1}{2}\pi$, $f'(x) > 0$ and $g'(x) > 0$ on $(0, \frac{1}{2}\pi)$. Hence f and g are both strictly increasing on $(0, \frac{1}{2}\pi)$. In particular, $f(x) > f(0) = 0$ and $g(x) > g(0) = 0$. Hence $\sin x < x < \tan x$ for $0 < x < \frac{1}{2}\pi$.

7. (a) $$\frac{3x^2}{x^3 - 1} - \frac{1}{x - 1} = \frac{3x^2 - (x^2 + x + 1)}{x^3 - 1} = \frac{2x^2 - x - 1}{x^3 - 1} \quad \text{for } x \neq 1$$

Therefore

$$\lim_{x \to 1} \left(\frac{3x^2}{x^3 - 1} - \frac{1}{x - 1} \right) = \lim_{x \to 1} \left(\frac{2x^2 - x - 1}{x^3 - 1} \right)$$

$$=^* \lim_{x \to 1} \frac{4x - 1}{3x^2} = 1$$

where * denotes an application of L'Hôpital's rule (version A). Incidentally, L'Hôpital's rule can be avoided here since, for $x \neq 1$,

$$\frac{2x^2 - x - 1}{x^3 - 1} = \frac{2x + 1}{x^2 + x + 1} \to 1 \quad \text{as } x \to 1$$

(b) Now

$$\operatorname{cosec} x - \cot x = \left(\frac{1}{\sin x} - \frac{\cos x}{\sin c} \right) = \frac{1 - \cos x}{\sin x}$$

and

$$\lim_{x \to 0} \frac{1 - \cos x}{\sin x} =^* \lim_{x \to 0} \frac{\sin x}{\cos x} = 0$$

Hence $\lim_{x \to 0} (\operatorname{cosec} x - \cot x) = 0$.

(c) $\log_e \left[\lim_{x \to \pi/2} (\sin x)^{\cosec 2x} \right] = \lim_{x \to \pi/2} \log_e \left[(\sin x)^{\cosec 2x} \right]$

$$= \lim_{x \to \pi/2} \cosec 2x \log_e (\sin x)$$

$$= \lim_{x \to \pi/2} \frac{\log_e (\sin x)}{\sin 2x}$$

$$\overset{*}{=} \lim_{x \to \pi/2} \frac{\cot x}{2 \cos 2x} = 0$$

Hence $\lim_{x \to \pi/2} (\sin x)^{\cosec 2x} = 1$.

(d) $\lim_{x \to 0} \dfrac{x^2}{\cosh x - 1} \overset{*}{=} \lim_{x \to 0} \dfrac{2x}{\sinh x} \overset{*}{=} \lim_{x \to 0} \dfrac{2}{\cosh x} = 2$

(e) $\lim_{x \to 0} \dfrac{\sqrt{1 + x} - 1}{\sqrt{1 - x} - 1} \overset{*}{=} \lim_{x \to 0} \dfrac{-\sqrt{1 - x}}{\sqrt{1 + x}} = -1.$

(f) Now

$$\log_e \left[\lim_{x \to 0} (\cosh x)^{1/x^2} \right] = \lim_{x \to 0} \log_e \left[(\cosh x)^{1/x^2} \right]$$

$$= \lim_{x \to 0} \frac{\log_e \cosh x}{x^2}$$

$$\overset{*}{=} \lim_{x \to 0} \frac{\tanh x}{2x} \overset{*}{=} \lim_{x \to 0} \frac{\sech^2 x}{2} = \frac{1}{2}$$

Hence $\lim_{x \to 0} (\cosh x)^{1/x^2} = \sqrt{e}$.

8. (a) If $f(x) = x^2$ then $f'(x) = 2x$, $f''(x) = 2$ and $f^{(r)}(x) = 0$ for $r \geq 3$. If $g(x) = e^{-x}$ then

$$g^{(r)}(x) = (-1)^r e^{-x} \quad \text{for all } r \geq 1$$

Hence

$$(f \cdot g)^{(4)} = \sum_{r=0}^{4} \binom{4}{r} f^{(r)}(x) g^{(4-r)}(x)$$

$$= \binom{4}{0} x^2 (-1)^4 e^{-x} + \binom{4}{1} 2x (-1)^3 e^{-x} + \binom{4}{2} 2 (-1)^2 e^{-x}$$

$$= e^{-x} (x^2 - 8x + 12)$$

(b) If $f(x) = x^3$ then $f'(x) = 3x^2$, $f''(x) = 6x$, $f^{(3)}(x) = 6$ and $f^{(r)}(x) = 0$ for $r \geq 4$. If $g(x) = \log_e x$ then

$$g^{(r)}(x) = \frac{(-1)^{r-1}(r - 1)!}{x^r} \quad \text{for all } r \geq 1$$

Hence

$$(f \cdot g)^{(6)} = \sum_{r=0}^{6} \binom{6}{r} f^{(r)}(x) g^{(6-r)}(x)$$

$$= \binom{6}{0} x^3 \frac{(-1)^5 5!}{x^6} + \binom{6}{1} 3x^2 \frac{(-1)^4 4!}{x^5}$$

$$+ \binom{6}{2} 6x \frac{(-1)^3 3!}{x^4} + \binom{6}{3} 6 \frac{(-1)^2 2!}{x^3}$$

$$= \frac{12}{x^3}$$

Exercises 6.3

1.   If $f(x) = \log_e (1 + x)$ then

$$f^{(r)}(x) = \frac{(-1)^{r-1}(r-1)!}{(1+x)^r} \quad \text{for } r \geqslant 1$$

Hence $f(0) = 0$ and

$$f^{(r)}(0) = (-1)^{r-1}(r-1)! \quad \text{for } r \geqslant 1$$

Thus the Taylor polynomial of degree 5 is

$$T_5 f(x) = x - \tfrac{1}{2}x^2 + \tfrac{1}{3}x^3 - \tfrac{1}{4}x^4 + \tfrac{1}{5}x^5$$

The remainder term is

$$R_5 f(b) = \frac{f^{(6)}(c)b^6}{6!} = \frac{-5!b^6}{(1+c)^6 6!} = -\frac{1}{6}\left(\frac{b}{1+c}\right)^6$$

for some c between 0 and b.

(a) $R_5 f(1) = -\frac{1}{6}[1/(1+c)]^6$, where $0 < c < 1$, and so

$$|R_5 f(1)| = \frac{1}{6}\left(\frac{1}{1+c}\right)^6 < \tfrac{1}{6}$$

(b) $R_5 f(0.1) = -\frac{1}{6}[0.1/(1+c)]^6$, where $0 < c < 0.1$, and so

$$|R_5 f(0.1)| = \frac{10^{-6}}{6}\left(\frac{1}{1+c}\right)^6 < 1.7 \times 10^{-7}$$

Now

$$\log_e 1.1 = \log_e (1 + 0.1) = T_5 f(0.1) + R_5 f(0.1)$$

Since $R_5f(0.1)$ lies between $\pm 10^{-6}$, $T_5f(0.1)$ provides an estimate for $\log_e 1.1$ that is certainly accurate to four decimal places. In fact, $\log_e 1.1 = 0.0953$ to four decimal places.

2. If $f(x) = \cos x$ then $f^{(r)}(x) = \cos(x + \tfrac{1}{2}r\pi)$, and so $f^{(r)}(0) = \cos \tfrac{1}{2}r\pi$ for $r \geqslant 1$. Hence

$$T_5f(x) = 1 - \frac{x^2}{2!} + \frac{x^4}{4!}$$

and this gives $T_5f(0.1) = 0.995\,004\,167$ to nine decimal places. Now

$$R_5f(0.1) = \frac{-\cos c(0.1)^6}{6!}$$

and hence $|R_5f(0.1)| < 1.4 \times 10^{-9}$. Hence

$$\cos 0.1 = T_5f(0.1) + R_5f(0.1)$$
$$= 0.995\,004\,17$$

correct to eight decimal places.

3. $$\sum_{r=n+1}^{\infty} \frac{1}{r!} = \frac{1}{(n+1)!} + \frac{1}{(n+2)!} + \cdots$$

$$= \frac{1}{(n+1)n!} + \frac{1}{(n+2)(n+1)n!} + \cdots$$

$$= \frac{1}{n!}\left[\frac{1}{n+1} + \frac{1}{(n+2)(n+1)} + \cdots\right]$$

$$< \frac{1}{n!}\left[\frac{1}{(n+1)} + \frac{1}{(n+1)^2} + \cdots\right]$$

Since $\sum_{r=0}^{\infty} x^r = 1/(1-x)$ for $|x| < 1$,

$$\sum_{r=1}^{\infty} \left(\frac{1}{n+1}\right)^r = \frac{1}{1 - 1/(n+1)} - 1 = \frac{1}{n}$$

Therefore

$$\sum_{r=n+1}^{\infty} \frac{1}{r!} < \frac{1}{nn!}$$

Now

$$e = \sum_{r=0}^{\infty} \frac{1}{r!}$$

and so

$$e - \sum_{r=0}^{n} \frac{1}{r!} = \sum_{r=n+1}^{\infty} \frac{1}{r!} < \frac{1}{nn!}$$

Assume that $e = p/q$, where $p, q \in \mathbb{N}$ and $q \neq 0$. Let $n = q$. Then

$$\frac{p}{q} - \sum_{r=0}^{q} \frac{1}{r!} < \frac{1}{qq!}$$

Hence

$$q! \left(\frac{p}{q} - \sum_{r=0}^{q} \frac{1}{r!} \right) < \frac{1}{q} \leqslant 1$$

But

$$q! \left(\frac{p}{q} - \sum_{r=0}^{q} \frac{1}{r!} \right)$$

is a non-zero positive integer. This contradiction shows that e is an irrational number.

4.   Since

$$e^x = 1 + x + \frac{x^2}{2!} + \frac{x^3}{3!} + \frac{x^4}{4!} + \frac{x^5}{5!} + \frac{x^6}{6!} + \dots \quad \text{for all } x$$

it follows that

$$e^{-x} = 1 - x + \frac{x^2}{2!} - \frac{x^3}{3!} + \frac{x^4}{4!} - \frac{x^5}{5!} + \frac{x^6}{6!} + \dots \quad \text{for all } x$$

Adding and dividing by two gives

$$\cosh x = 1 + \frac{x^2}{2!} + \frac{x^4}{4!} + \frac{x^6}{6!} + \dots$$

Since

$$\log_e (1 + x) = x - \tfrac{1}{2}x^2 + \tfrac{1}{3}x^3 + \dots \quad \text{for all } x$$

we have

$$\log_e (1 + x^2) = x^2 - \tfrac{1}{2}x^4 + \tfrac{1}{3}x^6 + \dots$$

Hence, ignoring terms higher than x^6,

$$\cosh x \log_e (1 + x^2) = \left(1 + \frac{x^2}{2!} + \frac{x^4}{4!} + \frac{x^6}{6!} \right) \left(x^2 - \frac{x^4}{2} + \frac{x^6}{3} \right)$$

$$= x^2 + \tfrac{1}{8}x^6$$

Now

$$\lim_{x \to 0} \frac{\cosh x \log_e (1 + x^2) - x^2}{x^6} = \lim_{x \to 0} \frac{\frac{1}{8}x^6 + \text{higher terms}}{x^6} = \frac{1}{8}$$

5.   Now for $h > 0$, there exist a θ_1 such that

$$f(a + h) = f(a) + hf'(a) + \frac{h^2}{2!} f''(a + \theta_1 h), \qquad \text{where } 0 < \theta_1 < 1$$

and a θ_2 such that

$$f(a - h) = f(a) - hf'(a) + \frac{h^2}{2!} f''(a - \theta_2 h), \qquad \text{where } 0 < \theta_2 < 1$$

Therefore

$$f(a + h) + f(a - h) - 2f(a) = \tfrac{1}{2}h^2[f''(a + \theta_1 h) + f''(a - \theta_2 h)]$$

Since f'' is continuous in a neighbourhood of a,

$$f''(a \pm \theta_i h) \to f''(a) \quad \text{as } h \to 0 \quad \text{for } i = 1, 2$$

Hence

$$\lim_{h \to 0} \frac{f(a + h) + f(a - h) - 2f(a)}{h^2} = f''(a)$$

6.   (a) If $f(x) = x^4 e^{-x}$ then $f'(x) = x^3(4 - x)e^{-x}$, and so stationary points occur when $x = 0$ and $x = 4$. Now

$$f''(x) = (x^4 - 8x^3 + 12x^2)e^{-x}$$

and so $f''(0) = 0$ and $f''(4) = -64e^{-4} < 0$. Hence $x = 4$ gives a local maximum of f. Now

$$f^{(3)}(x) = -(x^4 - 12x^3 + 36x^2 - 24x)e^{-x}$$

and hence $f^{(3)}(0) = 0$. Finally,

$$f^{(4)}(x) = (x^4 - 16x^3 + 72x^2 - 96x + 24)e^{-x}$$

and hence $f^{(4)}(0) = 24 > 0$. Thus $x = 0$ gives a local minimum of f. See Figure S.5(a).

(b) If $f(x) = x^2 \log_e x$ then

$$f'(x) = x(1 + 2 \log_e x)$$

and, since f is only defined for $x > 0$, f has one stationary point at $x = e^{-1/2}$. Now

$$f''(x) = 3 + 2 \log_e x$$

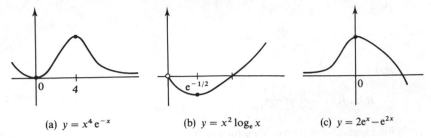

(a) $y = x^4 e^{-x}$ (b) $y = x^2 \log_e x$ (c) $y = 2e^x - e^{2x}$

Figure S.5

and so $f''(e^{-1/2}) = 2 > 0$. Hence f has a local minimum at $x = e^{-1/2}$. See Figure S.5(b).

(c) If $f(x) = 2e^x - e^{2x}$ then

$$f'(x) = 2e^x(1 - e^x)$$

and so f has a stationary point when $x = 0$. Since

$$f''(x) = 2e^x - 4e^{2x}$$

$f''(0) = -2 < 0$, and so f has a local maximum at $x = 0$. See Figure S.5(c).

Exercises 6.4

1.   If $f(x) = \sin x$ then $f^{(r)}(x) = \sin(x + \frac{1}{2}r\pi)$, $r \geq 0$. Hence $f^{(r)}(\frac{1}{6}\pi) = \sin(\frac{1}{6}\pi + \frac{1}{2}r\pi)$, $r \geq 0$. If r is even, $r = 2m$, and so

$$\sin(\tfrac{1}{6}\pi + \tfrac{1}{2}r\pi) = \sin(\tfrac{1}{6}\pi + m\pi)$$
$$= \sin\tfrac{1}{6}\pi\cos m\pi + \cos\tfrac{1}{6}\pi\sin m\pi$$
$$= (-1)^m \sin\tfrac{1}{6}\pi$$
$$= (-1)^{r/2}\tfrac{1}{2}$$

If r is odd, $r = 2m + 1$, and so

$$\sin(\tfrac{1}{6}\pi + \tfrac{1}{2}r\pi) = \sin(\tfrac{2}{3}\pi + m\pi)$$
$$= \sin\tfrac{2}{3}\pi\cos m\pi + \cos\tfrac{2}{3}\pi\sin m\pi$$
$$= (-1)^m \sin\tfrac{2}{3}\pi$$
$$= (-1)^{(r-1)/2}\sqrt{3}/2$$

Let

$$a_r = \begin{cases} (-1)^{r/2}\tfrac{1}{2} & \text{if } r \text{ is even} \\ (-1)^{(r-1)/2}\sqrt{3}/2 & \text{if } r \text{ is odd} \end{cases}$$

Then

$$T_{n,\pi/6}f(x) = \sum_{r=0}^{n} \frac{a_r}{r!} (x - \tfrac{1}{6}\pi)^r$$

and

$$R_{n,\pi/6}f(x) = \frac{(x - \tfrac{1}{6}\pi)^{n+1}}{(n+1)!} f^{(n+1)}(c)$$

for some c lying between x and $\frac{1}{6}\pi$. Now

$$|R_{n,\pi/6}f(x)| = \frac{|x - \tfrac{1}{6}\pi|^{n+1}}{(n+1)!} |f^{(n+1)}(c)|$$

$$= \frac{|x - \tfrac{1}{6}\pi|^{n+1}}{(n+1)!} |\sin[c + \tfrac{1}{2}(n+1)\pi]|$$

$$\leqslant \frac{|x - \tfrac{1}{6}\pi|^{n+1}}{(n+1)!}$$

For fixed x

$$|R_{n,\pi/6}f(x)| \leqslant \frac{K^{n+1}}{(n+1)!}, \quad \text{where} \quad K \geqslant 0 \text{ is a constant}$$

Since $(K^{n+1}/(n+1)!)$ is a null sequence (3.2.2(d)), $R_{n,\pi/6}f(x) \to 0$ as $n \to \infty$, by the sandwich rule for limits of sequences (3.1.3). Hence $\sum_{r=0}^{\infty}(a_r/r!)(x - \tfrac{1}{6}\pi)^r$ is the required series, which is absolutely convergent for all values of x, by a straightforward application of the ratio test (4.2.3).

2. From Question 1 above, the Taylor polynomial of degree four for $f(x) = \sin x$ at $\tfrac{1}{6}\pi$ is

$$T_{4,\pi/6}f(x) = \tfrac{1}{2} + \tfrac{1}{2}\sqrt{3}(x - \tfrac{1}{6}\pi) - \tfrac{1}{4}(x - \tfrac{1}{6}\pi)^2$$
$$- \tfrac{1}{12}\sqrt{3}(x - \tfrac{1}{6}\pi)^3$$

The corresponding remainder term is

$$R_{4,\pi/6}f(x) = \frac{(x - \tfrac{1}{6}\pi)^4}{4!} \sin(c + 2\pi)$$

$$= \frac{(x - \tfrac{1}{6}\pi)^4}{4!} \sin c$$

for some c lying between x and $\tfrac{1}{6}\pi$. Now $32° = 32\pi/180$ radians, and $32\pi/180 = \tfrac{1}{90}\pi + \tfrac{1}{6}\pi$. Hence

$$\sin 32° = \sin(\tfrac{1}{90}\pi + \tfrac{1}{6}\pi)$$
$$\approx T_{4,\pi/6}f(\tfrac{1}{90}\pi + \tfrac{1}{6}\pi)$$
$$= \tfrac{1}{2} + \tfrac{1}{2}\sqrt{3}(\tfrac{1}{90}\pi) - \tfrac{1}{4}(\tfrac{1}{90}\pi)^2 - \tfrac{1}{12}\sqrt{3}(\tfrac{1}{90}\pi)^3$$
$$= 0.529\,919\,23$$

working to eight decimal places. Since $|R_{4,\pi/6}f(32°)| \leq (\frac{1}{90}\pi)^4/4! \approx 0.000\,006\,19$, $\sin 32° = 0.5299$, accurate to four decimal places.

3.   (a) Example 1 in Section 6.4 gives

$$\log_e(1 + t) = \sum_{r=1}^{\infty} \frac{(-1)^{r-1}t^r}{r}$$

valid for $-1 < t \leq 1$. Let $t = x - 1$, so that $x = 1 + t$ and $0 < x \leq 2$. Then

$$\log_e x = \sum_{r=1}^{\infty} \frac{(-1)^{r-1}(x-1)^r}{r}$$

valid for $0 < x \leq 2$.

(b) The geometric series

$$\sum_{r=0}^{\infty} t^r = \frac{1}{1-t}$$

is valid for $-1 < t < 1$. Let $t = \frac{1}{2}(x+2)$, so that

$$\frac{1}{1-t} = \frac{1}{1 - \frac{1}{2}(x+2)} = -\frac{2}{x}$$

Then

$$\sum_{r=0}^{\infty} \left(\frac{x+2}{2}\right)^r = -\frac{2}{x}$$

where $-1 < \frac{1}{2}(x+2) < 1$. Hence

$$\frac{1}{x} = -\frac{1}{2}\sum_{r=0}^{\infty} \frac{(x+2)^r}{2^r}$$

valid for $-4 < x < 0$.

Chapter 7

Exercises 7.1

---

1.   (a) $F(x) = \frac{1}{3}x^3 + \frac{3}{2}x^2 - 2x$

(b) $F(x) = x + \frac{1}{3}\sin 3x$

(c) Since $e^x \cosh 2x = \frac{1}{2}e^x(e^{2x} + e^{-2x}) = \frac{1}{2}(e^{3x} + e^{-x})$, a suitable primitive is $F(x) = \frac{1}{6}(e^{3x} - 3e^{-x})$

(d) $F(x) = \sin^{-1}\frac{1}{3}x$

(e) $F(x) = \frac{1}{2}x^2$ if $x \geq 0$, and $F(x) = -\frac{1}{2}x^2$ if $x < 0$

2.

$$L(P_n) = \sum_{i=1}^{n} \frac{(i-1)^2}{n^2} \frac{1}{n}$$

$$= \frac{1}{n^3} \left[\tfrac{1}{6}(n-1)n(2n-1) \right] \quad \text{using Question 5(a) of Exercises 2.3}$$

$$= \frac{1}{6} \left(1 - \frac{1}{n} \right) \left(2 - \frac{1}{n} \right)$$

$$U(P_n) = \sum_{i=1}^{n} \frac{i^2}{n^2} \frac{1}{n}$$

$$= \frac{1}{n^3} \left[\tfrac{1}{6}n(n+1)(2n+1) \right]$$

$$= \frac{1}{6} \left(1 + \frac{1}{n} \right) \left(2 + \frac{1}{n} \right)$$

As $n \to \infty$, both $L(P_n)$ and $U(P_n)$ converge to $\frac{1}{3}$. Hence $\mathcal{L} = \mathcal{U} = \frac{1}{3}$, and so $f(x) = x^2$ is Riemann integrable on $[0, 1]$ and its value is $\frac{1}{3}$.

3.   First,

$$\int_1^2 \frac{1}{x} \, dx = [\log_e x]_1^2 = \log_e 2$$

Now

$$L(P_n) = \sum_{i=1}^{n} m_i(x_i - x_{i-1}) = \sum_{i=1}^{n} \frac{m_i}{n}$$

where

$$m_i = \inf \left\{ f(x) : 1 + \frac{i-1}{n} < x < 1 + \frac{i}{n} \right\}$$

Since f is decreasing on $[1, 2]$,

$$m_i = \frac{1}{1 + i/n}$$

Hence

$$L(P_n) = \sum_{i=1}^{n} \frac{1}{n+i}$$

Similarly,

$$U(P_n) = \sum_{i=1}^{n} \frac{1}{n + i - 1}$$

Hence

$$\frac{1}{n + 1} + \frac{1}{n + 2} + \ldots + \frac{1}{2n} \leq \int_1^2 \frac{1}{x}\, dx$$

$$\leq \frac{1}{n} + \frac{1}{n + 1} + \ldots + \frac{1}{2n - 1}$$

$$= \frac{1}{n + 1} + \frac{1}{n + 2} + \ldots + \frac{1}{2n}$$

$$+ \frac{1}{n} - \frac{1}{2n}$$

Thus

$$\log_e 2 - \frac{1}{2n} \leq \frac{1}{n + 1} + \frac{1}{n + 2} + \ldots + \frac{1}{2n} \leq \log_e 2$$

Now let $n \to \infty$ to deduce that

$$\frac{1}{n + 1} + \frac{1}{n + 2} + \ldots + \frac{1}{2n} \to \log_e 2$$

4.   Let $P = \{x_0, x_1, \ldots, x_n\}$ be a partition of $[a, b]$ and let $h(x) = g(x) - f(x)$, $x \in [a, b]$. Since $h(x) \geq 0$ for $x \in [a, b]$,

$$m_i = \inf\{h(x) : x_{i-1} < x < x_i\} \geq 0$$

Hence $L(P) \geq 0$ for h, and so $\int_a^b h(x)\, dx \geq 0$. By 7.1.10(1),

$$\int_a^b [g(x) - f(x)]\, dx = \int_a^b g(x)\, dx - \int_a^b f(x)\, dx \geq 0$$

Hence 7.1.10(3) follows.

In order to prove 7.1.10(4), we first need to show that $|f|$ is Riemann integrable on $[a, b]$. Let $U(P)$ and $\bar{U}(P)$ denote the upper sums of f and $|f|$ respectively relative to any partition P of $[a, b]$. Similarly, let $L(P)$ and $\bar{L}(P)$ denote the lower sums. The details are omitted here, but it can be shown that $0 \leq \bar{U}(P) - \bar{L}(P) \leq U(P) - L(P)$. For any $\varepsilon > 0$, P can be chosen so that $U(P) - L(P) < \varepsilon$. Therefore $\bar{U}(P) - \bar{L}(P) < \varepsilon$, and so, by Riemann's condition (7.1.5), it follows that $|f|$ is Riemann integrable on $[a, b]$.

To establish 7.1.10(4), note that

$$-|f(x)| \leq f(x) \leq |f(x)| \quad \text{for all } x \in [a, b]$$

Hence

$$-\int_a^b |f(x)|\, dx \leqslant \int_a^b f(x)\, dx \leqslant \int_a^b |f(x)|\, dx$$

It follows that

$$\left| \int_a^b f(x)\, dx \right| \leqslant \int_a^b |f(x)|\, dx$$

5. For a function f that is Riemann integrable on $[a, b]$ we have

$$\left| \int_a^b f(x)\, dx \right| \leqslant \int_a^b |f(x)|\, dx$$

Hence

$$|J_n| = \left| \frac{1}{n} \int_0^1 \frac{\sin nx}{1 + x^2}\, dx \right| \leqslant \frac{1}{n} \int_0^1 \left| \frac{\sin nx}{1 + x^2} \right|\, dx$$

Since $|\sin nx| \leqslant 1$,

$$|J_n| \leqslant \frac{1}{n} \int_0^1 \frac{1}{1 + x^2}\, dx = \frac{1}{n} [\tan^{-1} x]_0^1 = \frac{\pi}{4n}$$

Now J_n is sandwiched between $\pm \pi/4n$, and so $\lim_{n \to \infty} J_n = 0$.

6. The fundamental theorem of calculus does not apply to the function

$$f(x) = \frac{1}{(x - 1)^2}$$

on the interval $[0, 2]$ since f is undefined at $x = 1$ and hence f is not continuous on $[0, 2]$. The integral in this exercise is an example of a divergent improper integral – these are discussed in Section 7.3.

7. If $g(x) \geqslant 0$ on $[a, b]$, the area between the graphs of f and g and $x = a$ and $x = b$ is given by

$$\int_a^b f(x)\, dx - \int_a^b g(x)\, dx = \int_a^b [f(x) - g(x)]\, dx$$

as illustrated in Figure S.6(a). If, on the other hand, g has a minimum value of $c < 0$ on $[a, b]$ then consider the graphs of $f - c$ and $g - c$. (See Figures S.6(b, c).) The required area is then given by

$$\int_a^b [f(x) - c]\, dx - \int_a^b [g(x) - c]\, dx = \int_a^b [f(x) - g(x)]\, dx$$

as before.

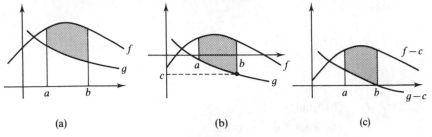

Figure S.6

(a) The given curves intersect when $2 + x - x^2 = 1 - x$, which gives $x^2 - 2x - 1 = 0$. This equation has roots $x = 1 \pm \sqrt{2}$. Hence the required area is

$$\int_{1-\sqrt{2}}^{1+\sqrt{2}} (2 + x - x^2 - 1 + x)\,dx = \int_{1-\sqrt{2}}^{1+\sqrt{2}} (1 + 2x - x^2)\,dx$$
$$= [x + x^2 - \tfrac{1}{3}x^3]_{1-\sqrt{2}}^{1+\sqrt{2}}$$
$$= \tfrac{8}{3}\sqrt{2}$$

(b) The given curves intersect when

$$\sin x = \cos x \quad (0 \leqslant x \leqslant \tfrac{3}{2}\pi)$$

This gives two solutions, namely $x = \tfrac{1}{4}\pi$ and $x = \tfrac{5}{4}\pi$. The required area is thus

$$\int_0^{\pi/4} (\cos x - \sin x)\,dx + \int_{\pi/4}^{5\pi/4} (\sin x - \cos x)\,dx$$
$$= [\sin x + \cos x]_0^{\pi/4} + [-\cos x - \sin x]_{\pi/4}^{5\pi/4}$$
$$= 3\sqrt{2} - 1$$

8.   Apply 7.1.12 to

$$\frac{nf(x)}{1 + n^2x^2}$$

on $[0, 1/\sqrt{n}]$ and on $[1/\sqrt{n}, 1]$. Hence there exist c_n and d_n as specified, with

$$J_1 = f(c_n)\int_0^{1/\sqrt{n}} \frac{n}{1 + n^2x^2}\,dx$$

and

$$J_2 = f(d_n)\int_{1/\sqrt{n}}^{1} \frac{n}{1 + n^2x^2}\,dx$$

Hence

$$J_1 = f(c_n)[\tan^{-1} nx]_0^{1/\sqrt{n}}$$
$$= f(c_n) \tan^{-1} \sqrt{n}$$

and

$$J_2 = f(d_n)[\tan^{-1} nx]_{1/\sqrt{n}}^1$$
$$= f(d_n)(\tan^{-1} n - \tan^{-1} \sqrt{n})$$

Let $n \to \infty$. Then $\tan^{-1} \sqrt{n} \to \frac{1}{2}\pi$ and $\tan^{-1} n \to \frac{1}{2}\pi$. Also, $c_n \to 0$ and, since f is continuous on $[0, 1]$, f is bounded on $[0, 1]$; therefore $f(d_n) \leq M$ for any n. Thus

$$J_1 \to \tfrac{1}{2}\pi f(0) \quad \text{and} \quad J_2 \to 0$$

Therefore

$$\int_0^1 \frac{nf(x)}{1 + n^2 x^2} \, dx \to \tfrac{1}{2}\pi f(0)$$

Exercises 7.2

1.   (a) $\displaystyle\int_1^2 (x^3 - x + 3) \, dx = [\tfrac{1}{4}x^4 - \tfrac{1}{2}x^2 + 3x]_1^2 = \tfrac{21}{4}$

     (b) $\displaystyle\int_0^{\pi/2} \sin 3x \, dx = [-\tfrac{1}{3} \cos 3x]_0^{\pi/2} = \tfrac{1}{3}$

     (c) If

$$\frac{3}{(1 + x)(2 - x)} = \frac{A}{1 + x} + \frac{B}{2 - x}$$

     then $A = B = 1$. Hence

$$\int_0^1 \frac{3}{(1 + x)(2 - x)} \, dx = \int_0^1 \frac{1}{1 + x} \, dx + \int_0^1 \frac{1}{2 - x} \, dx$$
$$= [\log_e (1 + x)]_0^1 + [-\log_e (2 - x)]_0^1$$
$$= \log_e 4$$

     (d) $\displaystyle\int_{-4}^4 \frac{1}{16 + x^2} \, dx = [\tfrac{1}{4} \tan^{-1} \tfrac{1}{4}x]_{-4}^4 = \tfrac{1}{8}\pi$

     (e) $\displaystyle\int_0^4 |2x - 3| \, dx = \int_0^{3/2} (3 - 2x) \, dx + \int_{3/2}^4 (2x - 3) \, dx$
$$= [3x - x^2]_0^{3/2} + [x^2 - 3x]_{3/2}^4 = \tfrac{17}{2}$$

(f) $\int_0^\pi x \cos x \, dx = [x \sin x]_0^\pi - \int_0^\pi \sin x \, dx$

$$= [\cos x]_0^\pi = -2$$

2. $\int_a^b f(x) \cos nx \, dx = \left[f(x) \frac{\sin nx}{n} \right]_a^b - \int_a^b f'(x) \frac{\sin nx}{n} \, dx$

$$= \frac{1}{n} [f(b) \sin nb - f(a) \sin na]$$

$$- \frac{1}{n} \int_a^b f'(x) \sin nx \, dx$$

Now

$$\left| \int_a^b f'(x) \sin nx \, dx \right| \leq \int_a^b |f'(x) \sin nx| \, dx \quad \text{by 7.1.10(4)}$$

$$\leq \int_a^b |f'(x)| \, dx \quad \text{since } |\sin \theta| \leq 1 \quad \text{for all } \theta$$

Since f' is continuous on $[a, b]$, f' is bounded on $[a, b]$, and hence there exists an M with $|f'(x)| \leq M$ for all $x \in [a, b]$. Therefore

$$\left| \int_a^b f(x) \cos nx \, dx \right| \leq \frac{1}{n} \left[|f(b) \sin nb| + |f(a) \sin na| + \int_a^b M \, dx \right]$$

$$\leq \frac{1}{n} [|f(b)| + |f(a)| + |b - a|M] = \frac{K}{n}$$

where K is a constant. Let $n \to \infty$ to see that $\int_a^b f(x) \cos nx \, dx \to 0$.

3. $\int_a^b f(x) g(x) \, dx = [f(x) G(x)]_a^b - \int_a^b f'(x) G(x) \, dx$

$$= f(b) G(b) - \int_a^b f'(x) G(x) \, dx \quad \text{since } G(a) = 0$$

By 7.1.12,

$$\int_a^b f'(x) G(x) \, dx = G(c) \int_a^b f'(x) \, dx \quad \text{for some } c, \ a < c < b$$

Hence $\int_a^b f'(x) G(x) \, dx = G(c)[f(b) - f(a)]$. Therefore

$$\int_a^b f(x) g(x) \, dx = f(b) G(b) - f(b) G(c) + f(a) G(c)$$

$$= f(b) \left[\int_a^b g(t) \, dt - \int_a^c g(t) \, dt \right] + f(a) \int_a^c g(t) \, dt$$

$$= f(a) \int_a^c g(x) \, dx + f(b) \int_c^b g(x) \, dx$$

$$\text{for some } c \in (a, b)$$

4. From Question 3 above,

$$\int_a^b \frac{1}{x} \sin x \, dx = \frac{1}{a} \int_a^c \sin x \, dx + \frac{1}{b} \int_c^b \sin x \, dx$$

Therefore

$$\left| \int_a^b \frac{1}{x} \sin x \, dx \right| = \left| \frac{1}{a} (\cos a - \cos c) + \frac{1}{b} (\cos c - \cos b) \right|$$

$$\leqslant \frac{2}{a} + \frac{2}{b} < \frac{4}{a}$$

Now

$$\left| \int_n^{n+1} \frac{\sin x}{x} \, dx \right| < \frac{4}{n} \to 0 \quad \text{as } n \to \infty$$

Hence

$$\lim_{n \to \infty} \int_n^{n+1} \frac{\sin x}{x} \, dx = 0$$

Exercises 7.3

1. (a) Now $x + [x] = x + n$ if $n \leqslant x < n + 1$. Hence

$$\int_{-2}^2 (x + [x]) \, dx = \int_{-2}^{-1} (x - 2) \, dx + \int_{-1}^0 (x - 1) \, dx + \int_0^1 x \, dx$$

$$+ \int_1^2 (x + 1) \, dx$$

$$= [\tfrac{1}{2}x^2 - 2x]_{-2}^{-1} + [\tfrac{1}{2}x^2 - x]_{-1}^0 + [\tfrac{1}{2}x^2]_0^1$$
$$+ [\tfrac{1}{2}x^2 + x]_1^2 = -2$$

(b) $\int_{-2}^2 f(x) \, dx = \int_{-2}^0 (-2x) \, dx + \int_0^2 x \, dx = [-x^2]_{-2}^0 + [\tfrac{1}{2}x^2]_0^2 = 6$

2. (a) $\int_1^n \frac{1}{x^3} \, dx = \left[\frac{1}{-2x^2} \right]_1^n = \frac{1}{2} - \frac{1}{2n^2} \to \frac{1}{2} \quad \text{as } n \to \infty$

Hence $\int_1^\infty (1/x^3) \, dx$ converges to $\tfrac{1}{2}$.

(b) Since $\sin^2 x = \tfrac{1}{2}(1 - \cos 2x)$,

$$\int_0^n \sin^2 x \, dx = [\tfrac{1}{2}(x - \tfrac{1}{2} \sin 2x)]_0^n = \tfrac{1}{2}(n - \tfrac{1}{2} \sin 2n)$$

which is unbounded as $n \to \infty$. Hence $\int_0^\infty \sin^2 x \, dx$ diverges.

(c) $\int_0^n e^{-2x} dx = [-\frac{1}{2}e^{-2x}]_0^n$

$$= \frac{1}{2}(1 - e^{-2n}) \to \frac{1}{2} \quad \text{as } n \to \infty$$

Hence $\int_0^\infty e^{-2x} dx$ converges to $\frac{1}{2}$.

(d) Now

$$\text{sech } x = \frac{1}{\cosh x} = \frac{2}{e^x + e^{-x}}$$

In order to find $\int \text{sech } x \, dx$, we try the substitution $u = e^x$. Then

$$\int \text{sech } x \, dx = \int \frac{2}{u + u^{-1}} \frac{1}{u} du$$

$$= \int \frac{2}{1 + u^2} du = 2\tan^{-1} u = 2\tan^{-1} e^x$$

Now

$$\int_{-n}^0 \text{sech } x \, dx = [2\tan^{-1} e^x]_{-n}^0$$

$$= 2\tan^{-1} 1 - 2\tan^{-1} e^{-n} \to \frac{1}{2}\pi \quad \text{as } n \to \infty$$

and

$$\int_0^n \text{sech } x \, dx = [2\tan^{-1} e^x]_0^n$$

$$= 2\tan^{-1} e^n - 2\tan^{-1} 1 \to \frac{1}{2}\pi \quad \text{as } n \to \infty$$

Hence $\int_{-\infty}^\infty \text{sech } x \, dx$ converges to π.

3.   (a) Since $\log_e x$ is undefined at $x = 0$, we consider

$$\int_\varepsilon^1 \log_e x \, dx = [x \log_e x - x]_\varepsilon^1 = -\varepsilon \log_e \varepsilon + \varepsilon - 1$$

Now

$$\lim_{\varepsilon \to 0+} \varepsilon \log_e \varepsilon = \lim_{\varepsilon \to 0+} \frac{\log_e \varepsilon}{\varepsilon^{-1}} = \lim_{\varepsilon \to 0+} \frac{\varepsilon^{-1}}{-\varepsilon^{-2}} = 0$$

by an obvious variant of L'Hôpital's rule. Hence

$$\int_0^1 \log_e x \, dx = \lim_{\varepsilon \to 0+} \int_\varepsilon^1 \log_e x \, dx = -1$$

(b) Since $1/(x-1)$ is undefined at $x = 1$, we consider

$$\int_{1+\varepsilon}^{2} \frac{1}{x-1}\,dx = [\log_e (x-1)]_{1+\varepsilon}^{2} = -\log_e \varepsilon$$

which is unbounded as $\varepsilon \to 0+$. Hence

$$\int_{1}^{2} \frac{1}{x-1}\,dx$$

diverges.

(c) Now

$$\int \sec x\,dx = \log_e (\sec x + \tan x)$$

and

$$\int \tan x\,dx = \log_e (\sec x)$$

and so

$$\int (\sec x - \tan x)\,dx = \log_e \left(\frac{\sec x + \tan x}{\sec x} \right)$$
$$= \log_e (1 + \sin x)$$

Since $\sec x - \tan x$ is undefined at $x = \frac{1}{2}\pi$, consider

$$\int_{0}^{\pi/2-\varepsilon} (\sec x - \tan x)\,dx = [\log_e (1 + \sin x)]_{0}^{\pi/2-\varepsilon}$$
$$= \log_e [1 + \sin (\tfrac{1}{2}\pi - \varepsilon)]$$
$$\to \log_e 2 \quad \text{as } x \to 0+$$

Hence

$$\int_{0}^{\pi/2} (\sec x - \tan x)\,dx = \lim_{\varepsilon \to 0+} \int_{0}^{\pi/2-\varepsilon} (\sec x - \tan x)\,dx$$
$$= \log_e 2$$

4.   First,

$$\int_{0}^{1} t^{x-1}\,dt = \left[\frac{t^x}{x} \right]_{0}^{1} = \frac{1}{x} \quad \text{for } x \geqslant 1$$

by the fundamental theorem of calculus (7.1.11). For $0 < x < 1$

$$\int_0^1 t^{x-1}\, dt = \lim_{\varepsilon \to 0+} \int_\varepsilon^1 t^{x-1}\, dt = \lim_{\varepsilon \to 0+} \left[\frac{t^x}{x}\right]_\varepsilon^1 = \lim_{\varepsilon \to 0+} \frac{1 - \varepsilon^x}{x}$$

Since $x > 0$, $\varepsilon^x \to 0$ as $\varepsilon \to 0+$. Hence

$$\int_0^1 t^{x-1}\, dt = \frac{1}{x} \quad \text{for } 0 < x < 1$$

5. $0 \le f(x) \le g(x)$ on $[a,\, b] \Rightarrow 0 \le \displaystyle\int_{a+\varepsilon}^b f(x)\, dx \le \int_{a+\varepsilon}^b g(x)\, dx$

Since $\int_a^b g(x)\, dx$ exists, $\int_{a+\varepsilon}^b f(x)\, dx$ is bounded as $\varepsilon \to 0+$. Also, $f(x) \ge 0$ means that $\int_{a+\varepsilon}^b f(x)\, dx$ is increasing as $\varepsilon \to 0+$. Hence $\lim_{\varepsilon \to 0+} \int_{a+\varepsilon}^b f(x)\, dx$ exists.

6.   (a) Now

$$0 \le \left(\frac{\sin x}{x}\right)^2 \le \frac{1}{x^2} \quad \text{for all } x \ge 1$$

Also

$$\int_1^\infty \frac{1}{x^2}\, dx = \lim_{n \to \infty} \int_1^n \frac{1}{x^2}\, dx = \lim_{n \to \infty} \left[\frac{-1}{x}\right]_1^n$$

$$= \lim_{n \to \infty} \left(1 - \frac{1}{n}\right) = 1$$

Hence

$$\int_1^\infty \frac{1}{x^2}\, dx$$

converges to 1. By the comparison test for integrals,

$$\int_1^\infty \left(\frac{\sin x}{x}\right)^2 dx$$

converges.

(b) $0 \le \dfrac{x}{1 + x^3} \le \dfrac{1}{x^2} \quad \text{for } x > 0$

Since $\int_1^\infty (1/x^2)\, dx$ converges, so too does

$$\int_1^\infty \frac{x}{1 + x^3}\, dx$$

(c) $0 \leqslant e^{-x^2} \leqslant e^{-x}$ for $x \geqslant 1$. Since

$$\int_1^\infty e^{-x}\, dx = \lim_{n \to \infty} \int_1^n e^{-x}\, dx$$

$$= \lim_{n \to \infty} [-e^{-x}]_1^n$$

$$= \lim_{n \to \infty} (e^{-1} - e^{-n}) = 1/e$$

$$\int_1^\infty e^{-x^2}\, dx \text{ converges}$$

7.   (a) For $0 < x < \frac{1}{2}\pi$, $0 < \sin x < x$, and so

$$\frac{1}{\sin x} > \frac{1}{x}$$

If

$$\int_0^{\pi/2} \frac{1}{\sin x}\, dx$$

is convergent then so too is

$$\int_0^{\pi/2} \frac{1}{x}\, dx$$

by Question 5 of Exercises 7.3. But

$$\int_0^{\pi/2} \frac{1}{x}\, dx$$

diverges (essentially Example 9 of Section 7.3), and so we have a contradiction. Hence

$$\int_0^{\pi/2} \frac{1}{\sin x}\, dx$$

is divergent.

(b) Since $e^x \geqslant 1 + x$ for all x, $e^{1/x} \geqslant 1 + 1/x$ for $x \neq 0$. Now

$$\int_0^1 \left(1 + \frac{1}{x}\right) dx = [x]_0^1 + \int_0^1 \frac{1}{x}\, dx = 1 + \int_0^1 \frac{1}{x}\, dx$$

which diverges. Hence $\int_0^1 e^{1/x}\, dx$ is divergent.

(c) For $0 < x < 1$, $0 < x < \tan x$, and so $0 < \sqrt{x} < \sqrt{\tan x}$. Hence

$$0 < \frac{1}{\sqrt{\tan x}} < \frac{1}{\sqrt{x}}$$

But $\int_0^1 (1/\sqrt{x})\,dx$ converges, by Example 7 of Section 7.3, and so

$$\int_0^1 \frac{1}{\sqrt{\tan x}}\,dx$$

is convergent.

8.   Since $\Gamma(\tfrac{1}{2}) = \sqrt{\pi}$, the statement is true for $n = 0$. Assume that

$$\Gamma(k + \tfrac{1}{2}) = \frac{(2k)!\sqrt{\pi}}{4^k k!}$$

for some $k \geqslant 1$. Now $\Gamma(x + 1) = x\Gamma(x)$, and so

$$
\begin{aligned}
\Gamma((k + 1) + \tfrac{1}{2}) &= \Gamma((k + \tfrac{1}{2}) + 1) \\
&= (k + \tfrac{1}{2})\Gamma(k + \tfrac{1}{2}) \\
&= \frac{(k + \tfrac{1}{2})(2k)!\sqrt{\pi}}{4^k k!} \\
&= \frac{(4k + 2)(2k)!\sqrt{\pi}}{4^{k+1} k!} \\
&= \frac{2(2k + 1)(2k)!\sqrt{\pi}}{4^{k+1} k!}\frac{(k + 1)}{(k + 1)} \\
&= \frac{(2k + 2)!\sqrt{\pi}}{4^{k+1}(k + 1)!}
\end{aligned}
$$

and so the statement also holds for $n = k + 1$. By induction on n,

$$\Gamma(n + \tfrac{1}{2}) = \frac{(2n)!\sqrt{\pi}}{4^n n!} \quad \text{for all } n \geqslant 0$$

Answers to problems

Chapter 1

2. (a) The statement is false (e.g. $m = n = 1$, $p = -2$).
       The converse is false (e.g. $m = -1$, $n = 2$, $p = 2$).

    (b) The statement is false (e.g. $m = -2$, $n = 1$).
       The converse is false (e.g. $m = 1$, $n = -2$).

    (c) The statement is false (e.g. $m = n = 1$).
       The converse is true.

    When m, n and p are positive:

    (a) remains false (e.g. $m = 4$, $n = 3$, $p = 6$), but its converse is true;

    (b) and its converse are both true;

    (c) remains false and its converse remains true.

4. Only the first statement is true.
       The second statement makes the false assertion that there is a number m exceeding $3n$ whatever the number n.
       The third statement claims falsely that there is a number n with $3n$ less than all numbers m; there is no such n when $m = 1$, 2 or 3.

5. Equality does not hold in part (c) when, for example, $A = \{0, 1\}$, $B = \emptyset$ and $C = \{1, 2\}$.

7. Equality does not hold in part (b) when, for example, $A = \{0, 1\}$, $B = \{1, 2\}$ and the subset $\{0, 2\}$, of $A \cup B$ is considered.

8. $$(f \circ g)(x) = \begin{cases} x^2 & \text{if } x \geqslant 0 \\ x + 1 & \text{if } -1 \leqslant x < 0 \\ x & \text{if } x < -1 \end{cases}$$

323

$$(g \circ f)(x) = \begin{cases} x^2 & \text{if } x \geqslant 0 \\ x & \text{if } x < 0 \end{cases}$$

$$(g \circ f)^{-1}(x) = \begin{cases} \sqrt{x} & \text{if } x \geqslant 0 \\ x & \text{if } x < 0 \end{cases}$$

9.   For $x \neq 0, 1$, $f^3(x) = x$ and $f^{-1}(x) = 1/(1-x)$

Chapter 2

3.   (a) All except A3, A4 and A8

         (b) All except A8

         (c) All except A4

4.   (a) $-1 < x < 2$ or $x > 4$

         (b) $-4 \leqslant x \leqslant -2$

7.   (a) $\sup S = 2$, $\inf S = -1$

         (b) $\sup S = \frac{9}{4}$, $\inf S = 0$

         (c) Since $|x + 1/x| \geqslant 2$, neither $\sup S$ nor $\inf S$ exist

         (d) $\sup S = 1$, $\inf S = \frac{1}{9}$

         (e) $\sup S = \frac{8}{9}$, $\inf S = 0$

9.   (a) No   (b) No   (c) Yes

Chapter 3

3.   (a) $-\frac{1}{3}$ (b) $\frac{1}{3}$ (c) divergent   (d) e^{-3} (e) 5

4.   (a) True   (b) False (e.g. $a_n = n + 1$, $b_n = -n$)

         (c) True   (d) False (e.g. $a_n = b = (-1)^n$)

6.   (a) Bounded, positive, increasing, convergent

         (b) Bounded below, eventually positive, increasing, divergent to ∞

         (c) Divergent

         (d) Bounded, divergent

         (e) Bounded, positive, oscillating, convergent

8. $\lim_{n \to \infty} a_n = \frac{1}{3}$ for all real values of α

9. $0, \frac{1}{2}, \frac{1}{4}, \frac{3}{8}, \frac{5}{16}, \frac{11}{32}$

         $b_{2n} = a_n$ (the sequence in Problem 8) when $a_1 = \frac{1}{2}$

         $b_{2n-1} = a_n$ when $a_1 = 0$

         Hence $\lim_{n \to \infty} b_n = \frac{1}{3}$

Chapter 4

3.   (a), (b), (c), (e), (f) and (g) are all convergent
(d) and (h) are divergent

4. $|x| < e$

6.   (a) is conditionally convergent
(b) and (c) are absolutely convergent
(d) is divergent

8.   (b) and (c) are convergent
(a) is divergent

9.   (a) 1   (b) 1   (c) $\frac{4}{27}$

Chapter 5

2.   (a) $\frac{1}{2}$ (b) $n/(2n-1)$ (c) $\frac{1}{2}$ (d) $\frac{1}{2}$

3.   (a) 2   (b) -2 (c) $\frac{1}{4}$ (d) 0   (e) $\frac{3}{4}$

4.   (b) One example is $f(x) = h(x) = x$, $g(x) = 1$ if $x \neq 0$ and $g(0) = 0$

6.   (a) Continuous for $x \neq n\pi$, n an integer
(b) Continuous for $x \neq 0$
(c) Continuous everywhere
(d) Continuous for $x \neq 0$ and $x \neq 1/n\pi$, n an integer

8.   (b) One example is $Q(x) = x$
(c) Two examples are $R(x) = x^2 + 1$ or $R(x) = 10$

9.   (a) $[-1, 1]$ (b) $[0, 1 + \sin^2 1]$
(c)$[0, \frac{3}{4}]$ (d) $[2, 4]$

Chapter 6

1.   (a) At $x = 0$, $f'_-(0) = -2$ and $f'_+(0) = 0$
At $x = 2$, $f'_-(2) = 0$ and $f'_+(2) = 2$
(b) At $x = n\pi$, n an integer, $f'_-(n\pi) = -1$ and $f'_+(n\pi) = 1$
(c) At $x = 0$, $f'_-(0) = 1$ and $f'_+(0) = -1$
(d) At $x = 0$, $f'_-(0) = -1$ and $f'_+(0) = 1$

7.   (a) The maximum value is $f(2) = 2$ and the minimum value is $f(0) = f(1) = 0$.

(b) The maximum value is $f(1) = 1/e$ and the minimum value is $f(0) = 0$.

(b) The maximum value is $f(\frac{1}{2}) = 4$ and the minimum value is $f(\frac{3}{2}) = 2$.

9. (a) 2   (b) $\frac{1}{6}$ (c) 0

12. (a) $1 + x + \frac{3}{2}x^2 + \frac{3}{2}x^3 + \frac{37}{24}x^4 + \frac{37}{24}x^5 + \frac{1111}{720}x^6$
   (b) $x + \frac{1}{2}x^2 - \frac{2}{3}x^3 + \frac{1}{4}x^4 + \frac{1}{5}x^4 - \frac{1}{3}x^6$

13. (a) Local minimum at $x = 0$
   (b) Horizontal points of inflection at $x = 2n\pi$, n an integer
   (c) Local minimum at $x = 0$
   (d) Local minimum at $x = \frac{2}{3}$ and local maximum at $x = 2$

14. (a) $\frac{4}{9} - \frac{16}{27}(x - \frac{1}{2}) + \frac{16}{27}(x - \frac{1}{2})^2 - \frac{128}{243}(x - \frac{1}{2})^3 + \frac{320}{729}(x - \frac{1}{2})^4$
   (b) $\frac{1}{2}\sqrt{3} + \frac{1}{2}(x - \frac{1}{3}\pi) - \frac{1}{4}\sqrt{3}(x - \frac{1}{3}\pi)^2 - \frac{1}{12}(x - \frac{1}{3}\pi)^3 + \frac{1}{48}\sqrt{3}(x - \frac{1}{3}\pi)^4$
   (c) $\log_e 3 + \frac{1}{3}(x - 2) - \frac{1}{18}(x - 2)^2 + \frac{1}{81}(x - 2)^3 - \frac{1}{324}(x - 2)^4$

Chapter 7

2. $L(P_n) = \frac{1}{2}\left(1 - \frac{1}{n}\right), \quad U(P_n) = \frac{1}{2}\left(1 + \frac{1}{n}\right) + \frac{2n - 1}{n^2}$

5. $L(P_n) = n^5 \sum_{i=1}^{n} \frac{1}{(n^2 + i^2)^3}$

$U(P_n) = n^5 \sum_{i=1}^{n} \frac{1}{[n^2 + (i - 1)^2]^3}$

The limit has the value $\frac{3}{32}\pi + \frac{1}{4}$.

7. $F(x) = \begin{cases} x & \text{if } 0 \leqslant x \leqslant 1 \\ 2x - 1 & \text{if } 1 \leqslant x \leqslant 2 \end{cases}$

$F'(1)$ does not exist!

9. (a) Divergent   (b) $\frac{1}{2}$ (c) $(1/2\sqrt{2})\tan^{-1}2\sqrt{2}$
   (d) Divergent   (e) Divergent   (f) 2

10. (b), (c), (d) and (f) are convergent
   (a) and (e) are divergent

Index of Symbols

Index

CHESTER COLLEGE LIBRARY